Twenty Years of Bialowieza:
A Mathematical Anthology

Aspects of Differential Geometric
Methods in Physics

World Scientific Monograph Series in Mathematics

Eds: Ron Donagi *(University of Pennsylvania)*, Rafael de la Llave *(University of Texas)* and Mikhail Shubin *(Northeastern University)*

Published

Vol. 1:   Almgren's Big Regularity Paper : Q-Valued Functions Minimizing Dirichlet's Integral and the Regularity of Area-Minimizing Rectifiable Currents up to Codimension 2
Eds. V. Scheffer and J. E. Taylor

Vol. 2:   Dynamics and Mission Design Near Libration Points
Vol. I Fundamentals: The Case of Collinear Libration Points
by G. Gómez, J. Llibre, R. Martinez and C. Simó

Vol. 3:   Dynamics and Mission Design Near Libration Points
Vol. II Fundamentals: The Case of Triangular Libration Points
by G. Gómez, J. Llibre, R. Martinez and C. Simó

Vol. 4:   Dynamics and Mission Design Near Libration Points
Vol. III Advanced Methods for Collinear Points
by G. Gómez, À. Jorba, J. Masdemont and C. Simó

Vol. 5:   Dynamics and Mission Design Near Libration Points
Vol. IV Advanced Methods for Triangular Points
by G. Gómez, À. Jorba, J. Masdemont and C. Simó

Vol. 6:   Hamiltonian Systems and Celestial Mechanics
Eds. J. Delgado, E. A. Lacomba, E. Pérez-Chavela and J. Llibre

Vol. 8:   Twenty Years of Bialowieza: A Mathematical Anthology
Aspects of Differential Geometry Methods in Physics
Eds. S. Twareque Ali, G. G. Emch, A. Odzijewicz, M. Schlichenmaier and S. L. Woronowicz

Forthcoming

Vol. 7:   Spectral Analysis of Differential Operators
by F. S. Rofe-Beketov, A. M. Kholkin and O. Milatovic

World Scientific Monograph Series
in Mathematics – Vol. 8

Twenty Years of Bialowieza:
A Mathematical Anthology

Aspects of Differential Geometric Methods in Physics

editors

S. Twareque Ali *(Concordia University, Canada)*
Gerard G. Emch *(University of Florida, USA)*
Anatol Odzijewicz *(University of Bialystok, Poland)*
Martin Schlichenmaier *(Université du Luxembourg, Luxembourg)*
Stanislaw L. Woronowicz *(University of Warsaw, Poland)*

World Scientific

NEW JERSEY • LONDON • SINGAPORE • BEIJING • SHANGHAI • HONG KONG • TAIPEI • CHENNAI

Published by

World Scientific Publishing Co. Pte. Ltd.

5 Toh Tuck Link, Singapore 596224

USA office: 27 Warren Street, Suite 401-402, Hackensack, NJ 07601

UK office: 57 Shelton Street, Covent Garden, London WC2H 9HE

British Library Cataloguing-in-Publication Data
A catalogue record for this book is available from the British Library.

TWENTY YEARS OF BIALOWIEZA: A MATHEMATICAL ANTHOLOGY
Aspects of Differential Geometry Methods in Physics

ISBN 981-256-146-3

Printed in Singapore by World Scientific Printers (S) Pte Ltd

Preface

Twenty Years of Białowieża: An Anthology

This volume marks the twentieth anniversary (which in fact passed three years ago) of the Białowieża series of meetings on Differential Geometric Methods in Physics. What once started out as a summer rendezvous for a few interested physicists and mathematicians from within Poland and some neighbouring countries, the Białowieża meetings have now grown into an annual pilgrimage for a devoted and growing international group, sharing scientific and professional experiences. Additionally, these meetings, held each year at the beginning of July, in the sylvan setting of the ageless Białowieża forests, have been the rite of passage for scores of graduate students and young researchers, starting out on their careers. The world of mathematical and theoretical physics knows no dearth of meetings, workshops, conferences, symposia ... each with its own series of proceedings volumes. What sets the Białowieża meetings apart is perhaps the "purity of expression", in mathematical terms and the close interaction between physicists and mathematicians that takes place during these week-long shared sojourns. More than at most mathematical physics meetings, the focus of attention here is rather on the mathematical structures than on the phenomenological aspects of the physical problems of interest. Added to that are the stimulating social ambience and camaraderie that have developed over the years.

The present collection is not a proceedings volume. Rather, it was conceived as a means to replicate the spirit of the Białowieża Workshops and to reflect its scientific tenor, as a tribute to the completion of two decades of a shared scientific experience. While the focus of the Białowieża workshops has naturally been on a small subset of areas within the broad gamut of mathematical physics – prominent among them being quantization techniques, coherent states, symplectic geometry, Poisson structures, infinite dimensional systems and new trends in the application of geometric methods to physics – yet even this narrow spectrum could not have been adequately represented in a single volume. Consequently, only a few topics could be chosen for inclusion in this collection. With this constraint in mind, a number of former participants and invited speakers were approached to contribute to this commemorative anthology, as it were, with papers that would represent a cross-section of the themes and topics that were, or could have been, discussed in these Workshops. Unfortunately, not everyone approached was able to contribute at this time. Nevertheless, in the opinion of the editors, what is presented here, does in many ways crystallize the Białowieża approach to mathematical physics.

In the course of its twenty-year history, the Workshop has offered a substantial core of mathematical lectures presented by mainstream mathematicians. Yet, as a result of the subject of the lectures and also of the personality of the lecturers, the audience has evolved to attract a number of mathematical physicists. The participation of the latter contributed different motivations and favored the raising of questions of intent. We all advertise to our students – and to our administrations – that mathematical techniques are demanded for the solution of problems in the physical sciences and in engineering; we know that this is no accident in the obvious sense that these techniques were often developed in direct response to questions from outside the ethereal realm of pure mathematics; and we all have our favorite examples. Nevertheless, what remains mysterious, as it did even to such creative practitioners as Wigner, is the "unreasonable effectiveness of mathematics". Mathematics that was developed as pure mathematics suddenly gets applied in fields ignored by the mathematical community. Flows running in these opposite directions have naturally surfaced in Białowieża also. While no external relevance is recognized as a precondition, the question of such relevance has been put forward as often as it has been brushed aside. These opposite pulls have contributed to the vitality of the Białowieża Workshops. We hope that some of the readers of the present collection will want to examine the various contributions from that angle, either demanding physical interpretation and motivation or insisting on mathematical purity and ingenuity. In this connection, we are reminded of the aphorism of the Polish mathematician Mark Kac who, upon considering the courtiers in attendance to the Queen of the Sciences, was moved to fear for her virtue. While Kac meant it as a barb against the purists nestled in the mathematical establishment, it seems to us that the aphorism could very well be double-edged. In this spirit, several of the contributions to this volume were solicited without concern for actual applications, but we believe none should be immune to this type of scrutiny.

A few words of introduction to the papers presented in this volume are in order. There is a cycle of four papers on different aspects of the theory of quantization – one of the most prominently represented areas of research at these workshops. The paper by N.P. Landsman entitled, "Functorial Quantization and the Guillemin-Sternberg Conjecture", addresses the issue of quantization in the presence of constraints; specifically, to find definitions for the arrows in following diagram that are reasonable and sharp enough to make sense of the question as to whether the diagram is commutative.

$$
\begin{array}{ccc}
Unconstrained & Q & Unconstrained \\
Classical\ system & \longrightarrow & Quantum\ system \\
& & \\
R \downarrow & & \downarrow R \\
& Q & \\
Classical\ system & \longrightarrow & Quantum\ system \\
with\ Constraints & & with\ Constraints
\end{array}
$$

Building up on the quantization approach now associated to the name of Raoul Bott, Landsman had recently proposed to view quantization as a functor between two categories, namely isomorphism classes of symplectic dual pairs and homotopy classes of Kasparov bimodules. To provide the definition of this functor between these categories is the main purpose of the present paper.

The two papers entitled, "Diffeomorphism Groups and Quantum Configurations as Mathematical Objects", by G.A. Goldin and "The Group of Volume Preserving Diffeomorphisms and the Lie Algebra of Unimodular Vector Fields: Survey of some Classical and not so Classical Results", by C. Roger, have a unifying aspect in that they both deal with the use of diffeomorphism groups as "receptacles" for several classical and quantum theories, notable among them being geometric and deformation quantization, local current algebras and their relation to quantized fields, vortex quantization in hydrodynamics, and anyons. The paper by Goldin has more of a physical flavour, with the mathematical structures introduced and explained in relatively non-technical terms. The author manages to survey an impressive spectrum of problems for which the diffeomorphism group could be used as an analyzing tool. The paper by C. Roger, bears distinctly a mathematician's stamp. It is a rigorous description of certain aspects of Lie algebras and "groups" of unimodular vector fields. The latter are vector fields that exponentiate to volume preserving diffeomorphisms. A number of mathematical results on the cohomology of the Lie algebra of vector fields is presented, one of the most important being the rigidity theorem, demonstrating that the Lie algebra of unimodular vector fields in dimensions greater or equal 3 admits no non-trivial formal deformations (Theorem 3.2). The theory is then extended to supermanifolds.

The fourth paper on the subject of quantization is, "Coherent State Method in Geometric Quantization" by A. Odzijewicz. Coherent states have been a constant theme running through the Białowieża meetings. The discussion in this paper is based on the construction of a *coherent state map*, between the classical phase space and a family of vectors on the quantized Hilbert space, which under conditions that are spelled out in the paper, yield a quantization of the original classical system. The situations where such a map yields the same results as the geometric quantization of Kostant-Souriau are worked out in some detail. A number of examples illustrate the theory.

There are two papers on the subject of symplectic and Poisson geometry. In the paper entitled, "Moduli Space of Germs of Symplectic Connections of Ricci type", by M. Cahen, the prototype for the structures studied arise, in the particular case of symplectic manifolds that are compact and simply connected, from complex projective spaces $P_n(C)$, equipped with their standard symplectic form and the Levi-Cevita connection associated to the Fubini-Study metric. The last section of the paper is devoted to the study of the moduli space of Ricci-type connections of $P_n(C)$ and it makes explicit the sense in which the Levi-Cevita connection for the Fubini-Study metric is an isolated point in this space. The general motivation for the paper is provided by an analogy with the moduli space of Einstein metrics, i.e. the space of Einstein metrics on a manifold M *modulo* the action of the diffeomorphism group of M. The other paper entitled, "Banach Lie-Poisson Spaces", by A. Odzijewicz and T.S. Ratiu, is a summary of the authors' work in this field spanning several years. The paper sets out the category of Banach Lie-Poisson manifolds, which are Poisson manifolds defined over Banach spaces, such that the duals of these spaces form Banach Lie-algebras under the Poisson bracket operation. The morphisms for the category of such spaces are the *linear Poisson maps*. The construction is illustrated with a number of examples of both mathematical and physical interest.

The paper by R. Picken, "A Cohomological Description of Abelian Bundles and Gerbes", stands by itself in that the geometric structures discussed there have potential applications to non-abelian gauge theories – a subject which has not featured in the mainstream of the Białowieża Workshops. Nevertheless, the paper adds a new flavour to the volume, even more so because of the easy pedagogical style of presentation. Gerbes with connection are higher order generalizations of abelian bundles with connection. The aim of this paper is to achieve a cohomological description of both gerbes and abelian bundles in which this generalization is seen to emerge in a natural way.

The paper by R.P. Langlands, entitled "The Renormalization Fixed Point as a Mathematical Object", also opens a new direction in the Białowieża repertoire, namely the delineation of universality classes in the critical phenomena of classical statistical mechanical systems exhibiting phase transitions. There is nevertheless a link to the main paths traditionally trod at the Workshop, namely the emergence of classical, or macroscopic, behaviour that are revealed in the course of asymptotic developments. Here, this process is provided by the successive iterations of rescaling known as renormalization (semi-)group techniques. Some of the immediacy of the present paper stems from its character as a review based mostly on numerical results obtained for well chosen models, specifically percolation and 2-D Ising.

In the paper by J.Hilgert entitled, "An Ergodic Arnold-Liouville Theorem for Locally Symmetric spaces", the main result is stated as the last theorem on the second last page of the paper, and it may be a good idea for the reader, before studying the paper in detail, to have a quick glance at this theorem. The author proposes to view it as an "ergodic Arnold-Liouville theorem" in analogy with the standard Arnold-Liouville theorem on integrable Hamiltonian systems. While the analysis is resolutely carried out within the realm of classical dynamical systems, the author also alludes to possible relevance to the quantization of some of the structures considered.

The paper entitled, "Spectra of Operators Associated with Dynamical Systems: from Ergodicity to the Duality Principle", by Antonevitch, *et al*, is of a functional analytic nature. The authors extend from reversible to irreversible shifts several aspects of the study of classical dynamical systems. Their main results for irreversible shifts is stated in Theorem 4.6 which establishes a duality – realized explicitly as a Legendre transform – between two quantities they define separately for the abstract version, *viz*, the spectral exponent of weighted shifts and a new dynamical entropy. To illustrate the theory, the last section of the paper is devoted to a discussion of the Perron-Frobenius systems and some of their properties.

In the paper by W. Pusz and S.L. Woronowicz entitled, "On Quantum Group of Unitary Operators: Quantum '$az + b$' Group", the authors discuss a theory whose inception and growth has in fact been coeval with the Białowieża meetings. The paper briefly surveys the present status of quantum group theory and introduces the concept of a quantum group of unitary operators, relevant to a study of non-compact, locally compact quantum groups. The theory is then illustrated by constructing a quantum "$az + b$" group.

As our commentaries should have amply demonstrated, we are happy to present this volume to our readers as a representative of the scientific variety of the Białowieża experience.

There remains only the pleasant task of expressing our debt of gratitude to our friend, colleague and co-sharer of the Białowieża experience, Mikhail Shubin, for his generous help with the editing of this volume and bringing it to fruition.

The Editors
(S.T. Ali, G.G. Emch,
A. Odzijewicz, M. Schlichenmaier
and S.L. Wronowicz)

Contents

Other Mathematical Methods

Chapter 1

Diffeomorphism Groups and Quantum Configurations

Gerald A. Goldin[1]

It is a special occasion to be writing for this 20th anniversary volume of the Workshop on Geometric Methods in Physics. The first of these conferences that I attended took place in 1992. It was already the 11th in the series. The special nature of the meeting intrigued me, and since then I haven't missed any. To pass one week each summer beneath the ancient trees in beautiful Białowieża Forest came to represent a fixed interval of peaceful scientific and personal reflection, about which the rest of the year turned. So I would like to begin by expressing gratitude to those who made this extraordinary series of meetings possible.

I want particularly to thank my friends S. Twareque Ali, Anatol Odzijewicz, and Aleksander Strasburger, who brought me to Białowieża and who contributed so much to the unique atmosphere of personal intimacy and creative scientific inquiry that has characterized the workshops.

This article discusses some issues in quantum theory from a very elementary but fundamental perspective. My purpose is to examine *why* the group of compactly supported diffeomorphisms of physical space, or one of several closely related infinite-dimensional groups, should play such a deep role in a unifying, geometric description of quantum mechanics. I hope to do so without presupposing prior familiarity with the subject; more specifics on various aspects may be found in earlier published articles, reviews, and the references therein [1; 2; 3; 4].

The idea of taking the diffeomorphism group as fundamental is related conceptually to other questions, especially the problem of quantization

---

[1]Departments of Mathematics and Physics, Rutgers University, New Brunswick, New Jersey 08903, USA, `gagoldin@dimacs.rutgers.edu`

1

(which has been a recurring theme of the Białowieża workshop series), and the relation between quantized fields, local current algebras, and particle statistics. After commenting on these issues, I mention very briefly some directions my collaborators and I are taking in the quantum theory of infinite-dimensional configuration spaces, motivated by the study of diffeomorphism group representations.

1.1   Diffeomorphism Groups and Physical Space

First let us define various diffeomorphism groups, and the corresponding Lie algebras of vector fields.

Let M and N be smooth, finite-dimensional Riemannian manifolds. A diffeomorphism is a one-to-one, onto, differentiable homeomorphism ϕ from M to N, whose inverse is likewise differentiable. We are interested here in diffeomorphisms that map M to itself. Any two such diffeomorphisms ϕ_1 and ϕ_2, acting successively, give a third diffeomorphism $\phi_2 \circ \phi_1$. The *support* of ϕ is the smallest closed set $K \subset M$ such that for all $\mathbf{x} \in M - K$, $\phi(\mathbf{x}) = \mathbf{x}$. Then the set of C^∞ diffeomorphisms of M having compact (but arbitrary) support forms a group under composition. We call this group $Diff^{\,c}(M)$, where the superscript c means "compact." It becomes a topological group when endowed with the topology of uniform convergence in all derivatives in compact sets. Of course if M itself is compact, this is just the full group of C^∞ diffeomorphisms $Diff(M)$. It is an infinite-dimensional group, whose unitary representations are of great interest for both mathematics and physics.

Associated with $Diff^{\,c}(M)$ is the infinite-dimensional Lie algebra $vect^c(M)$, consisting of the C^∞ (tangent) vector fields on M having compact support, endowed with the Lie bracket. The relation of $vect^c(M)$ to $Diff^{\,c}(M)$ is as follows. Let $s \to \phi_s$ $(s \in \mathbf{R})$ be a one-parameter group of diffeomorphisms of M, differentiable in s. Such a group defines a vector field $\mathbf{g}$ on M, whose value at $\mathbf{x} \in M$ is just the tangent vector to the curve $\phi_s(\mathbf{x})$ at $s = 0$. Thus we have $\partial_s \phi_s(\mathbf{x}) = \mathbf{g}(\phi_s(\mathbf{x}))$, with $\phi_{s=0}(\mathbf{x}) = \mathbf{x}$. We call ϕ_s the *flow* generated by the vector field $\mathbf{g}$. Evidently if the ϕ_s have support in a compact region K, then $\mathbf{g}$ vanishes outside K.

Now an arbitrary smooth vector field on a non-compact manifold can be exponentiated locally; but it does not necessarily exponentiate to a one-parameter group. The differentiable maps that are obtained by exponentiation may fail to be defined for all $\mathbf{x}$, or for all s. However, a *compactly-*

supported smooth vector field $\mathbf{g}$ on M always exponentiates to a flow in $Diff^c(M)$, which for specificity we label $\phi_s^{\mathbf{g}}$ (see for example [5]). If $\mathbf{g}_1$ and $\mathbf{g}_2$ are two such vector fields, their *Lie bracket* $[\mathbf{g}_1, \mathbf{g}_2]$ is the vector field that corresponds to the (infinitesimal) outcome of flowing (infinitesimally) by each of the two vector fields, in succession, and then flowing backward (infinitesimally) by each of the two vector fields. In local coordinates, $[\mathbf{g}_1, \mathbf{g}_2](\mathbf{x}) = \mathbf{g}_1(\mathbf{x}) \cdot \nabla \mathbf{g}_2(\mathbf{x}) - \mathbf{g}_2(\mathbf{x}) \cdot \nabla \mathbf{g}_1(\mathbf{x})$. The Lie bracket is again a C^∞ vector field having compact support. It satisfies the Jacobi identity and defines the Lie algebra structure on $vect^c(M)$. Of course the Lie algebra of vector fields is infinite dimensional, as is the diffeomorphism group.

Note that the C^k diffeomorphisms, $k = 1, 2, 3, ...,$ also form a group, as the product to two C^k diffeomorphisms is again C^k. It might thus seem that our restriction to C^∞ diffeomorphisms is overly strong. However, the Lie bracket of C^k vector fields involves a derivative, so in general it is only C^{k-1}. The condition that we have a bona fide Lie algebra, closed under the Lie bracket, therefore mandates the choice of C^∞ vector fields (for which we require C^∞ diffeomorphisms).

Let us now talk about d-dimensional physical space, $M = \mathbf{R}^d$.

An important subgroup of $Diff^c(\mathbf{R}^d)$ is the group of area- or volume-preserving diffeomorphisms $SDiff^c(\mathbf{R}^d)$, $d > 1$, where S stands for "special". (When $d = 2$, this coincides with the group of symplectic diffeomorphisms of the plane; for $d = 1$, however, the group is trivial.) The corresponding Lie subalgebra is $svect^c(\mathbf{R}^d)$ the algebra of *divergenceless* compactly-supported vector fields. The group $SDiff^c(\mathbf{R}^d)$ and the algebra $svect^c(\mathbf{R}^d)$ are important to the quantum theory of an ideal, incompressible fluid in $\mathbf{R}^d$, $d > 1$.

We also have some natural ways to enlarge $Diff^c(\mathbf{R}^d)$ that may be useful, without losing the correspondence between the resulting group and a Lie algebra of C^∞ vector fields. We may relax the condition of compact support, and modify correspondingly the topology of the group, in such a manner as to maintain the association with an algebra of vector fields that generate global flows. One possibility is to include diffeomorphisms that, in the limit as $|\mathbf{x}| \to \infty$, approach the identity map rapidly in all derivatives (faster than any polynomial). This group can be given the topology of uniform rapid convergence in all derivatives, and has been called $\mathcal{K}(\mathbf{R}^d)$. The Lie algebra corresponding to $\mathcal{K}(\mathbf{R}^d)$ consists of vector fields with components in Schwartz' space $\mathcal{S}(\mathbf{R}^d)$.

Alternatively, consider all C^∞ diffeomorphisms of $\mathbf{R}^d$ that coincide with some (uniform) translation outside of an arbitrary compact region $K \in \mathbf{R}^d$. These form a group that we may call $Diff^{\,\mathrm{trans}}(\mathbf{R}^d)$. We can obtain any such diffeomorphism by composing an element of $Diff^{\,c}(\mathbf{R}^d)$ with an element T of the translation group $\mathcal{T}(\mathbf{R}^d)$. Note further that there is a natural homomorphism from $\mathcal{T}(\mathbf{R}^d)$ to the group of automorphisms of $Diff^{\,c}(\mathbf{R}^d)$: for each translation T, we have the automorphism $\phi \to T \circ \phi \circ T^{-1}$. This lets us write $Diff^{\,\mathrm{trans}}(\mathbf{R}^d)$ as a semidirect product of $Diff^{\,c}(\mathbf{R}^d)$ with $\mathcal{T}(\mathbf{R}^d)$.

Similarly we may define the group $Diff^{\,\mathrm{rot}}(\mathbf{R}^d)$ [or respectively, the groups $Diff^{\,\mathrm{Eucl}}(\mathbf{R}^d)$, $Diff^{\,\mathrm{lin}}(\mathbf{R}^d)$, $Diff^{\,\mathrm{slin}}(\mathbf{R}^d)$, $Diff^{\,\mathrm{dil}}(\mathbf{R}^d)$, $Diff^{\,\mathrm{aff}}(\mathbf{R}^d)$, or $Diff^{\,\mathrm{saff}}(\mathbf{R}^d)$] to consist of all C^∞ diffeomorphisms of $\mathbf{R}^d$ that coincide, outside of some compact region, with some rotation (for $d > 1$) [or respectively, some Euclidean transformation, linear transformation, special linear transformation, dilatation, affine transformation, or special affine transformation]. For each such extension of $Diff^{\,c}(\mathbf{R}^d)$ we have a corresponding infinite-dimensional Lie algebra of vector fields on $\mathbf{R}^d$, where the vector fields coincide, outside of some compact region, with the infinitesimal generators of a finite-dimensional Lie group acting globally on $\mathbf{R}^d$. It is likewise natural to consider the extension of $SDiff^{\,c}(\mathbf{R}^d)$ by either the Euclidean group or the group of special linear or special affine transformations.

Again, we may enlarge each of the above groups to include diffeomorphisms which, in the limit as $|\mathbf{x}| \to \infty$, approach a translation [resp. Euclidean transformation, linear transformation, etc.] rapidly in all derivatives.

Finally, consider the natural semidirect product of a diffeomorphism group with an additive group of scalar functions on $\mathbf{R}^d$. Let $\mathcal{D}(\mathbf{R}^d)$ consist of the compactly supported C^∞ real-valued functions on $\mathbf{R}^d$, with its usual topology of uniform convergence in all derivatives in compact regions. A diffeomorphism $\phi \in Diff^{\,c}(\mathbf{R}^d)$ acts naturally on $\mathcal{D}(\mathbf{R}^d)$ by transforming the argument of each function; i.e., for $f \in \mathcal{D}(\mathbf{R}^d)$, $\phi : f \to f \circ \phi$. The map $(f, \phi) \to f \circ \phi$ is jointly continuous in f and ϕ. So we have the semidirect product group $\mathcal{D} \times Diff^{\,c}(\mathbf{R}^d)$, with $(f_1, \phi_1)(f_2, \phi_2) = (f_1 + f_2 \circ \phi_1, \phi_2 \circ \phi_1)$. Likewise, we may extend the other diffeomorphism groups mentioned above. In particular, it is natural also to consider the semidirect product group $\mathcal{S} \times \mathcal{K}(\mathbf{R}^d)$.

1.2 Diffeomorphism Group Representations and Quantum Mechanics

Why are the diffeomorphisms of $\mathbf{R}^d$, or of a more general manifold M, so important for quantum mechanics?

Let us think first of a diffeomorphism ϕ of M actively, as actually taking whatever might be located in a neighborhood $\mathcal{O}$ of a point $\mathbf{x}_0$, and moving it (while smoothly turning and distorting it) to a new neighborhood $\phi(\mathcal{O})$ containing $\phi(\mathbf{x}_0)$. Just as we identify the self-adjoint momentum operator P_x in quantum mechanics with the infinitesimal generator of the group of translations in the x-direction, or the self-adjoint angular momentum operator L_z with the infinitesimal generator of the group of rotations about the z-axis, we now consider a self-adjoint operator $J(\mathbf{g})$ for the *momentum density averaged with* $\mathbf{g}$ (also called a *local current*) and identify it with the infinitesimal generator of the flow that the vector field $\mathbf{g}$ generates.

That is, for each flow $\phi_s^{\mathbf{g}}$ there will be a continuous unitary group representation $V(\phi_s^{\mathbf{g}})$ in the Hilbert space $\mathcal{H}$ of our quantum theory, with $J(\mathbf{g}) = \lim_{s \to 0}(\hbar/is)[V(\phi_s^{\mathbf{g}}) - I]$ on its domain of definition. We expect that in the limit as $\mathbf{g}$ approaches a constant vector field in some direction (i.e., as $\mathbf{g}$ coincides with such a vector field in larger and larger regions), the matrix elements of $J(\mathbf{g})$ approach those of the total momentum operator in the specified direction.

Note that these identifications are not dynamical but *kinematical.* The existence of the self-adjoint operators P_x or L_z (as generators of group actions on the spatial manifold) does not depend on the Hamiltonian operator H being translation- or rotation-invariant (although it might be). It depends merely on the fact that the translations or rotations act smoothly as a group on the physical space. Likewise the existence of self-adjoint operators $J(\mathbf{g})$, representing the Lie algebra $vect^c(\mathbf{R}^d)$, depends not on the dynamics associated with a specific choice of H but merely on the existence of a continuous group action of $Diff^{\,c}(\mathbf{R}^d)$ on $\mathbf{R}^d$ (along with appropriate technical assumptions). Generic diffeomorphisms are not symmetries of the motion, but allow us to describe the kinematics locally.

In my opinion, there is also something to learn by seeing how far we can go in thinking of the diffeomorphism ϕ passively—as defining a *general coordinate transformation.* That is, ϕ provides us with a smooth way to modify our *description* of the locations of objects in space. In this interpretation, the metric in the N-particle configuration space is not invariant. Since the Schrödinger equation makes use of this metric by means of the bilinear form entering the kinetic energy term, we should set the time dimension aside for the moment, and apply our "passive" interpretation only to the fixed-time situation. In conventional quantum mechanics, the probability amplitude for a system in state Ψ_1 to be observed in state Ψ_2, given as usual by the inner product (Ψ_2, Ψ_1), is however a fixed-time notion. The "collapse of the wave packet" does not occur dynamically. Hence this inner product should remain unchanged by such a change of description, and we have reason to expect the modification of coordinates to be implemented in $\mathcal{H}$ by a unitary operator $V(\phi)$.

Plausibly, we expect the correspondence $\phi \to V(\phi)$ to be smooth and to respect the composition of diffeomorphisms. Thus we should have a continuous unitary representation (CUR), or at least a projective representation, of $Diff^{\,c}(\mathbf{R}^d)$ in $\mathcal{H}$. And given such a representation, the self-adjoint local currents $J(\mathbf{g})$ can be recovered as the infinitesimal generators of the continuous one-parameter unitary groups representing the flows.

Such a way of thinking could be extended to include the time dimension in theories that, unlike nonrelativistic quantum mechanics, are generally covariant—i.e., where there are no background structures and all the fields (including the space-time metric) are dynamical. In the context of quantum gravity, the reader is referred to an interesting discussion by Isham [6].

Now the interpretation of the local currents as averaged momentum densities gives us physical information about the quantum system. In nonrelativistic physics, momentum describes the flux of mass. Thus this interpretation suggests also the inclusion of a system of self-adjoint operators for the mass density, with the momentum density being the mass flux density. In other words, for each $f \in \mathcal{D}$ we should have a self-adjoint operator $\rho(f)$, so that $\rho(f_1)$ and $\rho(f_2)$ commute ($\forall f_1, f_2 \in \mathcal{D}$), and for which

$$V(\phi)\rho(f)V(\phi^{-1}) = \rho(f \circ \phi) \tag{1.1}$$

$[\forall \phi \in Diff^{\,c}(\mathbf{R}^d)]$. Furthermore we expect that when f is very near to the indicator function of a Borel region B in $\mathbf{R}^d$, then the matrix elements of $\rho(f)$ should approach those of the operator for total mass in B.

Eq. (1.1) exponentiates to a unitary representation of the semidirect product group $\mathcal{D} \times Diff^{\,c}(\mathbf{R}^d)$ or, with alternate conditions at $|\mathbf{x}| \to \infty$, $\mathcal{S} \times \mathcal{K}(\mathbf{R}^d)$. The operators $\rho(f)$ and $J(\mathbf{g})$ thus satisfy the nonrelativistic local current algebra [7; 8; 9],

$$[\rho(f_1),\, \rho(f_2)] = 0,$$

$$[\rho(f),\, J(\mathbf{g})] = i\hbar\rho(\mathbf{g}\cdot\nabla f),$$

$$[J(\mathbf{g}_1),\, J(\mathbf{g}_2)] = -i\hbar J([\mathbf{g}_1,\, \mathbf{g}_2])\,. \tag{1.2}$$

So far, the discussion parallels to a certain extent the standard way that the ordinary, global Euclidean symmetries of physical space show up in nonrelativistic quantum mechanics [10]. Following Mackey, we localize a particle by means of a "system of imprimitivity" (associating projection-valued measures with indicator functions on Borel sets in $\mathbf{R}^d$), we consider the stability subgroup of a point in $\mathbf{R}^d$, and we induce unitary representations of the global symmetry group from representations of the stability subgroup. But we immediately see some important differences. The group $\mathit{Diff}^{\,c}(\mathbf{R}^d)$ that we are here representing unitarily is infinite-dimensional, not a compact or locally compact Lie group. It also acts in an important sense *locally* rather than globally. That is, for any open region $\mathcal{O}$, there is a subgroup $\mathit{Diff}_{\mathcal{O}}(\mathbf{R}^d)$ of diffeomorphisms having support in $\mathcal{O}$, and a corresponding unitary group $V_{\mathcal{O}} = V[\mathit{Diff}^{\,c}(\mathbf{R}^d)]$ acting in $\mathcal{H}$. If O_1 and O_2 are disjoint, then $V_{\mathcal{O}_1}$ and $V_{\mathcal{O}_2}$ commute, so that we have a kind of Galilean locality built into the kinematics. We thus think of $\mathit{Diff}^{\,c}(\mathbf{R}^d)$ as a "local symmetry group." We shall shortly see that these features provide some important advantages, that are not present when we represent finite-dimensional Lie groups.

1.3   Causal Diffeomorphisms and Space-Time

But before exploring such issues, let us consider briefly the idea of general coordinate transformations of the space-time manifold $\mathbf{R}^{d+1}$, rather than just of $\mathbf{R}^d$. The natural transformations to consider are those diffeomorphisms $\Phi : \mathbf{R}^{d+1} \to \mathbf{R}^{d+1}$ that respect the *causal structure* of space-time.

In Galilean space-time, this means Φ must be such that $(t_1, \mathbf{x}_1)$ precedes $(t_2, \mathbf{x}_2)$ if and only if $\Phi(t_1, \mathbf{x}_1)$ precedes $\Phi(t_2, \mathbf{x}_2)$, while $(t_1, \mathbf{x}_1)$ and $(t_2, \mathbf{x}_2)$ are simultaneous if and only if $\Phi(t_1, \mathbf{x}_1)$ and $\Phi(t_2, \mathbf{x}_2)$ are simultaneous. Let us call these *causal diffeomorphisms*. Evidently the identity map has this property, while a diffeomorphism Φ is causal if and only if Φ^{-1} is causal; so we again have a group. We can write a general causal diffeomorphism in the form $\Phi(t, \mathbf{x}) = (t', \mathbf{x}') = (\tau(t), \phi_t(\mathbf{x}))$; where $\tau : \mathbf{R} \to \mathbf{R}$ is a diffeomorphism of the time dimension only, but $\phi_t(\mathbf{x})$ is a system of diffeomorphisms of $\mathbf{R}^d$ depending smoothly on t. In effect, we consider $\mathbf{R}^{d+1}$ as a bundle over $\mathbf{R}$ (the time axis), and take the group of bundle diffeomorphisms. Clearly the Galilean boosts ($t' = t$, $\mathbf{x}' = \mathbf{x} - \mathbf{v}t$) belong to this group, as do the time translations; and we have also the natural embedding of $Diff\,^c(\mathbf{R}^d)$ and its extensions given by ($t' = t$, $\mathbf{x}' = \phi(\mathbf{x})$). Representation of additional elements of this group would appear to be potentially interesting for the description of quantum mechanics in *nonuniformly moving or accelerating reference frames.* I will not discuss this idea further now, though it has been the subject of some interesting conversations in Białowieża [11].

The special relativistic situation is different. In Minkowskian space-time, causal diffeomorphisms must be such that if the relation of $(ct_1, \mathbf{x}_1)$ to $(ct_2, \mathbf{x}_2)$ is space-like [respectively, time-like preceding, time-like following, or light-like], the relation of $\Phi(ct_1, \mathbf{x}_1)$ to $\Phi(ct_2, \mathbf{x}_2)$ is also. In two-dimensional space-time, a diffeomorphism Φ of the Minkowskian plane that has this property may be characterized as acting independently on light cone coordinates. That is, write the general point (ct, x) in the form $(\chi_1, -\chi_1) + (\chi_2, \chi_2)$, where $\chi_1 = (ct - x)/2$ and $\chi_2 = (ct + x)/2$. Then there are two diffeomorphisms ϕ_1 and ϕ_2 of two different real lines (the left and the right light cone through the origin) such that with $\chi_1' = \phi_1(\chi_1)$ and $\chi_2' = \phi_2(\chi_2)$, $\Phi(ct, x) = (\chi_1', -\chi_1') + (\chi_2', \chi_2')$. We can thus realize a group of causal diffeomorphisms as the product $Diff\,^c(\mathbf{R}) \times Diff\,^c(\mathbf{R})$. Note, however, that even when ϕ_1 and ϕ_2 are compactly supported, Φ is not.

The appropriate local currents here are light cone currents, not fixed-time currents. The appropriate representations turn out to be projective representations of the local current algebra—so that we are dealing not just with two copies of the algebra of vector fields on $\mathbf{R}$, but two copies of the Virasoro algebra (the central extension of the algebra of vector fields on the real line or the circle). This leads us directly into conformal field theory in $1 + 1$ dimensions. However, the process of taking a nonrelativistic limit to recover the nonrelativistic local current algebra (1.2) and the group of diffeomorphisms of physical space is nontrivial.

In Minkowskian space-time of dimension greater than two, the group of causal diffeomorphisms is finite-dimensional (as is the conformal group). Indeed causal bijections, without further continuity assumptions, must be formed from Poincaré transformations together with dilatations [12]; we have lost the ability to deform the space-time *locally.* In a sense, special relativity in three or more space-time dimensions has a causal structure that is just "too rigid" for the diffeomorphism group.

Despite these facts, the semidirect product group $\mathcal{D} \times Diff^{\,c}(\mathbf{R}^d)$ still seems to play an interesting, mostly unexplored role in relativistic quantum field theory (QFT). While the physical world is relativistic, nonrelativistic quantum mechanics of course provides good approximations for observations at low energies. Thus, although local, relativistic algebras of observables necessarily connect subspaces of Hilbert space with different numbers of particles, if the local particle number makes sense there should exist mathematically (in a given reference frame) a system of operators for measuring the spatial locations of the particles and the flux of the particles. This is indeed the case, for instance, in the Fock representation of a relativistic canonical scalar field. Such operators are nonlocal and noncovariant, so they do not qualify as "local observables" for relativistic QFT. However, this means that the unitary representations of $\mathcal{D} \times Diff^{\,c}(\mathbf{R}^d)$ and the "nonrelativistic" local current algebra *can* exist at a fixed time even in relativistic models, and generally do. At low energies in particle theories, it is this current algebra that (approximately) describes the kinematics.

The situation changes again when we move from special to general relativity. Here the group of diffeomorphisms of a spacelike surface enters explicitly again, playing the role of a gauge group in the superspace formulation of quantum gravity. But that is a different subject.

1.4 Diffeomorphism Groups and Particle Configurations

Now let us return to our discussion of nonrelativistic quantum mechanics. Ordinarily an elementary introduction to the quantum kinematics of a single particle in $\mathbf{R}$ might begin with a representation of the Heisenberg algebra $[\,Q, P\,] = i\,C$ (where C is a central element) by (necessarily unbounded) self-adjoint operators. For the representation to be irreducible C must be a multiple of the identity, so we write $C = \hbar\,I$. Then, up to unitary equivalence, von Neumann showed the *uniqueness* of the irreducible representation realized in $\mathcal{H} = L^2(\mathbf{R})$ by $Q\,\Psi(q) = q\,\Psi(q),\ P\,\Psi(q) = -i\hbar\,\partial\Psi(q)/\partial q$. This uniqueness was long regarded as an advantage of the usual, simple prescription for quantization based on position and momentum operators. Another advantage is that the Heisenberg algebra treats Q and P equivalently, so that the symplectic structure of the qp-plane can be taken as a fundamental geometric starting point. In d dimensions, of course, we use $[Q^j, P^k] = i\hbar\,\delta^{jk}\,I\ (k = 1,...d)$, and $Q^j\,\Psi = q^j\,\Psi,\ P^k\,\Psi = -i\hbar\,\partial\Psi/\partial q^k$, while for systems of N particles, we make use of N distinct position and momentum (vector) coordinates.

In contrast, the self-adjoint representations of the local current algebra are *not* unique. Because the algebra is infinite-dimensional, its irreducible representations can encode much more information. Such a representation has the potential—all by itself—to describe multiparticle systems, even infinite-particle systems, and more. This is not surprising when we focus on the fact that the mass density operators can describe different regions of physical space independently of each other, or that diffeomorphisms can move different regions of physical space around independently. Suppose, for example, we have a system of N particles in $\mathbf{R}^3$ distinguishable by their masses $m_1, ..., m_N$. Let the configuration space C_N be the set of ordered N-tuples $(\mathbf{x}_1, ..., \mathbf{x}_N)$, $\mathbf{x}_j \in \mathbf{R}^3$, with $\mathbf{x}_j \neq \mathbf{x}_k$ for $j \neq k$. Then we have for each N, and each set of distinct masses, an inequivalent representation of the local current algebra: with $\Psi \in \mathcal{H}_\mathcal{N} = L^2(C_N)$,

$$\rho(f)\,\Psi(\mathbf{x}_1, ..., \mathbf{x}_N) = \Sigma_{j=1}^{N}\, m_j\, f(\mathbf{x}_j)\Psi(\mathbf{x}_1, ..., \mathbf{x}_N),$$

$$J(\mathbf{g})\,\Psi = -\,i\hbar\,\Sigma_{j=1}^{N}\,\frac{1}{2}\{\,\mathbf{g}(\mathbf{x}_j)\cdot\nabla_j\,\Psi\,+\,\nabla_j\,\cdot[\,\mathbf{g}(\mathbf{x}_j)\Psi\,]\}. \qquad (1.3)$$

The corresponding unitary representation $U(f)V(\phi)$ of the semidirect product group $\mathcal{D} \times \textit{Diff}^{\,c}(\mathbf{R}^d)$ can be written

$$U(f)\Psi(\mathbf{x}_1, ..., \mathbf{x}_N) = \exp[\,i\,\Sigma_{j=1}^{N}\,m_j f(\mathbf{x}_j)\,]\,\Psi(\mathbf{x}_1, ..., \mathbf{x}_N)\,,$$

$$V(\phi)\Psi(\mathbf{x}_1, ..., \mathbf{x}_N) = \Psi(\phi(\mathbf{x}_1), ..., \phi(\mathbf{x}_N))\sqrt{\Pi_{j=1}^{N}\mathcal{J}_\phi(\mathbf{x}_j)}\,, \qquad (1.4)$$

where $\mathcal{J}_\phi(\mathbf{x}_j)$ is the Jacobian of ϕ at $\mathbf{x}_j$.

Let us remark on some properties of these representations. Note that the operators $\rho(f)$ are bounded (and thus defined for all Ψ)—this is a consequence of the finiteness of the total mass, and does not hold for more general representations. The operators $J(\mathbf{g})$ are unbounded (and thus defined only on a dense domain in $\mathcal{H}_\mathcal{N}$). There exists a common, dense invariant domain of essential self-adjointness on which the current algebra is satisfied. Note the rather delicate interplay between the physical space $\mathbf{R}^3$ and the configuration space C_N: when f becomes (approximately) the indicator function of a region $B \subset \mathbf{R}^3$, the operator $\rho(f)$ approximately multiplies $\Psi(\mathbf{x}_1, ..., \mathbf{x}_N)$ by the sum of those masses m_j for which $\mathbf{x}_j \in B$, just as desired. Note also that when $\mathbf{g}$ becomes (approximately) a constant vector field on $\mathbf{R}^3$ in some direction—let us say, the unit vector field in the z-direction—then with $\mathbf{x}_j = (x_j, y_j, z_j)$, $J(\mathbf{g})\Psi$ becomes (approximately) $-i\hbar \Sigma_{j=1}^{N} \partial\Psi/\partial z_j$; i.e., it is the sum of the one-particle momentum operators in the z-direction, as desired.

Now the representation J of the vector fields alone given by (1.3) is reducible for $N > 1$, since all the operators $J(\mathbf{g})$ commute with permutations of the particle coordinates. Likewise the representation V of the diffeomorphism group alone given by (1.4) is reducible. Invariant subspaces of $\mathcal{H}_N$ for J and V are those having specified permutation symmetry; but of course the representation U does not respect this invariance when the particle masses are distinct. When the masses are identical, however, the particles are indistinguishable. In this case we obtain distinct (unitarily inequivalent) irreducible representations of the full current algebra on the fixed-symmetry subspaces of $\mathcal{H}_N$. In particular, the subspace $\mathcal{H}_N^s$ of totally symmetric wave functions carries the N-particle Bose representation, and the subspace $\mathcal{H}_N^a$ of totally antisymmetric wave functions carries a representation that we shall call (suppressing the discussion of spin) the N-particle Fermi representation.

We begin to see how general is the description of quantum mechanics by means of unitary representations of the group of diffeomorphisms of physical space. The unitarily inequivalent representations of this group correspond not only to quantum systems describing different numbers of particles, but also to different particle statistics. We have obtained the usual N-particle sectors of the nonrelativistic Fock space, without (yet) introducing canonical fields. An important remark, however, is that particle permutations here refer not to the indices j, but to the coordinate values $\mathbf{x}_j$. For $d = 1$, representations of the group of compactly supported diffeomorphisms of $\mathbf{R}^1$ *cannot* distinguish N-particle systems having different exchange statistics. This is because there are no compactly supported diffeomorphisms that actually exchange two distinct positions in $\mathbf{R}^1$. The distinction between bosons and fermions in one-dimensional space must be regarded as dynamical—encoded in the choice of a self-adjoint Hamiltonian.

In an important sense, the diffeomorphism group of physical space serves as a kind of "universal group" for quantum mechanics. Classification of its unitary representations amounts to classification of theoretically possible quantum-mechanical systems. If we did not know about N-particle configuration spaces, we would *discover* them by classifying the group representations.If we did not know about bosons and fermions, we would likewise discover them. Further, "exotic" statistics are also obtained this way. Distinct unitary representations of $Diff\,^c(\mathbf{R}^d)$ are induced by higher-dimensional representations of the symmetric group S_N, providing a description of parastatistics. The case $d = 2$ is especially interesting. Here the statistics of anyons and plektons were actually predicted in part from the study of representations of $Diff\,^c(\mathbf{R}^2)$ induced from unitary representations of the braid group B_N [13; 14; 15; 16; 17; 18; 19; 20]. Of course, additional considerations may restrict the possibilities; for example, we do not have a spin-statistics theorem merely from the kinematics of nonrelativistic quantum mechanics described here.

Note also that the condition $\mathbf{x}_j \neq \mathbf{x}_k$ for $j \neq k$ imposed on C_N is respected by diffeomorphisms. In removing the "diagonal" subspace from $\mathbf{R}^{3N}$, we are not merely excluding an arbitrary measure zero set, but a set invariant under diffeomorphisms—so that the diffeomorphism group acts transitively on the configurations that remain. This is of fundamental importance for the existence of the inequivalent representations describing particle statistics (see below), and does not depend on any assumption of a hard core repulsive potential between particles.

The configuration space C_N is also of considerable recent mathematical interest. Motivated by the investigation of Berry and Robbins into the spin-statistics connection [21], Atiyah and Bielawski study the existence, for each N, of a continuous map from C_N to the flag manifold $U(n)/T^n$ that respects the natural action of S_N on each space—a nontrivial problem with interesting generalizations [22; 23].

Another advantage of the diffeomorphism group approach to quantum theory is that because we have a local symmetry group, we are not restricted to $\mathbf{R}^d$ as the spatial manifold. We can represent the group of compactly-supported diffeomorphisms of a manifold, or a manifold with boundary, that lacks global translation- or rotation-invariance, that is not simply-connected, and so on. Thus, even when total momentum or angular momentum operators (the infinitesimal generators of spatial translations or rotations) do not exist or are not uniquely specified, we have a natural way to describe the quantum-mechanical possibilities, obtaining self-adjoint observables as the generators of one-parameter unitary groups.

1.5   Unitary Representations: Quasi-invariant Measures and Cocycles

Before going further, let me summarize (without too many technical details) a fairly general framework for realizing unitary representations of $Diff\,^c(\mathbf{R}^d)$

[14; 24; 25]. Consider a space Γ on which there is a continuous group action by $Diff^{\,c}(\mathbf{R}^d)$, written $\gamma \to \phi\gamma$. Let μ be a measure on Γ that is quasi-invariant under $Diff^{\,c}(\mathbf{R}^d)$—i.e., the group action respects the class of μ-measure zero sets. Then we write $\mathcal{H} = L^2_\mu(\Gamma, \mathcal{M})$, where $\mathcal{H}$ is the Hilbert space of μ-square integrable functions Ψ on Γ taking values in $\mathcal{M}$, and $\mathcal{M}$ is a complex inner product space (usually the complex numbers $\mathbf{C}$, but sometimes $\mathbf{C}^N$ or a separable Hilbert space). The inner product of Ψ_1 and Ψ_2 in $\mathcal{H}$ is given by

$$(\Psi_1, \Psi_2) = \int_\Gamma (\Psi_1(\gamma), \Psi_2(\gamma))_{\mathcal{M}} \, d\mu(\gamma), \qquad (1.5)$$

where the inner product in the integrand is taken in $\mathcal{M}$. Then we write

$$V(\phi)\Psi(\gamma) = \chi_\phi(\gamma)\Psi(\phi\gamma)\sqrt{\frac{d\mu_\phi}{d\mu}(\gamma)}, \qquad (1.6)$$

where: μ_ϕ is the measure obtained from μ by transforming by ϕ, $d\mu_\phi/d\mu$ is the Radon-Nikodym derivative (which exists by virtue of the quasi-invariance of μ), and $\chi_\phi(\gamma)$ is defined for each diffeomorphism ϕ (almost everywhere with respect to μ), as a unitary 1-cocycle acting in $\mathcal{M}$. That is, $\chi_\phi(\gamma) : \mathcal{M} \to \mathcal{M}$ is unitary, and ($\forall \phi_1, \phi_2$) these operators obey the cocycle equation

$$\chi_{\phi_1}(\gamma)\chi_{\phi_2}(\phi_1\gamma) = \chi_{\phi_2 \circ \phi_1}(\gamma) \qquad (1.7)$$

(μ-almost everywhere). Here Γ plays the role of the quantum configuration space. One way to obtain a suitable space Γ, following the method of semidirect products, is to consider the continuous dual space $\mathcal{D}'$ to $\mathcal{D}$. For $\gamma \in \mathcal{D}'$, the definition of $\phi\gamma$ by $\langle \phi\gamma, f \rangle = \langle \gamma, f \circ \phi \rangle$ gives us a group action, and consequently an *orbit* structure in $\mathcal{D}'$. This framework also allows us to represent the semidirect product group easily: for $f \in \mathcal{D}$, set $U(f)\Psi(\gamma) = \exp\left[i\langle \gamma, f \rangle \right]\Psi(\gamma)$.

The N-particle representations described above for $d = 3$ then correspond to particular orbits, as follows. Denote by $\delta_{\mathbf{x}} \in \mathcal{D}'$ the evaluation functional $\langle \delta_{\mathbf{x}}, f \rangle = f(\mathbf{x})$, and let $m_1, ... m_N$ be distinct positive real numbers (particle masses). Then the set $\{ \gamma = m_1 \delta_{\mathbf{x}_1} + ... + m_N \delta_{\mathbf{x}_N} \,|\, (\mathbf{x}_1, ... \mathbf{x}_N) \in C_N \} \subset \mathcal{D}'$ is such an orbit, and is straightforwardly identified with the configuration space C_N. Note that the indices here are artificial and not used; it is actually the distinct values of the masses that provide an ordering to the particle locations in the physical space. To complete the picture, μ

is (locally) the Lebesgue measure, $\mathcal{M}$ is $\mathbf{C}$, the cocycle $\chi_\phi(\gamma) \equiv 1$, and $d\mu_\phi/d\mu$ is the product of Jacobians that occurred in (1.4).

Alternatively, if the particles are indistinguishable with mass m, we have an orbit $\Gamma_N = \{ \gamma = m \Sigma_{j=1}^N \delta_{\mathbf{x}_j} | \mathbf{x}_j \neq \mathbf{x}_k \in \mathbf{R}^3 \text{ for } j \neq k \}$. This is just the collection of N-point subsets of $\mathbf{R}^3$, which can be identified with C_N/S_N. Again we may choose μ to be locally equivalent to Lebesgue measure. With $\chi_\phi(\gamma) \equiv 1$, we obtain a representation unitarily equivalent to the N-particle bosonic representation described above; this is easily seen by noting that C_N is the *universal covering space* of Γ_N, identifying Γ_N with a fundamental domain in C_N, and extending each wave functions on Γ_N to the corresponding totally symmetric wave functions on C_N. A *inequivalent* cocycle now describes the N-particle fermionic representation. In fact, the general principal here is that inequivalent representations of the diffeomorphism group describing different particle statistics are obtained as *induced representations* from inequivalent representations of the fundamental group of the configuration space. The stability subgroup in the diffeomorphism group (associated with a point $\gamma \in \Gamma_N$) maps homomorphically onto this fundamental group. The nontrivial fundamental group S_N of Γ_N is a consequence of the absence of the "diagonal" (discussed above for C_N). The possibility of exotic statistics for $d = 2$ derives from the fact that the fundamental group of the manifold of N-point configurations in the plane is larger than S_N—it is Artin's braid group B_N.

A different way to obtain configuration spaces for the diffeomorphism group is by the method of coadjoint orbits. This is closely related to the geometric quantization program, which I shall not discuss at this point.

1.6 Comments on Quantization

Until now I have not said much at all about quantization as such, except to contrast the diffeomorphism group approach with the approach to quantizing a particle system by representing the Heisenberg algebra or a finite-dimensional group of global symmetries.

For the case of a single particle without spin, the configuration space Γ_1 can be identified with the physical space M (although, more precisely, Γ_1 is the set of one-point subsets of M). Representing the algebra of scalar functions and vector fields then corresponds closely to the ideas of "Borel quantization" [26; 27; 28]. In this program, quantization begins with a configuration space manifold Γ; the classical phase space $\mathcal{P}$ is the cotangent bundle $T^*(\Gamma)$. The quantized theory is realized by representing an algebra of scalar functions on Γ, together with a Lie algebra of (tangent) vector fields on Γ, by self-adjoint operators in a Hilbert space of complex-valued, square integrable functions on Γ.

Let the phase space coordinates be $(\mathbf{q}, \mathbf{p})$, where $\mathbf{q} \in \Gamma$. Note that in the usual Poisson algebra over $\mathcal{P}$, the subspace consisting of C^∞ functions $f(\mathbf{q})$ of $\mathbf{q}$ only, and C^∞ functions of the form $\mathbf{g}(\mathbf{q}) \cdot \mathbf{p}$, where f and g have compact support, form a Poisson subalgebra. This algebra is isomorphic to the Lie algebra of local currents on the physical space M when $\Gamma = \Gamma_1$, while for the case of more than one particle, it is larger. The decision to represent this subalgebra (which does not, of course, contain the kinetic energy Hamiltonian) is thus a prescription for quantization of the kinematics of the system; in fact, Doebner and Tolar call their method "quantum Borel kinematics." Note that already in the one-particle case, one sacrifices the usual mathematical symmetry between position and momentum—the quantization scheme represents general functions of position, but only functions linear in the momentum.

The self-adjoint representation of a Lie algebra of vector fields on Γ suggests also a CUR of the group $\mathit{Diff}^{\,c}(\Gamma)$, a larger group than $\mathit{Diff}^{\,c}(M)$. When we began with $\mathit{Diff}^{\,c}(M)$, we were pleased to see a hierarchy of unitary representations emerge automatically, describing N-particle systems. But having fixed the configuration space from the outset, we no longer want "N-configuration" systems; only the "1-configuration" system. In quantum Borel kinematics the choice of representation is therefore restricted by means of a system of imprimitivity. A further restriction is imposed by requiring scalar-valued wave functions, ruling out parastatistics. Under these assumptions the quantization scheme is almost, but not quite, unique. The remaining possibilities are those that led to my subsequent work with H.-D. Doebner, where we proposed certain nonlinear quantum time-evolution equations [4; 29].

The approach that begins with local current algebra, or diffeomorphisms of the physical space, is, in a sense, *not* the quantization of a given classical system. Rather we obtain directly the class of possible quantum theories in a physical space, and for each one we find a configuration space interpretation. Classical physics on these configuration spaces is then recovered as a limiting case of the quantum theory. However, I would here like propose a point of view from which we may understand the local current algebra approach as tacitly embodying a quantization scheme.

Suppose we are given a classical configuration space Γ. The first step is to characterize a configuration $\gamma \in \Gamma$ as a (parameterized or unparameterized) embedding of a (discrete or continuous) set N into physical space M; i.e., where we have $\gamma : N \to M$. For any diffeomorphism $\phi \in Diff^c(M)$, define $\phi\gamma \in \Gamma$ as the composition of ϕ with the embedding γ. Then we have a natural, continuous homomorphism from $Diff^c(M)$ into $Diff^c(\Gamma)$. Correspondingly, a compactly supported (tangent) vector field on M defines a (tangent) vector field on Γ, so that we have a *distinguished class* of them. Thus, regarding $Diff^c(M)$ as a closed subgroup of $Diff^c(\Gamma)$, it is immediately clear that any CUR of the latter gives us a CUR of the former. But it is not necessary to start with the larger group. If we merely write a CUR of $Diff^c(M)$ in the space of square integrable functions on Γ, then we have obtained a quantization of the classical theory, with phase space $T^*(\Gamma)$, for which a corresponding distinguished class of functions on the phase space is represented by self-adjoint operators. This idea can even be generalized further, to realizing configurations in Γ as maps from N to tangent or cotangent bundles over M. The action of $Diff^c(M)$ lifted naturally to the bundle then gives us the group action on Γ.

1.7   Deriving Brackets for Creation and Annihilation Fields

The local current algebra given by Eqs. (1.2) was obtained originally by starting with canonical, nonrelativistic, second-quantized fields $\psi(\mathbf{x})$, $\psi^*(\mathbf{x})$ at fixed time t. We define operator-valued distributions $\rho(\mathbf{x}) = \psi^*(\mathbf{x})\psi(\mathbf{x})$ and $\mathbf{J}(\mathbf{x}) = -(i\hbar/2)\{\psi^*(\mathbf{x})\nabla\psi(\mathbf{x}) - [\nabla\psi^*(\mathbf{x})]\psi(\mathbf{x})\}$, construct $\rho(f) = \int f(\mathbf{x})\rho(\mathbf{x})d^3x$ and $J(\mathbf{g}) = \int \mathbf{g}(\mathbf{x}) \cdot \mathbf{J}(\mathbf{x})d^3x$, and derive the resulting Lie algebra of currents. The same algebra results whether one begins with fields satisfying canonical commutation or anticommutation relations. But supposing we take the local current algebra and group of diffeomorphisms of space to be fundamental, how do we then recover the creation and annihilation field operators?

In particular, we have seen that CURs of the diffeomorphism group are associated with configuration spaces of N particles (among other possibilities). These occur in the Fock representations of the canonical fields. But without fields, noting that there is more than one sort of N-particle representation (the particles may have different masses, or different statistics), how do we know when a series of representations fits together to form a hierarchy of the same sort of particle?

Abstracting from the N-particle subspaces of the Fock representations for canonical fields, the following conditions are proposed [30]. Suppose we have, for $N = 1, 2, 3, ...$, continuous unitary representations $U_N(f)V_N(\phi)$ of the semidirect product group $\mathcal{D} \times \textit{Diff}^{\,c}(\mathbf{R}^d)$ in Hilbert spaces $\mathcal{H}_N$. Let $\psi^*(\mathbf{x})$ be an operator-valued distribution modeled on a test-function space dense in $\mathcal{H}_1$. Write $\psi^*(h) = \int h(\mathbf{x})\psi^*(\mathbf{x})d^3x$, and suppose $\psi^*(h) : \mathcal{H}_N \to \mathcal{H}_{N+1}$. Thus $\mathcal{H}_1$ describes the type of object or primitive excitation in the hierarchy, and $\psi^*(h)$ will be interpreted as acting on an N-object state to create one more new object in state h. For the $U_N(f)V_N(\phi)$ to be a hierarchy, we now require

$$
\begin{aligned}
U_{N+1}(f)\psi^*(h) &= \psi^*(U_{N=1}(f)h)U_N(f)\,, \\
V_{N+1}(\phi)\psi^*(h) &= \psi^*(V_{N=1}(\phi)h)V_N(\phi)\,.
\end{aligned}
\tag{1.8}
$$

These conditions are extremely natural geometrically. The second equation states, in effect, that creating an excitation in a certain state and then deforming (in the space of $N + 1$ excitations) gives exactly the same result as deforming first (in the space of N excitations) and then creating an excitation in a state that is deformed (the latter taking place in the space of 1 excitation).

Eqns. (1.8) correspond to the operator commutator brackets

$$
\begin{aligned}
[\,\rho(f), \psi^*(h)\,] &= \psi^*(\rho_{N=1}(f)h)\,, \\
[\,J(\mathbf{g}), \psi^*(h)\,] &= \psi^*(J_{N=1}(\mathbf{g})h)\,.
\end{aligned}
\tag{1.9}
$$

The equations for ψ are obtained from (1.8) or (1.9) by taking the adjoint. Thus we have commutator brackets among the currents, and commutator brackets between the currents and the fields.

But the bracket obeyed by the field operators with each other now depends on (and may be discovered from) the concrete choice of representations of the diffeomorphism group to make up the hierarchy! In particular, the fact that second-quantized bosonic fields satisfy commutation relations, and fermionic fields satisfy anticommutation relations, is a *consequence* of the framework. Perhaps more surprisingly, in two-dimensional space, the field operators intertwining the *anyonic* representations of the diffeomorphism group by (1.8) or (1.9) satisfy q-commutation relations. These are not put in "by hand" but are a consequence of the framework, though only ordinary commutator brackets are assumed initially.

1.8 Infinite Dimensional Configuration Spaces

Let me close the discussion with brief mention of ongoing work with infinite-dimensional configuration spaces. One nice thing about the diffeomorphism group approach is that it provides a way to connect quantum and statistical physics, as well as a way to set up the quantum theory of extend objects in physical space.

The configuration space of infinite but locally finite subsets of $\mathbf{R}^d$ is well known in both quantum and statistical physics (see for example [31] for a discussion of local current algebra). Poisson measures and Gibbs measures describing equilibrium states on this space have the desired property of quasi-invariance, providing unitary representations of $Diff^{\,c}(\mathbf{R}^d)$.

One important generalization is to infinite configurations of particles having internal structure (marked configurations) [32]. If the "mark space" is taken to be a manifold S (for example, S may be S^2, a compact Lie group, or a locally compact homogeneous space for a group action), we consider $\mathbf{R}^d \times S$ as a bundle over $\mathbf{R}^d$. The relevant diffeomorphism group is then the group of bundle diffeomorphisms compactly supported in $\mathbf{R}^d$.

A second generalization is to countably infinite configurations with accumulation points in physical space. Here quasi-invariant measures can be constructed from self-similar random processes (interestingly and fairly straightforwardly for $d = 1$), leading again to unitary representations of the group of compactly supported diffeomorphisms. For $d = 1$, one has not only "particle cloud" configurations with isolated accumulation points, but also fractal configurations (random Cantor sets). The generalization to $d > 1$ posed some difficulties, but the construction of measures quasi-invariant under diffeomorphisms through self-similarity has been recently accomplished. [33; 34].

Finally there is a substantial body of work descriptive of excitations having spatial extent—manifolds embedded in physical space. Examples include arcs, loops, or ribbons of vorticity [35], and closed or open strings and membranes. We see that the group of diffeomorphisms of physical space, and related infinite-dimensional groups, provide a unified way to understand a highly diverse set of quantum-mechanical situations.

Bibliography

[1] Goldin, G. A., Menikoff, R. and Sharp, D. H. (1983). Diffeomorphism groups, gauge groups, and quantum theory. *Phys. Rev. Lett.* **51**, 2246-2249.

[2] Goldin, G. A. and Sharp, D. H. (1989). Diffeomorphism groups and local symmetries: Some applications in quantum physics. In Gruber, B. and Iachello, F. (eds.), *Symmetries in Science III.* New York: Plenum, 181-205.

[3] Goldin, G. A. and Sharp, D. H. (1991). The diffeomorphism group approach to anyons. *Int'l. J. Mod. Phys.* B **5**, 2625-2640.

[4] Goldin, G. A. (1992). The diffeomorphism group approach to nonlinear quantum systems. *Int'l. J. Mod. Phys.* B **6**, 1905-1916.

[5] Choquet-Bruhat, Y., DeWitt-Morette, C., with Dillard-Bleick, M. (1982). *Analysis, Manifolds and Physics, Part I* (revised edition). Amsterdam: Elsevier (North Holland).

[6] Isham, C. J. (1984). Topological and global aspects of quantum theory. In DeWitt, B. and Stora, R. (eds.), *Relativity, Groups and Topology II, Procs. 1983 Les Houches Summer School.* Amsterdam: North Holland, 1059-1290.

[7] Dashen, R. and Sharp, D. H. (1968). Currents as coordinates for hadron dynamics. *Phys. Rev.***165**, 1857.

[8] Goldin, G. A. and Sharp, D. H. (1970). Lie algebras of local currents and their representations. In Bargmann, V. (ed.), *Group Representations in Mathematics and Physics: Battelle-Seattle 1969 Rencontres.* Lecture Notes in Physics **6**, Berlin: Springer, 300-310.

[9] Goldin, G. A. (1971). Non-relativistic current algebras as unitary representations of groups. *J. Math. Phys.* **12**, 462-487.

[10] Mackey, G. W. (1968). *Induced Representations of Groups and Quantum Mechanics.* New York: W. A. Benjamin.

[11] Klink, W. (personal discussions and communications).

[12] Alexandrov, A. D. (1975). Mappings of spaces with families of cones and space-time transformations. *Annali di Matematica* (Bologna) **103**, 229-257.

[13] Leinaas, J. M. and Myrheim, J. (1977). On the theory of identical particles. *Nuovo Cimento* **37B**, 1.

[14] Goldin, G. A., Menikoff, R. and Sharp, D. H. (1980). Particle statistics from induced representations of a local current group. *J. Math. Phys.* **21**, 650-664. — (1981). Representations of a local current algebra in non-simply connected space and the Aharonov-Bohm effect. *J. Math. Phys.* **22**, 1664-1668.

[15] Wilczek, F. (1982). *Phys. Rev. Lett.* **48**, 1144.

[16] Goldin, G. A. and Sharp, D. H. (1983). Rotation generators in two-dimensional space and particles obeying unusual statistics. *Phys. Rev. D* **28**, 830-832.

[17] Goldin, G. A., Menikoff, R. and Sharp, D. H. (1985). Comment on 'General theory for quantum statistics in two dimensions'. *Phys. Rev. Lett.* **54**, 603.

[18] Goldin, G. A. (1987). Parastatistics, θ-statistics, and topological quantum mechanics from unitary representations of diffeomorphism groups. In Doebner, H.-D. and Hennig, J.-D. (eds.), *Procs. of the XV. Int'l. Conf. on Differential Geometric Methods in Theoretical Physics*. Singapore: World Scientific, 197-207.

[19] Wilczek, F. (1990). *Fractional Statistics and Anyon Superconductivity*. Singapore: World Scientific.

[20] Khare, A. (1997). *Fractional Statistics and Quantum Theory*. Singapore: World Scientific.

[21] Berry, M. V. and Robbins, J. M. (1997). Indistinguishability for quantum particles: spin, statistics and the geometric phase. *Proc. R. Soc. Lond.* A **453**, 1771-1790.

[22] Atiyah, M. (2001). Configurations of points. *Phil. Trans. R. Soc. Lond.* A **359**, 1375-1387.

[23] Atiyah, M. and Bielawski, R. (2002). Nahm's equations, configuration spaces and flag manifolds. *Bull. Braz. Math. Soc., New Series* **33**(2), 157-176.

[24] Goldin, G. A., Grodnik, J., Powers, R. T. and Sharp, D. H. (1974). Nonrelativistic current algebra in the 'N/V' limit. *J. Math. Phys.* **15**, 88-100.

[25] Ismagilov, R. S. (1996). *Representations of Infinite-Dimensional Groups*. Translations of Mathematical Monographs, v. 152. Providence, RI: Am. Math. Soc.

[26] Segal, I. E. (1960). Quantization of nonlinear systems. *J. Math. Phys.* **1**, 468-488.

[27] Doebner, H.-D. and Tolar, J. (1980). In Gruber, B. and Millman, R. S. (eds.), *Symposium on Symmetries in Science*. New York: Plenum, 475.

[28] Ali, S. T. (2002). Quantization techniques: A quick overview. In Govaerts, J., Hounkonnou, M. N. and Msezane, A. Z. (eds.), *Procs. of the Second Int'l. Workshop on Contemporary Problems in Mathematical Physics, Cotonou, Benin*. Singapore: World Scientific.

[29] Doebner, H.-D. and Goldin, G. A. (1992). On a general nonlinear Schrödinger equation admitting diffusion currents. *Physics Letters* A **162**, 397-401. — (1996). Introducing nonlinear gauge transformations in a family of nonlinear Schrödinger equations. *Phys. Rev.* A **54**, 3764-3771.

[30] Goldin, G. A. and Sharp, D. H. (1996). Diffeomorphism groups, anyon fields, and q-commutators. *Phys. Rev. Lett.* **76**, 1183-1187.

[31] Albeverio, S., Kondratiev, Yu. G. and Röckner, M. (1998). Analysis and geometry on configuration spaces. *J. Func. Anal.* **154**, 444-500 and bf 157, 242-291. — (1999). Diffeomorphism groups and current algebras: configuration space analysis in quantum theory. *Rev. Math. Phys.* **11**, 1-23.

[32] Goldin, G. A., Kondratiev, Yu. G., Kuna, T. and Silva, J. L. (research in preparation).

[33] Goldin, G. A. and Moschella, U. (1996). Quantum phase transitions from a new class of representations of a diffeomorphism group. *J. Phys. A: Math. Gen.* **28**, L475-L481. —. (2001). Random Cantor sets and measures quasi-invariant for diffeomorphism groups. In Schlichenmaier, M., Strasburger,

A., Ali, S. T. and Odzijewicz, A. (eds.), *Coherent States, Quantization and Gravity: Procs. of the XVII Workshop on Geometric Methods in Physics, Białowieża, July 3-9, 1998.* Warsaw: Wydawnictwa Uniwersytetu Warszawskiego, 263-272.

[34] Sakuraba, T. (2002). Unpublished Ph.D. thesis, Rutgers University Department of Mathematics. Goldin, G. A., Moschella, U. and Sakuraba, T. (research in preparation).

[35] Goldin, G. A., Owczarek, R. and Sharp, D. H. (2003), Quantum kinematics of bosonic vortex loops. Los Alamos National Laboratory preprint (to be submitted for publication).

Chapter 2

Functorial Quantization and the Guillemin–Sternberg Conjecture

N. P. Landsman[1]

Dedicated to Alan Weinstein, at his 60th birthday

Abstract: We propose that geometric quantization of symplectic manifolds is the arrow part of a functor, whose object part is deformation quantization of Poisson manifolds. The 'quantization commutes with reduction' conjecture of Guillemin and Sternberg then becomes a special case of the functoriality of quantization. In fact, our formulation yields almost unlimited generalizations of the Guillemin–Sternberg conjecture, extending it, for example, to arbitrary Lie groups or even Lie groupoids. Technically, this involves symplectic reduction and Weinstein's dual pairs on the classical side, and Kasparov's bivariant K-theory for C^*-algebras (KK-theory) on the quantum side.

2.1   Introduction

The theory of constraints and reduction in mechanics and field theory is important for physics, because the fundamental theories describing Nature (viz. electrodynamics, Yang–Mills theory, general relativity, and possibly also string theory) are a priori formulated as constrained systems (cf. [Sundermeyer (1982)]). The systematic investigation of classical constrained

[1] Korteweg–De Vries Institute for Mathematics, University of Amsterdam, Plantage Muidergracht 24, 1018 TV Amsterdam, The Netherlands, `npl@science.uva.nl`, `http://remote.science.uva.nl/~npl/`

systems was initiated by Dirac, whose ideas were reformulated mathematically as the theory of symplectic reduction (see, e.g., [Binz, Śniatycki and Fischer (1988); Landsman (1998)]). The procedure known as Marsden–Weinstein reduction [Meyer (1973); Marsden and Weinstein (1974); Abraham and Marsden (1985)] is a special case of this theory, which is easy to formulate, yet very rich in mathematical and physical applications. According to this procedure, a suitable action $G \circlearrowright M$ of a Lie group G on a symplectic manifold M produces another symplectic manifold, the reduced space M^0, which is a certain subspace of M/G. See [Marsden and Weinstein (2001)] for a recent overview.

In general, the traditional idea of quantization has always been that a phase space, i.e., a symplectic space M, should be quantized by a Hilbert space $H(M)$, and that the classical observables, viz. the (real-valued) smooth functions on M should be quantized by (self-adjoint) operators on H, which after all play the role of observables in quantum theory. What is the relationship between the quantization of M and the quantization of the reduced space M^0?

The quantization of constrained systems was first analyzed in a general setting in [Dirac (1964)], but there still exists no complete and satisfactory mathematical theory. Given some notion of quantization Q, the basic problem in such a theory would be to formulate a possible quantum analogue of the classical reduction procedure R, and compare the result of applying this procedure to the quantization of the unconstrained classical systems with the quantization of the classically reduced system. One would then hope that the order of quantization and reduction does not matter; this hope is symbolically expressed by '$[Q, R] = 0$.' This 'quantization commutes with reduction' principle can be turned into a mathematical conjecture once a precise meaning has been assigned to the operations Q and R. See [Guillemin, Ginzburg and Karshon (2002)] for a survey of the literature on this problem in the context of geometric quantization, and cf. [Landsman (1998)] for references (pre 1998) on other approaches.

In the context of (what we now call) Marsden–Weinstein reduction, Dirac proposed that the G action on M should be quantized by a unitary representation U of G on $H(M)$, while the so-called weak observables act on $H(M)$ by operators commuting with $U(G)$. The quantized reduction operation RQ then consists in taking the G invariant part $H(M)^G$ of $H(M)$, on which the weak observables then act by restriction. This idea makes rigorous mathematical sense in general only when G is compact. If, in addition, M is compact, one expects $H(M)$ to be finite-dimensional, and similarly for $H(M^0)$, so that the weakest possible form of the '$[Q, R] = 0$' conjecture, in which the action of observables is ignored, would be

$$H(M)^G \cong H(M^0). \qquad (2.1)$$

Here the $\cong$ sign stands for unitary isomorphism, and since the dimension is the only such invariant of a Hilbert space, one really is talking about a simple equality between numbers, i.e., $\dim(H(M)^G) = \dim(H(M^0))$.

Despite various refinements [Guillemin, Ginzburg and Karshon (2002)], some of which will be discussed below, (2.1) is basically the form in which the conjecture has been studied in the mathematical literature. This literature started with the seminal paper [Guillemin and Sternberg (1982)], after which the conjecture in any form resembling (2.1) is usually named. Using geometric quantization, they proved (2.1) under certain assumptions, among which the compactness of M and G are crucial. It is hard to think of a more favourable situation for quantization theory then the one assumed in [Guillemin and Sternberg (1982)]. Partly in order to generalize the Guillemin–Sternberg conjecture, in the mid-1990s a novel notion of quantization came up, which seems to incorporate all good features of geometric quantization whilst circumventing a number of its pitfalls; see [Sjamaar (1996); Guillemin, Ginzburg and Karshon (2002)], and references therein. This definition of quantization is sometimes attributed to Raoul Bott.

In this approach, quantization is simply defined as the index of a suitable Dirac operator $\slashed{D}$ naturally associated to M; when M carries a G action $G \circlearrowright M$, this index is understood in the equivariant sense, so that the quantization of M, or rather of $G \circlearrowright M$, is an element of the representation ring $R(G)$ of G. The quantization of the reduced space M^0 remains an integer. Taking the G invariant part of a representation induces a map $R(G) \to \mathbb{Z}$, in terms of which the Guillemin–Sternberg conjecture can then be stated in a very elegant form. In that form, it was proved in [Meinrenken (1998); Meinrenken and Sjamaar (1999)]; also see [Guillemin, Ginzburg and Karshon (2002); Paradan (2001a)] for other proofs and further references.

These ideas still only apply to the situation where M and G are compact (though cf. [Paradan (2001b)] for a special case where at least M is noncompact), which is highly undesirable for applications to both physics and mathematics. Furthermore, it would be welcome to have some direct motivation for the notion of quantization as an (equivariant) index, and if possible also to incorporate some extra structure. For example, when no G action is around, Bott's definition of quantization merely produces a number, and the entire idea of quantizing functions on M by operators is lost.

These problems can be addressed by combining geometric quantization with deformation quantization. In the latter, a Poisson manifold is quantized by an associative algebra, subject to a number of conditions. In the 'formal' setting, this should be an algebra over the commutative ring $\mathbb{C}[[\hbar]]$ of formal power series in one real variable [Bayen et al. (1978)], whereas in the 'strict' setting this should be a C^*-algebra over the commutative C^*-algebra $C(I)$ of continuous functions on the interval $I = [0, 1]$ [Landsman (2002)]. As in the entire context of relating classical to quantum mechanics [Landsman (1998)], the language of C^*-algebras is particularly attractive here. For our present purposes, it is sufficient to work with ordinary C^*-algebras (instead of C^*-algebras over $C(I)$); this amounts to quantizing at a fixed value of $\hbar$, as is usual also in geometric quantization. This simplification entails the need to impose prequantizability conditions on the symplectic manifolds in question.

In [Landsman (2002)] we proposed that quantization should be seen as a functor between categories whose arrows are equivalence classes of bimodules. What this means is rather different in the classical and in the quantum case [Landsman (2001); Landsman (2002)]. In the former, the arrows between Poisson manifolds are isomorphism classes of symplectic dual pairs [Karasev (1989); Weinstein (1983)]. In the latter, the arrows between (separable) C^*-algebras are homotopy classes of Kasparov bimodules [Kasparov (1981)]. Such bimodules are generalized Hilbert spaces equipped with a generalized Fredholm operator, such as (a bounded version of) a Dirac operator $\not{D}$.

In other words, quantization should map (isomorphism classes of) symplectic dual pairs into (homotopy classes of) Kasparov bimodules. More precisely, if Poisson manifolds P_1 and P_2 are quantized by (separable) C^*-algebras $Q(P_1)$ and $Q(P_2)$, respectively, then a symplectic dual pair $P_1 \leftarrow M \rightarrow P_2$ should be quantized by an element of the Kasparov group $KK(Q(P_1), Q(P_2))$. In the special case of a symplectic dual pair $pt \leftarrow M \rightarrow pt$, quantization should therefore produce an element of $KK\mathbb{C}, \mathbb{C}) \cong \mathbb{Z}$, i.e., an integer. This is precisely what Bott's index-theoretic definition of quantization does.

In [Landsman (2002)] we had no idea what the quantization functor should look like, and therefore missed the connection between the envisaged functoriality of quantization and the Guillemin–Sternberg conjecture. We now propose that a suitable generalization of Bott's definition of quantization will do the job, and will check that the Guillemin–Sternberg conjecture is actually a special case of functoriality. Conversely, requiring the functoriality of quantization on suitable symplectic dual pairs leads to almost unlimited generalizations of the Guillemin–Sternberg conjecture. For example, one can now remove the restriction that M and G have to be compact, in which case the quantization functor constructs the quantization of a canonical G action $G \circlearrowright M$ as a generalized equivariant index as defined in the K-theory of group C^*-algebras [Connes (1994)]. This relates the Guillemin–Sternberg conjecture to the Baum–Connes conjecture in noncommutative geometry [Baum, Connes and Higson (1994); Connes (1994)], in which it is postulated that the K-theory of a group C^*-algebra is exhausted by such indices. Moreover, techniques that have been developed in the context of the Baum–Connes conjecture [Connes (1994); Le Gall (1999); Paterson (2002)] enable one to state a generalized Guillemin–Sternberg conjecture even for Lie groupoid actions. Finally, our approach incorporates and illuminates the use of shriek maps in K-theory [Atiyah and Singer (1968); Connes (1982); Connes and Skandalis (1984); Hilsum and Skandalis (1987); Connes (1994)], whose functoriality turns out to be a special case of the functoriality of quantization.

Since this paper relates two different areas of mathematics, we have tried to make it largely self-contained. Following a brief review of classical reduction, we recall the idea of looking at symplectic dual pairs as arrows between Poisson manifolds. We then review the Guillemin–Sternberg conjecture in its original form, and subsequently, following a recapitulation of Spin^c structures and Dirac operators, in its modern form based on Bott's definition of quantization. We then explain how the quantization context naturally leads to KK-theory, including the idea of interpreting homotopy classes of Kasparov bimodules as arrows between C^*-algebras. We then show that the Guillemin–Sternberg conjecture is a special case of the functoriality of quantization. In the final two sections we consider generalizations of the Guillemin–Sternberg conjecture by relating quantization to K-homology and to foliation theory, respectively.

2.2   Classical Reduction

A Poisson manifold M is a manifold equipped with a Lie bracket $\{\,,\,\}$ on $C^\infty(M)$ with the property that for each $f \in C^\infty(M)$ the map $g \mapsto$

$\{f, g\}$ defines a derivation of the commutative algebra structure of $C^\infty(M)$ given by pointwise multiplication. Hence this map is given by a vector field ξ_f, called the Hamiltonian vector field of f. Symplectic manifolds are special instances of Poisson manifolds, characterized by the property that the Hamiltonian vector fields exhaust the tangent bundle. In that case, the Poisson bracket comes from a symplectic form ω on M in the usual way [Abraham and Marsden (1985)].

Suppose a Lie algebra $\mathfrak{g}$ acts on a Poisson manifold M in strongly Hamiltonian fashion. This means that there exist Lie algebra homomorphisms $X \mapsto X^M$ from $\mathfrak{g}$ to the space $\Gamma(M, TM)$ of vector fields on M and $X \mapsto J_X$ from $\mathfrak{g}$ to $C^\infty(M)$, with the property $X^M = \xi_{J_X}$. The functions J_X may be assembled into a so-called momentum map $J : S \to \mathfrak{g}^*$, defined by $\langle J(\sigma), X \rangle = J_X(\sigma)$. Here $\mathfrak{g}^*$ is the dual vector space of the Lie algebra $\mathfrak{g}$. This $\mathfrak{g}^*$ is canonically a Poisson manifold under the Lie–Poisson bracket, defined on linear functions (hence elements of $\mathfrak{g}^{**} = \mathfrak{g}$) by the Lie bracket. It follows that J is a Poisson map. Note that a smooth map between two Poisson manifolds is called Poisson when its pullback is a Lie algebra homomorphism (and anti-Poisson when it is an anti homomorphism). It may happen that the $\mathfrak{g}$ action comes from a G action $G \circlearrowright M$, where G is a Lie group with Lie algebra $\mathfrak{g}$: in that case, one has $X^M f(\sigma) = df(\exp(-tX)\sigma)/dt|t = 0$. The G action is called strongly Hamiltonian whenever the associated $\mathfrak{g}$ action is.

We now specialize to the case where M is symplectic. The symplectic quotient or reduced space defined by the G action, or physically by the constraint $J = 0$, is $M^0 = J^{-1}(0)/G$. In case that 0 is a regular value of J and the G action is proper and free on $J^{-1}(0)$, M^0 is a manifold, which moreover carries a unique symplectic form ω^0 with the property $i^*\omega = \pi^*\omega^0$. Here $i : J^{-1}(0) \hookrightarrow M$ is the inclusion and $\pi : J^{-1}(0) \to M^0$ is the projection map. Thus Marsden–Weinstein reduction produces a new symplectic manifold (M^0, ω^0) from a given symplectic manifold (M, ω) equipped with a strongly Hamiltonian G action [Meyer (1973); Marsden and Weinstein (1974); Abraham and Marsden (1985); Marsden and Weinstein (2001)]. If the stated assumptions are not met, singularities may arise in the reduced space (cf. [Sjamaar and Lerman (1991); Landsman, Pflaum and Schlichenmaier (2001); Pflaum (2001)]).

2.3 Symplectic Dual Pairs as Arrows

On the classical side, a bimodule over a pair P, Q of Poisson manifolds is by definition a so-called symplectic dual pair [Karasev (1989); Weinstein (1983)] $Q \leftarrow M \rightarrow P$, simply called a dual pair in what follows. Here M is a symplectic manifold, the map $Q \leftarrow M$ is Poisson, and $M \rightarrow P$ is anti-Poisson. Furthermore, the pullback of any function on P should Poisson-commute on M with the pullback of any function on Q. One of the motivating examples of a dual pair is $G\backslash M \leftarrow M \xrightarrow{J} \mathfrak{g}^*_-$, obtained from a strongly Hamiltonian G action on M with momentum map J.[2] Similarly, $\mathfrak{g}^*_- \leftarrow M^- \rightarrow G\backslash M$ is a dual pair.

Two Q-P dual pairs $Q \xleftarrow{q_i} \tilde{M}_i \xrightarrow{p_i} P$, $i = 1, 2$, are said to be isomorphic when there is a symplectomorphism $\varphi : \tilde{M}_1 \rightarrow \tilde{M}_2$ for which $q_2\varphi = q_1$ and $p_2\varphi = p_1$. We now interpret the equivalence class of a dual pair $Q \leftarrow M \rightarrow P$ as an arrow from Q to P. Two compatible dual pairs $Q \leftarrow M_1 \rightarrow P$ and $P \leftarrow M_2 \rightarrow R$ can be composed when firstly $M_1 \times_P M_2$ is a coisotropic submanifold of $M_1 \times M_2$, and secondly the associated symplectic quotient of $M_1 \times_P M_2$ by its canonical foliation is a manifold. We then denote the product of the dual pairs in question by $P \leftarrow M_1 \circledcirc_P M_2 \rightarrow R$. This product is well defined on equivalence classes, where it is associative, since the operation $\circledcirc$ is associative up to isomorphism. For example, when G is connected the product of the dual pairs $G\backslash M \leftarrow M \rightarrow \mathfrak{g}^*_-$ and $\mathfrak{g}^*_- \hookleftarrow 0 \rightarrow pt$ is $G\backslash M \leftarrow M^0 \rightarrow pt$, where M^0 is the Marsden–Weinstein quotient $J^{-1}(0)/G$ as before.

As explained in [Landsman (2001)] (see also [Bursztyn and Weinstein (2003)]), one can impose certain regularity conditions on both Poisson manifolds and dual pairs, which guarantee that all products exist and that one has identity arrows from P to P. Thus one obtains a category **Poisson** whose objects are (regular) Poisson manifolds and whose arrows are equivalence classes of (regular) dual pairs. The regularity condition on Poisson manifolds is very mild, and it is actually quite hard to construct an example that fails to satisfy it. On the other hand, many dual pairs one would like to use are not regular, such as $pt \leftarrow M \rightarrow pt$, where pt is the space consisting of a point. Also, although $G\backslash M \leftarrow M \rightarrow \mathfrak{g}^*_-$ is regular, $pt \leftarrow M \rightarrow \mathfrak{g}^*_-$ is not. Nonetheless, the product of $pt \leftarrow M \rightarrow \mathfrak{g}^*$ and $\mathfrak{g}^* \hookleftarrow 0 \rightarrow pt$ is well

---

[2]In general, we write P^- for a Poisson manifold P equipped with minus its Poisson bracket, but we write $\mathfrak{g}^*_-$ for $(\mathfrak{g}^*)^-$.

defined, and equal to

$$(pt \leftarrow M \rightarrow \mathfrak{g}_-^*) \circledcirc_{\mathfrak{g}_-^*} (\mathfrak{g}_-^* \hookleftarrow 0 \rightarrow pt) \cong pt \leftarrow M^0 \rightarrow pt. \qquad (2.2)$$

Another example is the dual pair $X \xleftarrow{\pi} T^*X \xrightarrow{f \circ \pi} Y$ defined by a smooth map $X \xrightarrow{f} Y$. Here X and Y are manifolds with zero Poisson bracket, f is smooth, and T^*X has the canonical Poisson structure. The product of $X \leftarrow T^*X \rightarrow Y$ with the dual pair $Y \leftarrow T^*Y \rightarrow Z$ induced by $Y \xrightarrow{g} Z$ is

$$(X \leftarrow T^*X \rightarrow Y) \circledcirc_Y (Y \leftarrow T^*Y \rightarrow Z) \cong X \leftarrow T^*X \xrightarrow{g \circ f \circ \pi} Z. \qquad (2.3)$$

Note that the dual pairs defined by a G action $G \circlearrowright M$ and by a map $X \xrightarrow{f} Y$ are both special cases of a very general functorial construction involving Lie groupoids [Landsman (2000)]. Such examples indicate that products of dual pairs lying in a certain class often make sense when the regularity condition is not satisfied. Thus in the present paper we shall not impose the regularity conditions on dual pairs, refraining from a complete categorical structure. It will still be possible to map arrows of the above type into arrows in the category KK defined below, and to check functoriality of this map, interpreted as quantization, with respect to the product $\circledcirc$. It is in this rather pragmatic sense that the notion of functoriality will be understood in what follows.

2.4 The Guillemin–Sternberg Conjecture

Guillemin and Sternberg [Guillemin and Sternberg (1982)] considered the case in which the symplectic manifold M is compact, prequantizable, and equipped with a positive-definite complex polarization $\mathcal{J}$. Recall that a symplectic manifold (M, ω) is called prequantizable when the cohomology class $[\omega]/2\pi$ in $H^2(M, \mathbb{R})$ is integral, i.e., lies in the image of $H^2(M, \mathbb{Z})$ under the natural homomorphism $H^2(M, \mathbb{Z}) \rightarrow H^2(M, \mathbb{R})$. In that case, there exists a line bundle L_ω over M whose first Chern class $c_1(L_\omega)$ maps to $[\omega]/2\pi$ under this homomorphism; L_ω is called the prequantization line bundle over M. In general, this bundle is not unique.

Under these circumstances, the quantization operation Q is well-defined through geometric quantization [Guillemin, Ginzburg and Karshon (2002)]: one picks a connection ∇ on L_ω whose curvature is ω, and defines the Hilbert space $H(M)$ as the space $H = H^0(M, L_\omega)$ of polarized sections of L_ω (i.e., of sections annihilated by all ∇_X, $X \in \mathcal{J}$).

Now suppose that M carries a strongly Hamiltonian action $G \circlearrowleft M$ of a compact Lie group G that leaves $\mathcal{J}$ invariant. The Hilbert space $H(M)$ then carries a natural unitary representation of G determined by the classical data, as polarized sections of L_ω are mapped into each other by the pullback of the G action. Moreover, it turns out that the reduced space M^0 inherits all relevant structures on M (except, of course, the G action), so that it is quantizable as well, in the same fashion. Thus (2.1) becomes, in obvious notation, $H^0(M, L_\omega)^G \cong H^0(M^0, L_\omega^0)$, which Guillemin and Sternberg indeed managed to prove. The idea of the proof is to define a map from $H^0(M, L_\omega)^G$ to $H^0(M^0, L_\omega^0)$ by simply restricting a G invariant polarized section of L_ω to $J^{-1}(0)$; this map is then shown to be an isomorphism [Guillemin and Sternberg (1982)].

2.5 Spin^c Structures and Dirac Operators

The new approach to geometric quantization mentioned in the Introduction is based on the notion of a Spin^c structure on M, which we briefly recall.[3] A large number of approaches to Spin^c structures exist, of which the ones relating this concept to K-theory [Atiyah, Bott and Shapiro (1964); Lawson and Michelsohn (1989)], to K-homology [Baum and Douglas (1982); Higson and Roe (2000)], to KK-theory [Connes and Skandalis (1984)], to E-theory [Connes (1994)] (all these approaches are, in turn, closely linked to index theory), and to Morita equivalence of C^*-algebras [Plymen (1986); Gracia-Bondía, Várilly and Figueroa (2001)] are particularly relevant to our theme. We will return to some of these in due course, but for the moment a purely differential-geometric approach is appropriate [Duistermaat (1996); Guillemin, Ginzburg and Karshon (2002)].

Firstly, the compact Lie group $\mathrm{Spin}^c(n)$ is a nontrivial central extension of $SO(n)$ by $U(1)$, defined as $\mathrm{Spin}^c(n) = \mathrm{Spin}(n) \times_{Z_2} U(1)$, where $\mathrm{Spin}(n)$ is the usual twofold cover of $SO(n)$, and Z_2 is seen as the subgroup $\{(1,1), (-1,-1)\}$ of $\mathrm{Spin}(n) \times U(1)$. Thus one has the obvious homomorphisms $\pi : \mathrm{Spin}^c(n) \to SO(n) \cong \mathrm{Spin}(n)/Z_2$, given by projection on the first factor, and $\det : \mathrm{Spin}^c(n) \to U(1)$, defined by $[x, z] \mapsto z^2$.

Let $n = \dim(M)$. A Spin^c structure $(P, \cong)$ on M is by definition a principal $\mathrm{Spin}^c(n)$-bundle P over M with an isomorphism $P \times_\pi \mathbb{R}^n \cong TM$ of vector bundles. Here the bundle on the left-hand side is the bundle as-

---

[3]Such a structure may more generally be defined on a real vector bundle E over M; when E is the tangent bundle TM we obtain the special case discussed in the main text.

sociated to P by the defining representation of $SO(n)$. Various structures on M canonically induce a Spin^c structure on M, such as a Spin structure or an almost complex structure. Note that a Spin^c structure on M, when it exists, is not unique: up to homotopy, the class of possible Spin^c structures on M (with given orientation) is parametrized by the Picard group $H^2(M,\mathbb{Z})$ [Guillemin, Ginzburg and Karshon (2002)].

A Spin^c structure defines a number of vector bundles over M associated to P by various representations of $\mathrm{Spin}^c(n)$. The first of these, which is isomorphic to the bundle TM, has just been mentioned.[4] The second is the canonical line bundle $L = P \times_{\det} \mathbb{C}$ associated to P by the defining representation of $U(1)$. Thirdly, $\mathrm{Spin}^c(n)$ has a canonical unitary representation Δ_n on a finite-dimensional Hilbert space S, the so-called (complex) spin representation, which for odd n is irreducible, and for even n decomposes into two irreducibles $\Delta_n = \Delta_n^+ \oplus \Delta_n^-$ on $S = S^+ \oplus S^-$. Thus one has an associated spinor bundle $\mathcal{S} = P \times_{\Delta_n} S$, which for even n decomposes into the direct sum $\mathcal{S}^\pm = P \times_{\Delta_n} S^\pm$. Thus the physical interpretation of Spin^c structures involves gravity, electromagnetism, and fermions.

A Spin^c structure on M defines a vector bundle action $TM \to \mathrm{End}(\mathcal{S})$ by Clifford multiplication, since both TM and $\mathcal{S}$ are subspaces of the Clifford bundle $Cl(TM)$ over M. This action may be seen as a map $c : \Gamma(TM \otimes_M \mathcal{S}) \to \Gamma(\mathcal{S})$. Furthermore, a connection on P induces one on $\mathcal{S}$, which amounts to a covariant derivative $\nabla : \Gamma(\mathcal{S}) \to \Gamma(T^*M \otimes_M \mathcal{S})$. Identifying T^*M with TM through the Riemannian metric g determined by the Spin^c structure and composing these maps yields the Dirac operator

$$\slashed{D} : \Gamma(\mathcal{S}) \xrightarrow{\nabla} \Gamma(T^*M \otimes_M \mathcal{S}) \xrightarrow{g \otimes \mathrm{id}} \Gamma(TM \otimes_M \mathcal{S}) \xrightarrow{c} \Gamma(\mathcal{S}).$$

This elliptic first-order linear differential operator is formally self-adjoint, and can be turned into a bounded self-adjoint operator $\tilde{\slashed{D}} = \slashed{D}/\sqrt{1 + \slashed{D}^* \slashed{D}} : L^2(\mathcal{S}) \to L^2(\mathcal{S})$, where $L^2(\mathcal{S})$ stands for the Hilbert space of L^2-sections of the vector bundle $\mathcal{S}$. When M is even-dimensional, $\slashed{D}$ is odd with respect to the decomposition $\mathcal{S} = \mathcal{S}^+ \oplus \mathcal{S}^-$, so that one obtains the chiral Dirac operator $\slashed{D}^+ : \Gamma(\mathcal{S}^+) \to \Gamma(\mathcal{S}^-)$, with formal adjoint $\slashed{D}^- : \Gamma(\mathcal{S}^-) \to \Gamma(\mathcal{S}^+)$, by restriction. Similarly, one has $\tilde{\slashed{D}}^\pm : L^2(\mathcal{S}^\pm) \to L^2(\mathcal{S}^\mp)$.

---

[4]It induces both an orientation and a Riemannian metric on M, by transferring the standard orientation and metric on $\mathbb{R}^n$ to E. Conversely, given an orientation and a Riemannian metric on M, one should require a Spin^c structure on M to be compatible with these.

2.6   Bott's Definition of Quantization

The first step in Bott's definition of quantization is to canonically associate a Spin^c structure to a given symplectic and prequantizable manifold (M, ω) [Guillemin, Ginzburg and Karshon (2002); Meinrenken (1998)]. First, one picks an almost complex structure $\mathcal{J}$ on M that is compatible with ω (in that $\omega(-, \mathcal{J}-)$ is positive definite and symmetric, i.e., a metric). This J canonically induces a Spin^c structure $P_{\mathcal{J}}$ on TM [Duistermaat (1996); Guillemin, Ginzburg and Karshon (2002)], but this is not the right one to use here. The Spin^c structure P needed to quantize M is the one obtained by twisting $P_{\mathcal{J}}$ with the prequantization line bundle L_ω. This means (cf. [Guillemin, Ginzburg and Karshon (2002)], App. D.2.7) that $P = P_{\mathcal{J}} \times_{\ker(\pi)} U(L_\omega)$, where $\pi : \mathrm{Spin}^c(n) \to SO(n)$ was defined in the preceding section (note that $\ker(\pi) \cong U(1)$), and $U(L_\omega) \subset L_\omega$ is the unit circle bundle.[5]

When M is compact, the operators $\tilde{\slashed{D}}^{\pm}$ determined by the Spin^c structure $(P, \cong)$ have finite-dimensional kernels, whose dimensions define the quantization of (M, ω) as

$$Q(M, \omega) = \mathrm{index}(\slashed{D}^{+}) = \dim \ker(\slashed{D}^{+}) - \dim \ker(\slashed{D}^{-}). \qquad (2.4)$$

In fact, the corresponding Hilbert space operators $\tilde{\slashed{D}}^{\pm}$ are Fredholm, and by elliptic regularity $\mathrm{index}(\slashed{D}^{+})$ coincides with the Fredholm index $\dim \ker(\tilde{\slashed{D}}^{+}) - \dim \ker(\tilde{\slashed{D}}^{-})$ of $\tilde{\slashed{D}}^{+}$. This notion of quantization just associates an integer to (M, ω). This number turns out to be independent of the choice of the Spin^c structure on M, as long as it satisfies the above requirement, and is entirely determined by the cohomology class $[\omega]$ (as remarked earlier, this is not true for the Spin^c structure and the associated Dirac operator itself) [Guillemin, Ginzburg and Karshon (2002)].

This definition of quantization gains in substance when a compact Lie group G acts on M in strongly Hamiltonian fashion. In that case, the pertinent Spin^c structure may be chosen to be G invariant, and the spaces $\ker(\slashed{D}^{\pm})$ are finite-dimensional complex G modules. Hence

$$G\text{-index}(\slashed{D}^{+}) = [\ker(\slashed{D}^{+})] - [\ker(\slashed{D}^{-})] \qquad (2.5)$$

defines an element of the representation ring $R(G)$ of G.[6] Thus, the quan-

---

[5]In fact, this construction needs to be corrected in some cases [Guillemin, Ginzburg and Karshon (2002); Paradan (2001b)], but this correction complicates the statement of the Guillemin–Sternberg conjecture, and will not be discussed here.

[6]$R(G)$ is defined as the abelian group with one generator $[L]$ for each finite-

tization of (M, ω) with associated G action may be defined as

$$Q(G \circlearrowright M, \omega) = G\text{-index}(\slashed{D}^+) \in R(G). \qquad (2.6)$$

As before, this element only depends on $[\omega]$ (and on the G action). The same definition arises from the Hilbert space setting: the Hilbert spaces $L^2(\mathcal{S}^\pm)$ carry unitary representations $U^\pm$ of G in the obvious way, and the bounded Dirac operators $\tilde{\slashed{D}}^\pm$ are equivariant under these, so that $\ker(\tilde{\slashed{D}}^\pm)$ are unitary G modules. Replacing $\slashed{D}^\pm$ in (2.5) by $\tilde{\slashed{D}}^\pm$ then yields an element of the ring of unitary finite-dimensional representations of G, which for a compact group is the same as $R(G)$. When G is trivial, one may identify $R(e)$ with $\mathbb{Z}$ through the map $[V] - [W] \mapsto \dim(V) - \dim(W)$, so that (2.4) emerges as a special case of (2.6).

In this setting, the Guillemin–Sternberg conjecture makes sense as long as M and G are compact. The Hilbert space $H^0(M, L_\omega)^G$ in the original version of the conjecture is now replaced by the image $Q(G \circlearrowright M, \omega)_0$ of $Q(G \circlearrowright M, \omega)$ in $\mathbb{Z}$ under the map $[V] - [W] \mapsto \dim(V_0) - \dim(W_0)$, where V_0 is the G invariant part of V, etc. The right-hand side of the conjecture is the quantization of the reduced space M^0 (which inherits a Spin^c structure from M) according to (2.4). Denoting the pertinent Dirac operator on M^0 by $\slashed{D}_0$, the Guillemin–Sternberg conjecture in the setting of Bott's definiton of quantization is therefore simply

$$G\text{-index}(\slashed{D}^+)_0 = \text{index}(\slashed{D}_0^+). \qquad (2.7)$$

In this form, the conjecture was proved in [Meinrenken (1998)]; it even holds when 0 fails to be a regular value of J [Meinrenken and Sjamaar (1999)]. Also see [Guillemin, Ginzburg and Karshon (2002); Paradan (2001a)] for other proofs and further references.

Bott's definition of quantization (2.6) or (2.4) isn't actually all that far removed from the traditional idea of associating a group representation on a Hilbert space with a strongly Hamiltonian action on a symplectic manifold. In fact, when the symplectic form ω is sufficiently large, the space $\ker(\slashed{D}^-)$ tends to vanish [Braverman (1998)], so that $Q(G \circlearrowright M, \omega)$ is really a representation of G, up to isomorphism. This is relevant in the semiclassical regime, where one quantizes $(M, \omega/\hbar)$ for small values of $\hbar$.

---

dimensional complex representation L of G, and relations $[L] = [M]$ when L and M are equivalent and $[L] + [M] = [L \oplus M]$. The tensor product of representations defines a ring structure on $R(G)$.

2.7 From Quantization to KK-Theory

To motivate the use of Kasparov's bivariant K-theory, or KK-theory, in the light of the Guillemin–Sternberg conjecture and Bott's definition of quantization, let us recall a result from functional analysis (see, e.g., [Douglas (1998)]). Recall that a bounded operator $F : H^+ \to H^-$ between two Hilbert spaces is called Fredholm when it is invertible up to compact operators, that is, when there exists a bounded operator $F' : H^- \to H^+$, called a parametrix of F, such that $FF' - 1$ and $F'F - 1$ are compact operators on H^- and H^+, respectively. A key result is then that the space $\mathcal{F}(H^+, H^-)/ \overset{h}{\sim}$ of homotopy equivalence classes $[F]$ of Fredholm operators F (where the notion of homotopy is defined with respect to operator-norm continuous paths in the space of all Fredholm operators) is homeomorphic to $\mathbb{Z}$, where the pertinent homeomorphism is given by $[F] \mapsto \mathrm{index}(F)$.

Hence in Bott's definition of quantization (2.4) we may work with $[\tilde{\slashed{D}}^+]$ instead of with $\mathrm{index}(\slashed{D}^+)$ ($= \mathrm{index}(\tilde{\slashed{D}}^+)$). Thus we put

$$Q(pt \leftarrow M \to pt) = [\tilde{\slashed{D}}^+]. \tag{2.8}$$

As indicated by the notation, we regard the right-hand side of (2.8) as the quantization of (the isomorphism class of) the dual pair on the left-hand side. It will become clear shortly that this homotopy class is an element of the Kasparov group $KK(\mathbb{C}, \mathbb{C})$, where we regard $\mathbb{C}$ as the C^*-algebra that quantizes the Poisson manifold pt. This group is isomorphic to $\mathbb{Z}$, and the image of $[F]^7$ under the isomorphism $KK(\mathbb{C}, \mathbb{C}) \to \mathbb{Z}$ is precisely $\mathrm{index}(F)$. Clearly, this isomorphism links (2.8) to (2.4).

2.8 Kasparov Bimodules as Arrows

To generalize this idea to more complicated dual pairs, we need Kasparov's theory [Kasparov (1981)] (see also [Blackadar (1999)] for a full treatment and [Higson (1990); Skandalis (1991); Connes (1994)] for very useful introductions), which is a systematic machinery for dealing with homotopy classes of generalized Fredholm operators. The first step is to generalize the notion of a Hilbert space, which we here regard as a Hilbert $\mathbb{C}$-$\mathbb{C}$ bimodule, to the concept of a Hilbert A-B bimodule, where A and B are separable C^*-algebras (which in our setting emerge as the quantizations of Poisson

---

[7]More precisely, of the homotopy class $[F, H^+, H^-]$, where $H^\pm$ are $\mathbb{C}$-$\mathbb{C}$ Hilbert bimodules under the action $z \mapsto z1$, $z \in \mathbb{C}$.

manifolds P and Q). The correct generalization was introduced by Rieffel in a different context [Rieffel (1974)], and has already been used in the theory of constrained quantization in [Landsman (1998)].

An A-B Hilbert bimodule is an algebraic A-B bimodule E (where A and B are seen as complex algebras, so that E is a complex linear space) with a compatible B-valued inner product. This is a sesquilinear map $\langle\,,\,\rangle : E \times E \to B$, linear in the second and antilinear in the first entry, satisfying $\langle x, y\rangle^* = \langle y, x\rangle$, $\langle x, x\rangle \geq 0$, and $\langle x, x\rangle = 0$ iff $x = 0$. The compatibility of the inner product with the remaining structures means that firstly E has to be complete in the norm $\|x\|^2 = \|\langle x, x\rangle\|$, secondly that $\langle x, yb\rangle = \langle x, y\rangle b$, and thirdly that $\langle a^*x, y\rangle = \langle x, ay\rangle$ for all $x, y \in E$, $b \in B$, and $a \in A$. The latter condition may be expressed by saying that a is adjointable, with adjoint a^*; this is a nontrivial condition even when a is bounded (note that an adjointable operator is automatically bounded). The best example of all this is the A-A Hilbert bimodule $E = A$, with the obvious actions and the inner product $\langle a, b\rangle = a^*b$.

An A-$\mathbb{C}$ Hilbert bimodule is simply a Hilbert space equipped with a representation of A. A $\mathbb{C}$-B Hilbert bimodule is called a Hilbert B module, or Hilbert C^*-module over B.

Adjointable operators on an A-B Hilbert bimodule E are the analogues of bounded operators on a Hilbert space; the collection of all adjointable operators indeed forms a C^*-algebra. The role of compact operators on E is played by operators that can be approximated in norm by linear combinations of rank one operators of the form $z \mapsto x\langle y, z\rangle$ for $x, y \in E$ (such operators are automatically adjointable). Again, as for Hilbert spaces, the space of all compact operators on E is a C^*-algebra. In the example ending the preceding paragraph, the left A action turns out to be by compact operators. A Fredholm operator, then, is an adjointable operator that is invertible up to compact operators.

Now an A-B Kasparov bimodule is a pair of countably generated A-B Hilbert bimodules (E^+, E^-) with an 'almost' Fredholm operator $F :$ $E^+ \to E^-$ that 'almost' intertwines the A actions on E^+ and E^-. The first condition means that there is an adjointable operator $F' : H^- \to H^+$ such that $a(FF' - 1)$ and $a(F'F - 1)$ are compact for all $a \in A$, and the second states that $aF - Fa$ is compact for all $a \in A$. With the structure of $E^\pm$ as A-B Hilbert bimodules understood, we denote such a Kasparov bimodule simply by (F, E^+, E^-).

For $B = \mathbb{C}$ this is sometimes called a Fredholm module [Connes (1994)]. A key example of a Fredholm module is given by $E^\pm = L^2(\mathcal{S}^\pm)$, and $F =$

$\tilde{\not{D}}^+$. When M is compact, this works for both $A = \mathbb{C}$ and $A = C(M)$, but when M isn't one must take $A = C_0(M)$. For general A and B, it follows from the definitions that if A acts on E by compact operators, then the choice $F = 0$ yields a Kasparov bimodule. This applies, for instance, to the A-A Hilbert bimodule $(E^+ = A, E^- = 0)$.

A homotopy of A-B Kasparov bimodules is an A-$C([0, 1], B)$ Kasparov bimodule. The ensuing set $KK(A, B)$ of homotopy classes of A-B Kasparov bimodules may more conveniently be described as the quotient of the set of all A-B Kasparov bimodules by the equivalence relation generated by unitary equivalence, translation of F along norm-continuous paths (of almost intertwining almost Fredholm operators), and the addition of degenerate Kasparov bimodules. The latter are those for which the operators $aF - Fa$, $a(FF' - 1)$ and $a(F'F - 1)$ are not merely compact but zero for all $a \in A$. Using the polar decomposition, one may always choose representatives for which all $(F' - F^*)a$ are compact (so that F is almost unitary), and this is often included in the definition of a Kasparov bimodules. In that case, the condition that $(F' - F^*)a = 0$ is added to the definition of a degenerate Kasparov bimodule.

It is not difficult to see that $KK(A, B)$ is an abelian group; the group operation is the direct sum of both bimodules and operators F, and the inverse of the class of a Kasparov bimodule is found by swapping E^+ and E^- and replacing $F : E^+ \to E^-$ by its parametrix $F' : E^- \to E^+$. Moreover, with respect to *-homomorphisms between C^*-algebras the association $(A, B) \mapsto KK(A, B)$ is contravariant in the first entry, and covariant in the second.

Let us note that for any C^*-algebra A the group $KK(\mathbb{C}, A)$ is naturally isomorphic to the algebraic K-theory group $K_0(A)$.[8] Hence as far as K_0 is concerned, K-theory is a special case of KK-theory. Explicitly, the isomorphism $KK(\mathbb{C}, A) \to K_0(A)$ is the generalized index map[9]

$$[F, E^+, E^-] \mapsto [\ker(F)] - [\ker(F')]. \tag{2.9}$$

---

[8]When A has a unit, $K_0(A)$ may be defined as the abelian group with one generator $[E]$ for each finitely generated projective (f.g.p.) right module over A, and relations $[E] = [E']$ when E and E' are isomorphic, and $[E] + [E'] = [E \oplus E']$. For example, when X is a compact Hausdorff space one has $K_0(C(X)) = K^0(X)$, the topological K-theory of Atiyah and Hirzebruch [Karoubi (1978)]. When A has no unit, $K_0(A)$ is defined as the kernel of the canonical map $K_0(\tilde{A}) \to K_0(\mathbb{C})$, where $\tilde{A} = A \oplus \mathbb{C}$ is the unitization of A.

[9]The representatives F and F' of their respective homotopy classes have to be chosen such that their kernels in the A modules E^- and E^+ are indeed f.g.p.

A remarkable aspect of Kasparov's theory is the existence of a product

$$KK(A, B) \times KK(B, C) \to KK(A, C),$$

which is functorial in all conceivable ways. Disregarding F, this would be easy to define, since one feature of algebraic bimodules that survives in the Hilbert case is the existence of a bimodule tensor product [Rieffel (1974)]: from an A-B Hilbert bimodule E and a B-C Hilbert bimodule $\tilde{E}$ one can form an A-C Hilbert bimodule $E \hat{\otimes}_B \tilde{E}$, called the interior tensor product of E and $\tilde{E}$. However, the composition of the almost Fredholm operators in question is too complicated to be explained here (see [Connes and Skandalis (1984); Higson (1990); Skandalis (1991); Connes (1994); Kucerovsky (1997); Blackadar (1999)]). In any case, this product leads to the category KK, whose objects are separable C^*-algebras, and whose arrows are Kasparov's KK-groups.

To close this section, let us mention that we only use the 'even' part of KK-theory; in general, each KK group is $\mathbb{Z}_2$ graded, and what we have called $KK(A, B)$ is really $KK_0(A, B)$. This restriction is possible because symplectic manifolds happen to be even-dimensional.

2.9 The Guillemin–Sternberg Conjecture Revisited

Let us return to a strongly Hamiltonian group action $G \circlearrowright M$, with associated dual pair $pt \leftarrow M \to \mathfrak{g}^*_-$. To quantize this dual pair, we first note that the quantization of the Poisson manifold $\mathfrak{g}^*$ is the group C^*-algebra $C^*(G)$ [Rieffel (1990); Landsman (1998)]; this is probably the best understood example in C^*-algebraic quantization theory.[10] Although this holds for any G with given Lie algebra, to obtain a unique functor we assume G to be connected and simply connected. Hence the quantization of the dual pair $pt \leftarrow M \to \mathfrak{g}^*_-$ should be an element of the Kasparov group $KK(\mathbb{C}, C^*(G)) \cong K_0(C^*(G))$.

When G is compact, which we assume throughout the remainder of this section, one may identify $K_0(C^*(G))$ with the representation ring $R(G)$; this is because finitely generated projective modules over $C^*(G)$ may be identified with finite-dimensional unitary representations of G. Now assume that M is compact as well. Seen as an element of $R(G)$, the quantization of $pt \leftarrow M \to \mathfrak{g}^*_-$ is given by G-index($\not{D}^+$), as in (2.5); this is just a

---

[10]Here $C^*(G)$ is a suitable completion of the convolution algebra on G determined by a Haar measure [Dixmier (1977); Landsman (1998)].

reinterpretation of Bott's definition (2.6) of quantization. It is slightly more involved to explain the quantization of $pt \leftarrow M \rightarrow \mathfrak{g}^*_-$ when it is seen as an element of $KK(\mathbb{C}, C^*(G))$. Firstly, one turns the Hilbert spaces $L^2(\mathcal{S}^\pm)$ into Hilbert $C^*(G)$ modules, as follows [Baum, Connes and Higson (1994); Valette (2002); Valette (2003)].

The canonical G actions $U^\pm$ on $L^2(\mathcal{S}^\pm)$ induce right actions $\pi^\pm$ of $C^*(G)$ by $\pi^\pm_-(f) = \int_G dx\, f(x) U^\pm(x^{-1})$, where $f \in C(G)$ (the action of a general element of $C^*(G)$ is then defined by continuity). Furthermore, one obtains a $C^*(G)$ valued inner product on $L^2(\mathcal{S}^\pm)$ by the formula

$$\langle \psi, \varphi \rangle : x \mapsto (\psi, U^\pm(x)\varphi), \tag{2.10}$$

which defines an element of $C(G) \subset C^*(G)$. Completing $L^2(\mathcal{S}^\pm)$ in the norm

$$\|\psi\|^2 = \|\langle \psi, \psi \rangle\|_{C^*(G)} \tag{2.11}$$

then yields Hilbert $C^*(G)$ modules $E^\pm(\mathcal{S})$. The operator $\tilde{\slashed{D}}^+ : L^2(\mathcal{S}^+) \rightarrow L^2(\mathcal{S}^-)$ extends to an adjointable operator $\hat{\slashed{D}}^+ : E^+(\mathcal{S}) \rightarrow E^-(\mathcal{S})$ by continuity, and the triple $(\hat{\slashed{D}}^+, E^+(\mathcal{S}), E^-(\mathcal{S}))$ defines a $\mathbb{C}$-$C^*(G)$ Kasparov bimodule, whose homotopy class is the desired element of $KK(\mathbb{C}, C^*(G))$, i.e.,

$$Q(pt \leftarrow M \rightarrow \mathfrak{g}^*_-) = [\hat{\slashed{D}}^+, E^+(\mathcal{S}), E^-(\mathcal{S})]. \tag{2.12}$$

The canonical isomorphism $KK(\mathbb{C}, C^*(G)) \rightarrow K_0(C^*(G)) = R(G)$ given by (2.9) indeed maps this element to G-index$(\slashed{D}^+)$.

Apart from the dual pair $pt \leftarrow M \rightarrow \mathfrak{g}^*_-$, the momentum map associated to the action $G \circlearrowright M$ equally well leads to a dual pair $\mathfrak{g}^*_- \leftarrow M^- \rightarrow pt$. This is to be quantized by an element of $KK(C^*(G), \mathbb{C}) \cong K^0(C^*(G))$, the so-called Kasparov representation ring of G (cf. [Higson and Roe (2000)]). This time, we interpret the Hilbert spaces $L^2(\mathcal{S}^\pm)$ as $C^*(G)$-$\mathbb{C}$ Hilbert bimodules, where the pertinent representations $\pi^\pm$ of $C^*(G)$ are given by a very slight adaptation of the procedure sketched in the preceding paragraph: to obtain left actions instead of right actions, we now put $\pi^\pm(f) = \int_G dx\, f(x) U^\pm(x)$. Since $\tilde{\slashed{D}}^+ U^+(x) = U^-(x)\tilde{\slashed{D}}^+$ for all $x \in G$, one now has $\tilde{\slashed{D}}^+ \pi^+(f) = \pi^-(f)\tilde{\slashed{D}}^+$ for all $f \in C^*(G)$. Since $\tilde{\slashed{D}}^+$ is Fredholm one thus obtains an element $[\tilde{\slashed{D}}^+, L^2(\mathcal{S}^+), L^2(\mathcal{S}^-)]$ of $KK(C^*(G), \mathbb{C})$, which we regard as the quantization of the dual pair $\mathfrak{g}^*_- \leftarrow S^- \rightarrow pt$.

The very simplest example is the dual pair $\mathfrak{g}_-^* \hookleftarrow 0 \to pt$, whose quantization is just

$$Q(g_-^* \hookleftarrow 0 \to pt) = [0, \mathbb{C}, \mathbb{C}], \tag{2.13}$$

where the $C^*(G)$-$\mathbb{C}$ Hilbert bimodules $\mathbb{C}$ carry the trivial representation of G. A simple computation of the Kasparov product

$$KK(\mathbb{C}, C^*(G)) \times KK(C^*(G)), \mathbb{C}) \to KK(\mathbb{C}, \mathbb{C}) \cong K_0(\mathbb{C}) \cong \mathbb{Z}$$

yields

$$[\hat{\not{D}}^+, E^+(\mathcal{S}), E^-(\mathcal{S})] \times [0, \mathbb{C}, \mathbb{C}] = G\text{-index}(\not{D}^+)_0, \tag{2.14}$$

cf. (2.7) and preceding text. In fact, $y \times [0, \mathbb{C}, \mathbb{C}]$ is just the image of y under the map $KK(\mathbb{C}, C^*(G)) \to KK(\mathbb{C}, \mathbb{C})$ functorially induced by the *-homomorphism $C^*(G) \to \mathbb{C}$ given by the trivial representation of G.

As explained around (2.8), if we identify $KK(\mathbb{C}, \mathbb{C})$ with $\mathbb{Z}$ as above, the reduced space M^0 is quantized by

$$Q(pt \leftarrow M^0 \to pt) = \text{index}(\not{D}_0^+). \tag{2.15}$$

Combining (2.2), (2.12), (2.13), (2.14), and (2.15), we see that the functoriality condition

$$Q(pt \leftarrow M \to \mathfrak{g}_-^*) \times Q(g_-^* \hookleftarrow 0 \to pt) =$$
$$Q((pt \leftarrow M \to \mathfrak{g}_-^*) \otimes_{\mathfrak{g}_-^*} (\mathfrak{g}_-^* \hookleftarrow 0 \to pt)) \tag{2.16}$$

is precisely the Guillemin–Sternberg conjecture (2.7).

2.10 Guillemin–Sternberg for Noncompact Groups

The above reformulation of the Guillemin–Sternberg conjecture as a special case of the functoriality of Bott's definition of quantization paves the way for far-reaching generalizations of this conjecture. Firstly, one can now consider noncompact G and M, as long as the G action on M is proper. It is convenient to use the language of K-homology (cf. [Higson and Roe (2000)]). The K-homology group of a manifold M is just defined as the Kasparov group $K_0(M) = KK(C_0(M), \mathbb{C})$. A Spin^c structure on M defines an element $[\tilde{\not{D}}^+]$ of $K_0(M)$ through its associated Dirac operator. This so-called fundamental class never vanishes. It is independent of the connection picked to define $\not{D}$, and is the analogue in K-homology of the fundamental class in ordinary homology defined by the orientation of M [Higson and Roe

(2000)]. From this point of view, Bott's quantization (2.4) of (M, ω), which in our setting is the quantization of the dual pair $pt \leftarrow M \rightarrow pt$, is the image of the fundamental class of M determined by the symplectic structure as explained, under the map $KK(C_0(M), \mathbb{C}) \rightarrow KK(\mathbb{C}, \mathbb{C})$ obtained by forgetting the $C_0(M)$ actions on $L^2(\mathcal{S}^\pm)$ (followed by the isomorphism $KK(\mathbb{C}, \mathbb{C}) \rightarrow \mathbb{Z}$).

In the presence of a proper G action, one uses the equivariant K-homology group $K_0^G(M) = KK^G(C_0(M), \mathbb{C})$, which is defined like $KK(C_0(M), \mathbb{C})$, but with the additional stipulation that the Hilbert spaces $H^\pm$ in the Kasparov bimodule (F, H^+, H^-) are unitary G modules, in such a way that F is equivariant, and the representations of $C_0(M)$ on $E^\pm$ are covariant under G [Kasparov (1988); Valette (2002)]. One now has a canonical map $K_0^G(M) \rightarrow K_0(C^*(G))$, called the analytic assembly map, which plays a key role in the Baum–Connes conjecture [Baum, Connes and Higson (1994)]. Replacing $K_0(C^*(G))$ with $KK(\mathbb{C}, C^*(G))$, this map is defined by a slight generalization of the construction of the element $[\hat{\slashed{D}}^+, E^+(\mathcal{S}), E^-(\mathcal{S})]$ of $KK(\mathbb{C}, C^*(G))$ explained prior to (2.12); cf. [Valette (2002)] for details. The basic idea is to define the $C_c(G)$-valued inner products (2.10) on the dense subspace $C_c(M)L^2(\mathcal{S}^\pm)$, completing these subspaces in the norm (2.11) to obtain the Hilbert $C^*(G)$ modules $E^\pm(\mathcal{S})$.[11]

It follows that the element of $KK(\mathbb{C}, C^*(G))$ that quantizes the dual pair $pt \leftarrow M \rightarrow \mathfrak{g}_-^*$ a la Bott is just the image of the pertinent fundamental class of M under the analytic assembly map.[12] The functoriality condition (2.16) remains well defined, but the computation (2.14) is invalid for noncompact groups, so that for noncompact G the left-hand side of the Guillemin–Sternberg conjecture is simply given by the left-hand side instead of the right-hand side of (2.14).[13] This yields a generalization of the Guillemin–Sternberg conjecture to noncompact groups, where G-index$(\slashed{D}^+)_0$ in (2.7) is now reinterpreted as the image of G-index$(\slashed{D}^+) \in$

---

[11]We here assume that G is unimodular, which guarantees that (2.10) is positive. This was shown for discrete G in Lemma 3 in [Valette (2003)], but the proof apparently works for unimodular groups in general. In general, the construction in the preceding section produces a Hilbert module over the reduced group C^*-algebra $C_r^*(G)$ [Baum, Connes and Higson (1994)]. This is sufficient for the Baum–Connes conjecture, but not for our generalized Guillemin–Sternberg conjecture.

[12]Cf. [Landsman (2003)] for an exposition of the link between the analytic assembly map and C^*-algebraic deformation quantization, following Connes's discussion of this map in E-theory [Connes (1994)].

[13]A complication arises when M does not admit a G invariant Spin^c structure. For techniques to overcome this cf. [Hilsum and Skandalis (1987); Paradan (2001b)].

$K_0(C^*(G))$ under the map $K_0(C^*(G)) \to \mathbb{Z}$ induced in K-theory by the *-homomorphism $f \mapsto \int_G dx\, f(x)$ from $C^*(G)$ to $\mathbb{C}$.

As a first example, consider the case where $G = \Gamma$ is discrete and infinite. One then simply has $M^0 = M/\Gamma$, and $D\!\!\!/_0^+$ is just the operator on M/Γ whose lift is $D\!\!\!/^+$. Using Atiyah's L^2-index theorem [Atiyah (1976)], our generalized Guillemin–Sternberg conjecture is equivalent to

$$G\text{-index}(D\!\!\!/^+)_0 = \mathrm{tr} \circ \pi_* \circ G\text{-index}(D\!\!\!/^+).$$

Here $\pi_* : K_0(C^*(\Gamma)) \to K_0(C_r^*(\Gamma))$ is the K-theory map functorially induced by the canonical projection $\pi : C^*(\Gamma) \to C_r^*(\Gamma)$, and $\mathrm{tr} : K_0(C^*(\Gamma)) \to \mathbb{C}$ is defined by the pairing of the trace $f \mapsto f(e)$ on $C_r^*(\Gamma)$ (seen as a cyclic cocycle) with K-theory [Connes (1994)].

2.11 Foliation Theory and Quantization

A second generalization of the Guillemin–Sternberg conjecture arises when one considers strongly Hamiltonian actions of Lie groupoids on symplectic manifolds; the pertinent symplectic reduction procedure was first studied in [Mikami and Weinstein (1988)], and is actually a special case of the product $\odot$ [Landsman (1998); Bursztyn and Weinstein (2003)]. Furthermore, the appropriate construction of elements of $K_0(C^*(G))$ has been given in [Connes (1994); Paterson (2002)]. A very interesting special case comes from foliation theory, as follows (cf. [Connes (1982); Hilsum and Skandalis (1987); Connes (1994); Moerdijk (1993); Mrčun (1996)]). Let (V_i, F_i), $i = 1, 2$, be foliations with associated holonomy groupoids $G(V_i, F_i)$ (assumed to be Hausdorff for simplicity). A smooth generalized map f between the leaf spaces V_1/F_1 and V_2/F_2 is defined as a smooth right principal bibundle M_f between the Lie groupoids $G(V_1, F_1)$ and $G(V_2, F_2)$. Classically, such a bibundle defines a dual pair $T^*F_1 \leftarrow T^*M_f \to T^*F_2$ [Landsman (2000)]. Here $TF_i \subset TV_i$ is the tangent bundle to the foliation (V_i, F_i), whose dual bundle T^*F_i has a canonical Poisson structure.[14] Quantum mechanically, f defines an element [Connes (1982); Hilsum and Skandalis (1987)]

$$f_! \in KK(C^*(G(V_1, F_1)), C^*(G(V_2, F_2))).$$

---

[14]The best way to see this is to interpret TF_i as the Lie algebroid of $G(V_i, F_i)$, and to pass to the canonical Poisson structure on the dual bundle $A^*(G)$ to the Lie algebroid $A(G)$ of any Lie groupoid G.

In our functorial approach to quantization, $f_!$ is interpreted as the quantization of the dual pair $T^*F_1 \leftarrow T^*M_f \rightarrow T^*F_2$. The functoriality of quantization among dual pairs of the same type then follows from the computations in [Hilsum and Skandalis (1987); Landsman (2000)]. The construction and functoriality of shriek maps in [Atiyah and Singer (1968); Connes (1982)] is a special case of this, in which the V_i are both trivially foliated.

Bibliography

Abraham, R. and Marsden, J.E. (1985). Foundations of Mechanics, 2nd ed. Addison Wesley, Redwood City.

Atiyah, M.F. (1976). Elliptic operators, discrete groups and von Neumann algebras. Astérisque 32-33, 43–72.

Atiyah, M.F., Bott, R. and Shapiro, A. (1964). Clifford modules. Topology 3 suppl. 1, 3–38.

Atiyah, M.F. and Singer, I.M. (1968). The index of elliptic operators I. Ann. Math. 87, 485–530.

Baum, P., Connes, A. and Higson, N. (1994). Classifying space for proper actions and K-theory of group C^*-algebras. Contemp. Math. 167, 241–291.

Baum, P. and Douglas, R.G. (1982). K homology and index theory. Proc. Sympos. Pure Math. 38, 117–173.

Bayen, F., Flato, M., Fronsdal, C., Lichnerowicz, A. and Sternheimer, D. (1978). Deformation theory and quantization. I, II. Ann. Phys. (N.Y.) 110, 61–110, 111–151.

Binz, E., J., Śniatycki, J. and Fischer, H. (1988). The Geometry of Classical Fields. North–Holland, Amsterdam

Blackadar, B. (1999). K-theory for Operator Algebras, 2nd ed. Cambridge University Press, Cambridge.

Braverman, M. (1998). Vanishing theorems for the kernel of a Dirac operator. arXiv:math.DG/9805127.

Bursztyn, H., and Weinstein, A. (2003). Picard groups in Poisson geometry. Moscow Math. J., to appear. arXiv:math.SG/0304048.

Connes, A. (1982). A survey of foliations and operator algebras. Proc. Sympos. Pure Math. 38, 521–628.

Connes, A. (1994). Noncommutative Geometry, Academic Press, San Diego.

Connes, A. and Skandalis, G. (1984). The longitudinal index theorem for foliations. Publ. Res. Inst. Math. Sci. 20, 1139–1183.

Dirac, P.A.M. (1964). Lectures on Quantum Mechanics. Belfer School of Science, Yeshiva University, New York.

Dixmier, J. (1977). C^*-Algebras, North–Holland, Amsterdam.

Duistermaat, J. J. (1996). The Heat Kernel Lefschetz Fixed Point Formula for the Spin-c Dirac Operator. Birkhäuser, Boston.

Douglas, R.G. (1998). Banach Algebra Techniques in Operator Theory. Second edition. Springer-Verlag, New York.

Gracia-Bondía, J.M., Várilly, J.C. and Figueroa, H. (2001). Elements of Noncommutative Geometry. Birkhäuser, Boston.

Guillemin, V., Ginzburg, V. and Karshon, Y. (2002). Moment Maps, Cobordisms, and Hamiltonian Group Actions. American Mathematical Society, Providence, RI.

Guillemin, V. and Sternberg, S. (1982). Geometric quantization and multiplicities of group representations. Inv. Math. 67, 515–538.

Higson, N. (1990). A primer on KK-theory. Proc. Sympos. Pure Math. 51, Part 1, 239–283.

Higson, N. and Roe, J. (2000). Analytic K-homology. Oxford University Press, Oxford.

Hilsum, M. and Skandalis, G. (1987). Morphismes K-orientés d'espaces de feuilles et fonctorialité en théorie de Kasparov (d'après une conjecture d'A. Connes). Ann. Sci. 'Ecole Norm. Sup. (4) 20, 325–390.

Karasev, M.V. (1989). The Maslov quantization conditions in higher cohomology and analogs of notions developed in Lie theory for canonical fibre bundles of symplectic manifolds. I, II. Selecta Math. Soviet. 8, 213–234, 235–258.

Karoubi, M. (1978). K-theory: An Introduction. Springer-Verlag, Heidelberg.

Kasparov, G.G. (1981). The operator K-functor and extensions of C^*-algebras. Math. USSR Izvestija SSSR 16, 513–572.

Kasparov, G.G. (1988). Equivariant KK-theory and the Novikov conjecture. Invent. Math. 91, 147–201.

Kucerovsky, D. (1997). The KK-product of unbounded modules. K-theory 11, 17–34.

Le Gall, P. (1999). Théorie de Kasparov équivariante et groupoïdes. I. K-Theory 16, 361–390.

Landsman, N.P. (1998). Mathematical Topics Between Classical and Quantum Mechanics. Springer, New York.

Landsman, N.P. (2000). The Muhly-Renault-Williams theorem for Lie groupoids and its classical counterpart. Lett. Math. Phys. 54, 43–59. arXiv:math-ph/0008005.

Landsman, N.P. (2001). Quantized reduction as a tensor product. Ref. [Landsman, Pflaum and Schlichenmaier (2001)], pp. 137–180. arXiv:math-ph/0008004.

Landsman, N.P. (2002). Quantization as a functor. Contemp. Math. 315, 9–24. arXiv:math-ph/0107023.

Landsman, N.P. (2003). Deformation quantization and the Baum-Connes conjecture. Comm. Math. Phys. 237, 87–103. arXiv:math-ph/0210015.

Landsman, N.P., Pflaum, M. and Schlichenmaier, M., eds. (2001). Quantization of Singular Symplectic Quotients. Birkhäuser, Basel.

Lawson, H. B. and Michelsohn, M.-L. (1989). Spin Geometry. Princeton University Press, Princeton, NJ.

Marsden, J.E. and Weinstein, A. (1974). Reduction of symplectic manifolds with symmetry. Rep. Math. Physics 5, 121-130.

Marsden, J.E. and Weinstein, A. (2001). Comments on the history, theory, and applications of symplectic reduction. Ref. [Landsman, Pflaum and Schlichenmaier (2001)], pp. 1–19.

Meinrenken, E. (1998). Symplectic surgery and the Spin^c-Dirac operator. Adv. Math. 134, 240–277.

Meinrenken, E. and Sjamaar, R. (1999). Singular reduction and quantization. Topology 38, 699–762.

Meyer, K. (1973). Symmetries and integrals in mechanics. Peixoto, M.M., ed., Dynamical systems, pp. 259–272. Academic Press, New York.

Mikami, K. and Weinstein, A. (1988). Moments and reduction for symplectic groupoids. Publ. RIMS Kyoto Univ. 24, 121–140.

Moerdijk, I. (1993). Foliations, groupoids and Grothendieck etendues. Rev. Academia Ciencias. Zaragoza 48, 5–33.

Mrčun, J. (1996). Stability and Invariants of Hilsum–Skandalis Maps, Ph.D. thesis, University of Utrecht.

Paradan, P.-E. (2001a). Localization of the Riemann-Roch character. J. Funct. Anal. 187, 442–509.

Paradan, P.-E. (2001b). Spin^c quantization and the K-multiplicities of the discrete series. arXiv:math.DG/0103222.

Paterson, A. (2002). The equivariant analytic index for proper group actions. Preprint.

Pflaum, M. (2001). Analytic and Geometric Study of Stratified Spaces. Lecture Notes in Mathematics 1768. Springer, Berlin.

Plymen, R. J. (1986). Strong Morita equivalence, spinors and symplectic spinors. J. Operator Theory 16, 305–324.

Rieffel, M.A. (1974). Induced representations of C^*-algebras. Adv. Math. 13, 176–257.

Rieffel, M.A. (1990). Lie group convolution algebras as deformation quantization of linear Poisson structures. Amer. J. Math. 112, 657–686.

Sjamaar, R. (1996). Symplectic reduction and Riemann-Roch formulas for multiplicities. Bull. Amer. Math. Soc. (N.S.) 33, 327–338.

Sjamaar, R. and Lerman, E. (1991). Stratified symplectic spaces and reduction. Ann. Math. 134, 375–422.

Skandalis, G. (1991). Kasparov's bivariant K-theory and applications. Exposition. Math. 9, 193–250.

Sundermeyer, K. (1982). Constrained Dynamics. Lecture Notes in Physics 169. Springer, Berlin.

Valette, A. (2002). Introduction to the Baum–Connes Conjecture. Birkhäuser, Basel.

Valette, A. (2003). On the Baum–Connes assembly map for discrete groups. With an appendix by D. Kucerovsky. Preprint.

Weinstein, A. (1983). The local structure of Poisson manifolds. J. Diff. Geom. 18, 523–557.

Chapter 3

Coherent State Method in Geometric Quantization

Anatol Odzijewicz[1]

Abstract: A notion of a mechanical system is introduced. On the basis of this idea the relationship of the coherent state method with the Kostant–Souriau geometric quantization and path integral quantization is described.

Contents

3.1 Introduction

Coherent states were discovered in 1926 by Schrödinger [22] and about forty years later, were used by Glauber [6] for the description of laser light coherence. Subsequently, their applications to a wide class of physical phenomena were found, see [8]. In mathematical physics, the coherent state

[1]Institute of Physics, University of Bialystok, Lipowa 41, PL-15424 Bialystok, Poland, `aodzijew@labfiz.uwb.edu.pl`

method was initiated by Klauder [8] and Perelomov [21]. Currently, the coherent state method is an intensively investigated part of mathematical physics, related among others directly to the theory of wavelets [1].

In our review, which is based on [13; 14], we shall demonstrate the considerable importance of the coherent state method in the theory of quantization. For this reason we will introduce the notion of the coherent state map and then we will demonstrate the relationship of the coherent state method with the Kostant-Souriau method of geometric quantization [9; 23] and Feynman's path integral quantization [3].

In Section 3.2 and Section 3.3 we define the coherent state map and introduce the concept of physical system.

In Section 3.4 and Section 3.5 we discuss the role of the coherent state map in the Kostant-Souriau geometric quantization. Besides other we show that the classical physical systems quantized in Kostant-Souriau sense are exactly those which admit the existence of a coherent state map. The relation between Kostant-Souriau geometric quantization and Feynman path integral quantization will also be exhibited.

Some examples of mechanical systems are presented in Section 3.6.

The quantum algebras related to the coherent states map were introduced in [17]. The structure of these algebras is inherited by the spaces of covariant symbols of their elements. In [17] and [15] it is shown that covariant symbols correspond to the classical observables and their product (the star product) is the natural generalization of the Moyal [10] product to the general symplectic manifold case. However, because of limitation of the space we do not discuss this topic here.

The author was motivated to write this contribution, not merely because of the importance of the subject of quantization, but also by the fact that coherent states have formed such an important theme in the Białowieza Workshops.

3.2   Coherent State Map

In the standard model of quantum mechanics the space of states consists of the density operators. In order to describe the geometric structure of this space we present some necessary facts and notations.

Let $\mathcal{H}$ be a separable complex Hilbert space. By $L^1(\mathcal{H})$ we denote the complex Banach space of trace class operators. The Banach space dual to $L^1(\mathcal{H})$ is isomorphic to the Banach space of bounded operators $L^\infty(\mathcal{H})$, e.g.

see [11]. Consistently to the above by $H^1(\mathcal{H})$ and $H^\infty(\mathcal{H})$ we shall denote the real Banach spaces of Hermitian trace class operators and Hermitian bounded operators. The space $iH^\infty(\mathcal{H})$ is the real Banach Lie algebra with the commutator taken as the Lie bracket $[\cdot,\cdot]$. The pairing

$$\langle \rho, x \rangle := i \operatorname{Tr}(\rho x) \tag{3.2.1}$$

between $\rho \in H^1(\mathcal{H})$ and $x \in iH^\infty(\mathcal{H})$ defines the Banach spaces isomorphism $H^1(\mathcal{H})^* \cong iH^\infty(\mathcal{H})$.

So, $H^1(\mathcal{H})$ is the predual of the Banach Lie algebra $iH^\infty(\mathcal{H})$ and according to [2; 19] it is a Banach Lie-Poisson space with Poisson bracket

$$\{f, g\}(\rho) := i \operatorname{Tr}\left(\rho[Df(\rho), Dg(\rho)]\right), \tag{3.2.2}$$

where $Df(\rho), Dg(\rho) \in H^1(\mathcal{H})^*$ are the derivatives of the functions $f, g \in C^\infty(H^1(\mathcal{H}))$ taken at the element $\rho \in H^1(\mathcal{H})$. Expressing the trace class operator ρ in the coordinates

$$\rho = \sum_{k,l=0}^{\infty} \rho_{kl} |k\rangle\langle l|, \tag{3.2.3}$$

where $\{|k\rangle\}_{k=0}^{\infty}$ is the orthonormal basis in $\mathcal{H}$, we obtain from (3.2.2) the coordinate expression

$$\{f, g\} = \sum_{k,l,n=0}^{\infty} \rho_{kl} \left(\frac{\partial f}{\partial \rho_{ln}} \frac{\partial g}{\partial \rho_{nk}} - \frac{\partial g}{\partial \rho_{ln}} \frac{\partial f}{\partial \rho_{nk}} \right) \tag{3.2.4}$$

for the Poisson bracket.

Let $GU^\infty(\mathcal{H})$ be the Banach group of all unitary operators. The coadjoint action of $GU^\infty(\mathcal{H})$ on $H^1(\mathcal{H}) \subset (iH^\infty(\mathcal{H}))^*$ is given by

$$\operatorname{Ad}_g^* \rho = g\rho g^+, \qquad g \in GU^\infty(\mathcal{H}) \tag{3.2.5}$$

and the symplectic leaves of the Banach Lie-Poisson space $H^1(\mathcal{H})$ are the orbits of this action, see [19]. Thus the states space $\mathcal{S} \subset H^1(\mathcal{H})$, which is the intersection of the cone of positive trace class operators with the unit sphere $\{\rho : \|\rho\|_1 = 1\}$, splits into the symplectic leaves. One can define the pure states $\rho \in \mathcal{S}$ as the values $\rho = \iota([\psi])$ of the map $\iota : \mathbb{CP}(\mathcal{H}) \to H^1(\mathcal{H})$ of the complex projective Hilbert space $\mathbb{CP}(\mathcal{H})$ into $H^1(\mathcal{H})$ defined by

$$\iota([\psi]) = \frac{|\psi\rangle\langle\psi|}{\langle\psi|\psi\rangle}, \tag{3.2.6}$$

where $[\psi] := \mathbb{C}\psi$, $0 \neq \psi \in \mathcal{H}$. The image $\iota(\mathbb{CP}(\mathcal{H})) \subset \mathcal{S}$ of $\mathbb{CP}(\mathcal{H})$ we identify with the $\mathrm{Ad}^*(GU^\infty(\mathcal{H}))$-orbit $\mathcal{O}$ of some element $\iota([\psi])$. One proves, e.g. see [2], that ι is an embedding and that the symplectic structure $\omega_{\mathcal{O}}$ of $\mathcal{O}$ defined by the Poisson structure of $H^1(\mathcal{H})$ coincides with the Fubini-Study form $\iota^* \omega_{\mathcal{O}} = \omega_{FS}$

$$\omega_{FS} = i\partial\overline{\partial}\log(1 + z^+z) = i \sum_{k,l=0}^{\infty} (1 + z^+z)^{-2}((1 + z^+z)\delta_{kl} + z_k\overline{z}_l)dz_k \wedge dz_l,$$

$$(3.2.7)$$

where $z_k = \frac{\psi_k}{\psi_0}$, $k \in \mathbb{N}$, $|\psi\rangle = \sum_{k=0}^{\infty} \psi_k|k\rangle$, $\psi_0 \neq 0$ and $z^+z = \sum_{k=0}^{\infty} \overline{z}_k z_k$.

A simple computation shows that the Hamilton equation takes the following form

$$\frac{d}{dt}\rho = i[Dh(\rho), \rho], \tag{3.2.8}$$

where $h \in C^\infty(H^1(\mathcal{H}))$ is some fixed Hamiltonian. If $h(\rho) = \mathrm{Tr}(\rho H)$, $H \in H^\infty(\mathcal{H})$, i.e. the Hamiltonian is a linear function, then (3.2.8) is reduced to the Heisenberg equation

$$\frac{d}{dt}\rho = i[H, \rho]. \tag{3.2.9}$$

We conclude the above considerations by the remark that quantum mechanics is an infinite dimensional Hamiltonian mechanics.

Now, we have the necessary background to introduce the notion of the coherent states map. One sees that the space of pure states is an infinite dimensional strong symplectic Banach manifold what implies the impossibility to control experimentally all pure states, since, for example, one needs for this purpose infinite time. Therefore, one is obligated to consider only a finite parameter family of pure states. Moreover, it is natural to assume that these parameters form a manifold M and the way of parameterization is given by a map of M into the space of pure states

$$\mathcal{K} : M \longrightarrow \mathbb{CP}(\mathcal{H}). \tag{3.2.10}$$

Definition 3.2.1 A map (3.2.10) such that $\mathcal{K}(M)$ is linearly dense in $\mathcal{H}$ and the pullback $\mathcal{K}^*\omega_{FS} =: \omega$ of the Fubini-Study form ω_{FS} is a symplectic form, we shall call **coherent state map**.

If (3.2.10) is a coherent state map then (M, ω) can be considered as the phase space of a system in the sense of classical Hamiltonian mechanics. The pure states $\mathcal{K}(m)$, $m \in M$, we shall call the **coherent states** of the system. Identifying $\mathcal{K}(m)$ with m we shall think about the coherent state $\mathcal{K}(m)$ as the classical mechanical state.

Motivated by the above we introduce the following definition (see [14]).

Definition 3.2.2 By a **mechanical system** we shall understand the following triple:

(1) a symplectic manifold (M, ω);
(2) a complex separable Hilbert space $\mathcal{H}$;
(3) a symplectic map $\mathcal{K}$ of M into $\mathbb{CP}(\mathcal{H})$.

Additionally let us mention that the energy operator H occurring in the Heisenberg equation is an unbounded self-adjoint operator in general. Thus in this case (3.2.9) makes only sense for such $\rho \in \mathcal{S}$ for which the image $\rho(\mathcal{H})$ is contained in the domain $D(H)$ of H. Hence, fixing the energy operator and other observables of the mechanical system we shall assume that their domains contain $\mathcal{K}(M)$.

Now we come back to the statistical interpretation of quantum mechanics and discuss the metric structure of $\mathbb{CP}(\mathcal{H})$ in this context. To this end, let us fix two pure states

$$\iota([\psi]) = \frac{|\psi\rangle\langle\psi|}{\langle\psi|\psi\rangle} \quad \& \quad \iota([\varphi]) = \frac{|\varphi\rangle\langle\varphi|}{\langle\varphi|\varphi\rangle} \,, \tag{3.2.11}$$

where $\varphi, \psi \in \mathcal{H}$. Since $H^1(\mathcal{H}) \subset H^\infty(\mathcal{H})$ one can consider, for example the state $\iota([\varphi])$ as an observable. Thus, according to the standard statistical model of quantum mechanics, one assumes that the probability of finding the system in the state $\iota([\varphi])$, when one knows that it is in the state $\iota([\psi])$, is given by

$$\mathrm{Tr}(\iota([\psi])\iota([\varphi])) = \frac{|\langle\psi|\varphi\rangle|^2}{\langle\psi|\psi\rangle\langle\varphi|\varphi\rangle} \,. \tag{3.2.12}$$

The complex valued quantity

$$a([\psi], [\varphi]) := \frac{\langle\psi|\varphi\rangle}{\sqrt{\langle\psi|\psi\rangle\langle\varphi|\varphi\rangle}} \,, \tag{3.2.13}$$

called the **transition amplitude** between the pure states $\iota([\psi])$ and $\iota([\varphi])$, plays a fundamental role in quantum mechanical considerations, see for example Feynman [3]. The following formula

$$\|\iota([\psi]) - \iota([\varphi])\|_1 = \sqrt{2}(1 - |a([\psi], [\varphi])|^2)^{\frac{1}{2}} \tag{3.2.14}$$

explains the relation between the $\|\cdot\|_1$-distance and the transition probability $|a([\psi],[\varphi])|^2$. One sees from (3.2.14) that the transition probability from $\iota([\psi])$ to $\iota([\varphi])$ is nearly 1 if these states are close in the sense of the $\|\cdot\|_1$-metric. The sequence of states $\{\iota([\psi_n])\}_{n=0}^{\infty}$ of the physical system is a Cauchy sequence if starting from some state $\iota([\psi_N])$ the probability $|a([\psi_n],[\psi_m])|^2$ of successive transitions $\iota([\psi_n]) \to \iota([\psi_m])$ is arbitrarily close to one for $m, n > N$.

The transition probability $|a([\psi],[\varphi])|^2$ is a quantity measurable in a direct way. So, it is natural to assume that the set $\mathrm{Mor}(\mathbb{CP}(\mathcal{H}_1), \mathbb{CP}(\mathcal{H}_2))$ of morphisms between $\mathbb{CP}(\mathcal{H}_1)$ and $\mathbb{CP}(\mathcal{H}_2))$ consists of maps $\Sigma : \mathbb{CP}(\mathcal{H}_1) \longrightarrow \mathbb{CP}(\mathcal{H}_2))$ which preserve the corresponding transition probabilities, i.e. $\Sigma \in \mathrm{Mor}(\mathbb{CP}(\mathcal{H}_1), \mathbb{CP}(\mathcal{H}_2))$ if

$$|a_2(\Sigma([\psi]), \Sigma([\varphi]))|^2 = |a_1([\psi],[\varphi])|^2 \tag{3.2.15}$$

for any $[\psi], [\varphi] \in \mathbb{CP}(\mathcal{H})$, or equivalently the maps which preserve the $\|\cdot\|_1$ metric.

For two physical systems $(M_1, \mathcal{H}_1, \mathcal{K}_1 : M_1 \to \mathbb{CP}(\mathcal{H}_1))$ and $(M_2, \mathcal{H}_2, \mathcal{K}_2 : M_2 \to \mathbb{CP}(\mathcal{H}_2))$ we shall define morphisms by the following commutative diagram

$$\begin{array}{ccc} M_1 & \xrightarrow{\ \mathcal{K}_1\ } & \mathbb{CP}(\mathcal{H}_1) \\ {\scriptstyle \sigma}\downarrow & & \downarrow{\scriptstyle \Sigma} \\ M_2 & \xrightarrow{\ \mathcal{K}_2\ } & \mathbb{CP}(\mathcal{H}_2) \end{array} \tag{3.2.16}$$

where $\sigma \in SpC^{\infty}(M_1, M_2)$ is a symplectomorphism and $\Sigma \in \mathrm{Mor}(\mathbb{CP}(\mathcal{H}_1), \mathbb{CP}(\mathcal{H}_2))$. The morphism Σ is univocally defined by σ. It is so by Wigner theorem [24] due to the assumption that $\mathcal{K}(M)$ is linearly dense in $\mathcal{H}$.

Therefore mechanical systems form a category. We shall denote this category by $\mathfrak{P}$.

3.3   Different Representations of Mechanical Systems

The mechanical system defined in the previous section can be described in terms of some analytic or geometric objects defined on the classical phase space M.

We shall start with the coordinate description of the coherent state map. Let us fix an atlas $\{\Omega_\alpha, \Phi_\alpha\}_{\alpha \in I}$, where Ω_α is the open domain of the chart $\Phi_\alpha : \Omega_\alpha \to \mathbb{R}^n$, with the property that for any $\alpha \in I$ there exists smooth map

$$K_\alpha : \Omega_\alpha \to \mathcal{H} \tag{3.3.1}$$

such that $K_\alpha(q) \neq 0$ for $q \in \Omega_\alpha$. One has the consistency condition

$$K_\beta(q) = g_{\beta\gamma}(q)K_\gamma(q) \tag{3.3.2}$$

for $q \in \Omega_\beta \cap \Omega_\gamma$, where the maps

$$g_{\beta\gamma} : \Omega_\beta \cap \Omega_\gamma \longrightarrow \mathbb{C} \setminus \{0\} \tag{3.3.3}$$

form a smooth cocycle, i.e.

$$g_{\beta\gamma}(p) = g_{\beta\delta}(p)g_{\delta\gamma}(p) \tag{3.3.4}$$

for $p \in \Omega_\beta \cap \Omega_\gamma \cap \Omega_\delta$. The system of maps $\{\mathcal{K}_\alpha\}_{\alpha \in I}$ we shall call a trivialization of the coherent states map $\mathcal{K}$ if one has

$$\mathcal{K}(q) = [K_\alpha(q)] = \mathbb{C}K_\alpha(q) \tag{3.3.5}$$

for $q \in \Omega_\alpha$.

The coordinate description of the coherent state map is directly related to the straightforward construction of the coherent state map in the experimental way. For two fixed points $q \in \Omega_\alpha$ and $p \in \Omega_\beta$ the transition amplitude $a_{\overline{\alpha}\beta}(q,p)$ from the coherent state $\iota([K_\alpha(q)])$ to the coherent state $\iota([K_\beta(p)])$ has the following properties:

(1)

$$a_{\overline{\alpha}\beta}(q,p) = u_{\beta\gamma}(p)a_{\overline{\alpha}\gamma}(q,p) \tag{3.3.6}$$

for $p \in \Omega_\beta \cap \Omega_\gamma$, where $u_{\alpha\beta} : \Omega_\beta \cap \Omega_\gamma \to \mathbb{S}^1$ is a unitary cocycle;

(2)

$$\overline{a_{\overline{\alpha}\beta}(q,p)} = a_{\overline{\beta}\alpha}(p,q) \tag{3.3.7}$$

$$a_{\overline{\alpha}\alpha}(q,q) = 1 \tag{3.3.8}$$

for $q \in \Omega_\alpha$ and $p \in \Omega_\beta$;

(3)

$$\det \begin{pmatrix} a_{\overline{\alpha}_1\alpha_1}(q_1,q_1) & \cdots & a_{\overline{\alpha}_1\alpha_N}(q_1,q_N) \\ \vdots & & \vdots \\ a_{\overline{\alpha}_N\alpha_1}(q_N,q_1) & \cdots & a_{\overline{\alpha}_N\alpha_N}(q_N,q_N) \end{pmatrix} \geqslant 0 \tag{3.3.9}$$

for any $q_1 \in \Omega_{\alpha_1}, \ldots, q_N \in \Omega_{\alpha_N}$ and any $N \in \mathbb{N}$.

The transition amplitude $\{a_{\overline{\alpha}\beta}(q,p)\}$ is the quantity which can be directly obtained by the measurement procedure. Let us recall for this reason that $|a_{\overline{\alpha}\beta}(q,p)|^2$ is the transition probability and the phase $|a_{\overline{\alpha}\beta}(q,p)|^{-1} a_{\overline{\alpha}\beta}(q,p)$ is responsible for the quantum interference effects.

Having experimentally given transition amplitude one could obtain the coherent state map. In order to do that we shall introduce the notion of Hermitian kernel. Therefore let us consider the complex line bundle

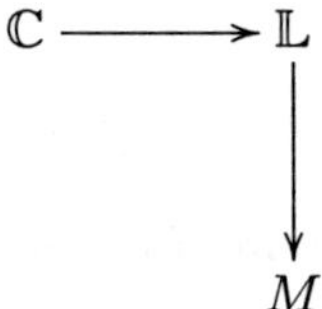

over the manifold M with a fixed local trivialization

$$s_\alpha : \Omega_\alpha \longrightarrow \mathbb{L} \tag{3.3.10}$$

$$g_{\alpha\beta} : \Omega_\alpha \cap \Omega_\beta \longrightarrow \mathbb{C} \setminus \{0\},$$

i.e. $s_\alpha(m) \neq 0$ for $m \in \Omega_\alpha$ and

$$s_\alpha = g_{\alpha\beta} s_\beta \quad \text{on } \Omega_\alpha \cap \Omega_\beta \tag{3.3.11}$$

and

$$g_{\alpha\beta} g_{\beta\gamma} = g_{\alpha\gamma} \quad \text{on } \Omega_\alpha \cap \Omega_\beta \cap \Omega_\gamma, \tag{3.3.12}$$

where $(\Omega_\alpha, \varphi_\alpha)_{\alpha \in I}$ forms an atlas of M.

Using the projections

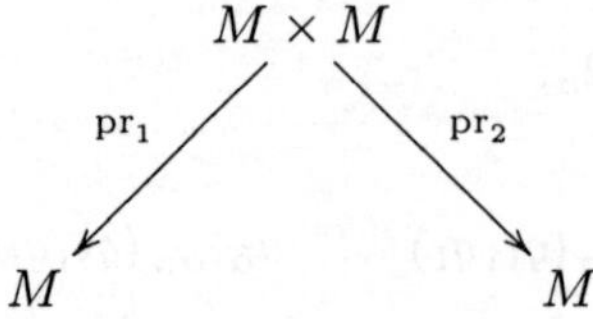

on the first and second components of the product $M \times M$ one can define the line bundle

$$\mathbb{C} \twoheadrightarrow \mathrm{pr}_1^* \overline{\mathbb{L}^*} \otimes \mathrm{pr}_2^* \mathbb{L}^*$$

$$\Big\downarrow \tag{3.3.13}$$

$$M \times M$$

with local trivialization defined by the tensor product

$$\mathrm{pr}_1^* \,\overline{s}_\alpha^* \otimes \mathrm{pr}_2^* \,\overline{s}_\beta^* : \Omega_\alpha \times \Omega_\beta \to \mathrm{pr}_1^* \overline{\mathbb{L}^*} \otimes \mathrm{pr}_2^* \mathbb{L}^* \tag{3.3.14}$$

of the pullbacks of the local frames given by (3.3.10).

Let us explain here that $\mathbb{L}^*$ is dual to $\mathbb{L}$ and $\overline{\mathbb{L}^*}$ is complex conjugation of $\mathbb{L}^*$. The line bundle (3.3.13) by definition is the tensor product of the pullbacks $\mathrm{pr}_1^* \overline{\mathbb{L}^*}$ and $\mathrm{pr}_2^* \mathbb{L}^*$ of $\overline{\mathbb{L}^*}$ and $\mathbb{L}^*$ respectively.

Definition 3.3.1 A section $K_{\mathbb{L}} \in C^\infty(M \times M, \mathrm{pr}_1^* \overline{\mathbb{L}^*} \otimes \mathrm{pr}_2^* \mathbb{L}^*)$ we shall call a positive Hermitian kernel iff

$$\overline{K_{\overline{\alpha}\beta}(q,p)} = K_{\overline{\beta}\alpha}(p,q),$$

$$K_{\overline{\alpha}\alpha}(q,q) > 0, \tag{3.3.15}$$

$$\sum_{k,j=1}^{N} K_{\overline{\alpha_j}\alpha_k}(q_j,q_k)\overline{v^j}v^k \geqslant 0$$

for any $q \in \Omega_\alpha$, $p \in \Omega_\beta$, $q_k \in \Omega_{\alpha_k}$, $v^1,\ldots,v^N \in \mathbb{C}$ and any set of indices $\alpha, \beta, \alpha_1, \ldots, \alpha_N$ resulting from a covering of M by open sets Ω_α, $\alpha \in I$, where

$$K_{\overline{\alpha}\beta} : \Omega_\alpha \times \Omega_\beta \longrightarrow \mathbb{C} \tag{3.3.16}$$

are the coordinate functions of $K_{\mathbb{L}}$ defined by

$$K_{\mathbb{L}} = K_{\overline{\alpha}\beta}(q,p) \,\mathrm{pr}_1^* \,\overline{s}_\alpha^*(q) \otimes \mathrm{pr}_2^* \,\overline{s}_\beta^*(p) \tag{3.3.17}$$

on $\Omega_\alpha \times \Omega_\beta$.

It follows immediately from the transformation rule

$$K_{\overline{\alpha}\beta}(q,p) = \overline{g_{\alpha\gamma}(q)}g_{\beta\delta}(p)K_{\overline{\gamma}\delta}(q,p) \tag{3.3.18}$$

for $q \in \Omega_\alpha \cap \Omega_\gamma$ and $p \in \Omega_\beta \cap \Omega_\delta$ that the conditions (3.3.15) are independent with respect to the choice of frame.

The relation of $K_{\mathbb{L}}$ to the transition amplitude on M is recognized by noticing that

$$a_{\overline{\alpha},\beta}(q,p) := \frac{K_{\overline{\alpha}\beta}(q,p)}{K_{\overline{\alpha}\alpha}(q,q)^{\frac{1}{2}} K_{\overline{\beta}\beta}(p,p)^{\frac{1}{2}}} \tag{3.3.19}$$

fulfills the properties (3.3.6)-(3.3.9). So, we shall identify transition amplitudes with the positive Hermitian kernels.

The line bundles with distinguished positive Hermitian kernels ($\mathbb{L} \to M, K_{\mathbb{L}}$) form a category $\mathfrak{K}$ for which the morphisms set $\mathrm{Mor}[(\mathbb{L}_1 \to M_1, K_{\mathbb{L}_1}), (\mathbb{L}_2 \to M_2, K_{\mathbb{L}_2})]$ is given by $f : M_2 \to M_1$ such that

$$\mathbb{L}_2 = f^*\mathbb{L}_1 = \{(m,\xi) \in M_2 \times \mathbb{L}_1 : f(m) = \pi_1(\xi)\} \tag{3.3.20}$$

and

$$K_{\mathbb{L}_2} = f^*K_{\mathbb{L}_1} = K^1_{\overline{\alpha}\beta}(f(q),f(p))\,\mathrm{pr}_1^* \overline{s^{1*}_{\alpha}(f(q))} \otimes \mathrm{pr}_2^* s^{1*}_{\beta}(f(p)) \tag{3.3.21}$$

i.e.

$$K^2_{\overline{\alpha}\beta}(q,p) = K^1_{\overline{\alpha}\beta}(f(q),f(p)) \tag{3.3.22}$$

for $q \in f^{-1}(\Omega_\alpha)$ and $p \in f^{-1}(\Omega_\beta)$.

The above expresses the covariant character of the transition amplitude and its independence of the choice of the coordinates.

There is a covariant functor

$$\mathcal{F}_{\mathcal{KP}} : \mathfrak{P} \longrightarrow \mathfrak{K} \tag{3.3.23}$$

between the category of physical systems $\mathfrak{P}$ and the category $\mathfrak{K}$ of positive Hermitian kernels, naturally defined by

$$\mathbb{L} = \mathcal{K}^*\mathbb{E} = \{(m,\xi) \in M \times \mathbb{E} : \mathcal{K}(m) = \pi(\xi)\} \tag{3.3.24}$$

and by $K_{\mathbb{L}} = \mathcal{K}^*K_{\mathbb{E}}$, i.e.

$$K_{\overline{\alpha}\beta}(q,p) = \langle K_\alpha(q)|K_\beta(p)\rangle, \tag{3.3.25}$$

where $K_\alpha : \Omega_\alpha \to \mathbb{C} \setminus \{0\}$ is given by (3.3.1)-(3.3.5). The functor $\mathcal{F}_{\mathfrak{K}\mathfrak{P}}$ maps $(\sigma,\Sigma) \in \mathrm{Mor}(M_1 \xrightarrow{\mathcal{K}_1} \mathbb{CP}(\mathcal{H}_1), M_2 \xrightarrow{\mathcal{K}_2} \mathbb{CP}(\mathcal{H}_2))$ to the morphism $f \in \mathrm{Mor}[(\mathbb{L}_1 \to M_1, K_{\mathbb{L}_1}), (\mathbb{L}_2 \to M_2, K_{\mathbb{L}_2})]$ by $f = \sigma$.

Let us recall that the tautological complex line bundle

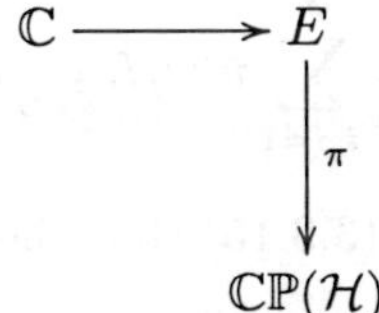

over $\mathbb{CP}(\mathcal{H})$ is defined by

$$\mathbb{E} := \{(\psi, l) \in \mathcal{H} \times \mathbb{CP}(\mathcal{H}) : \psi \in l\} \qquad (3.3.26)$$

and the bundle projection π is by definition the projection on the second component of the product $\mathcal{H} \times \mathbb{CP}(\mathcal{H})$. The bundle fibre $\pi^{-1}(l) =: \mathbb{E}_l$ is given by the complex line $l \subset \mathcal{H}$. With the use of the projection $\mu : \mathbb{E} \to \mathcal{H}$ on the first factor of the product $\mathcal{H} \times \mathbb{CP}(\mathcal{H})$ we define the Hermitian kernel $K_{\mathbb{E}}(l, k) : \pi^{-1}(l) \times \pi^{-1}(k) \to \mathbb{C}$ by

$$K_{\mathbb{E}}(l, k)(\xi, \eta) := \langle \mu(\xi) | \mu(\eta) \rangle, \qquad (3.3.27)$$

where $\xi \in \pi^{-1}(l)$ and $\eta \in \pi^{-1}(k)$. It follows directly from the definition that $K_{\mathbb{E}}$ is a smooth section of the bundle

$$\mathrm{pr}_1^* \overline{\mathbb{E}}^* \otimes \mathrm{pr}_2^* \mathbb{E}^* \longrightarrow \mathbb{CP}(\mathcal{H}) \times \mathbb{CP}(\mathcal{H}), \qquad (3.3.28)$$

where $\mathrm{pr}_1^* \overline{\mathbb{E}}^* \to \mathbb{CP}(\mathcal{H}) \times \mathbb{CP}(\mathcal{H})$ is the pull back of the dual of the complex conjugated bundle $\overline{\mathbb{E}}^*$ given by the projection $\mathrm{pr}_1 : \mathbb{CP}(\mathcal{H}) \times \mathbb{CP}(\mathcal{H}) \to \mathbb{CP}(\mathcal{H})$ on the first factor of the product and $\mathrm{pr}_2^* \mathbb{E}^* \to \mathbb{CP}(\mathcal{H}) \times \mathbb{CP}(\mathcal{H})$ is the pull back of $\mathbb{E}^*$ by the projection pr_2 on the second factor.

Therefore, the tautological bundle $\mathbb{E} \to \mathbb{CP}(\mathcal{H})$ has a canonically defined Hermitian kernel $K_{\mathbb{E}} \in \Gamma^\infty(\mathrm{pr}_1^* \overline{\mathbb{E}}^* \otimes \mathrm{pr}_2^* \mathbb{E}^*, \mathbb{CP}(\mathcal{H}) \times \mathbb{CP}(\mathcal{H}))$.

The positive Hermitian kernel $K_{\mathbb{L}}$ defines a complex separable Hilbert space $\mathcal{H}_{\mathbb{L}}$ realized as a vector subspace of the space $\Gamma(M, \overline{\mathbb{L}^*})$ of the sections of the bundle $\overline{\mathbb{L}}^* \to M$. One obtains $\mathcal{H}_{\mathbb{L}}$ in the following way. Let us take the vector space $V_{K,\mathbb{L}}$ of finite linear combinations

$$v = \sum_{i=1}^{N} v_i K_{\beta_i}(q_i) \qquad (3.3.29)$$

of sections

$$K_{\beta_i}(q_i) = K_{\overline{\beta}\beta_i}(p, q_i) \overline{s_\beta^*}(p) \in \Gamma(M, \overline{\mathbb{L}^*}), \qquad (3.3.30)$$

where $q_i \in \Omega_{\beta_i}$ and $p \in \Omega_\beta$, with the scalar product defined by

$$\langle v|w \rangle := \sum_{i,j=1}^{N} \overline{v_i} w_j K_{\overline{\beta}_i \beta_j}(q_i, q_j). \qquad (3.3.31)$$

It follows from the properties (3.3.15) that the pairing (3.3.31) is sesquilinear and that

$$\left| \sum_{i=1}^{N} \overline{v_i} K_{\beta_i \beta}(q_i, p) \right|^2 = |\langle v|K_\beta(p)\rangle|^2 \leqslant \langle v|v \rangle K_{\overline{\beta}\beta}(p, p), \qquad (3.3.32)$$

from which one has $v = 0$ iff $\langle v|v \rangle = 0$. Therefore (3.3.31) defines a positive definite scalar product on $V_{K,\mathbb{L}}$.

Proposition 3.3.2 *The unitary space $V_{K,\mathbb{L}}$ extends in a canonical and unique way to the Hilbert space $\mathcal{H}_{K,\mathbb{L}}$, which is a vector subspace of $\Gamma(M, \overline{L^*})$.*

Obviously, for $v \in \mathcal{H}_{K,\mathbb{L}}$ one has

$$v = v_\alpha(p)\overline{s_\alpha^*} = \langle K_\alpha(p)|v \rangle \overline{s_\alpha^*}, \qquad (3.3.33)$$

which shows that the evaluation functional $e_\alpha(p) : \mathcal{H}_{K,\mathbb{L}} \to \mathbb{C}$ defined by

$$e_\alpha(p)(v) := v_\alpha(p) \qquad (3.3.34)$$

is a continuous linear functional and $e_\alpha(p)$ depends smoothly on $p \in \Omega_\alpha$. Hence, we see that the Hilbert space $\mathcal{H}_{K,\mathbb{L}} \subset \Gamma(M, \overline{\mathbb{L}^*})$ possesses the property that the evaluation functionals $e_\alpha(p) : \mathcal{H}_{K,\mathbb{L}} \to \mathbb{C}$ are continuous and define smooth maps

$$e_\alpha : \Omega_\alpha \to \mathcal{H}_{K,\mathbb{L}}^* \setminus \{0\} \qquad (3.3.35)$$

for $\alpha \in I$. Since $e_\alpha(p)(K_\alpha(p)) = K_{\overline{\alpha}\alpha}(p, p) > 0$, e_α does not take zero value in $\mathcal{H}_{K,\mathbb{L}} \equiv \mathcal{H}_{K,\mathbb{L}}^*$.

In particular, the canonical Hermitian kernel $K_{\mathbb{E}} \in \Gamma^\infty(\mathbb{CP}(\mathcal{H}) \times \mathbb{CP}(\mathcal{H}), \mathrm{pr}_1^* \overline{\mathbb{E}}^* \otimes \mathrm{pr}_2^* \overline{\mathbb{E}}^*)$ defines the Hilbert space $\mathcal{H}_{\mathbb{E}} := I(\mathcal{H}) \subset \Gamma(\mathbb{CP}(\mathcal{H}), \overline{\mathbb{E}}^*)$, where the vector spaces monomorphism I is defined by

$$I : \mathcal{H} \ni \psi \longrightarrow \langle \mu(\cdot)|\psi \rangle \in \Gamma(\mathbb{CP}(\mathcal{H}), \overline{\mathbb{E}}^*). \qquad (3.3.36)$$

Motivated by the preceding construction let us introduce the category $\mathfrak{H}$ of line bundles $\mathbb{L} \to M$ with distinguished Hilbert space $H_{\mathbb{L}}$ which is realized as a vector subspace of $\Gamma(M, \overline{L^*})$ and has the property that the evaluation functionals $e_\alpha(p)$ are continuous, i.e. $\|e_\alpha(p)(v)\| \leqslant M_{\alpha,p}\|v\|$ for $v \in \mathcal{H}_\mathbb{L}$, $M_{\alpha,p} > 0$ and define smooth maps $e_\alpha : \Omega_\alpha \to \mathcal{H}_\mathbb{L}^* \setminus \{0\}$.

By definition, the morphisms set

$$\mathrm{Mor}[(\mathbb{L}_1 \to M_1, \mathcal{H}_{\mathbb{L}_1}), (\mathbb{L}_2 \to M_2, \mathcal{H}_{\mathbb{L}_2})] \tag{3.3.37}$$

will consist of maps $f : M_2 \to M_1$ which satisfy $f^*\mathbb{L}_1 = \mathbb{L}_2$ and $f^*\mathcal{H}_{\mathbb{L}_1} = \mathcal{H}_{\mathbb{L}_2}$. In order to prove the correctness of the definition let us show that the vector space

$$f^*\mathcal{H}_{\mathbb{L}_1} = \{f^*v | v \in \mathcal{H}_{\mathbb{L}_1}\} \tag{3.3.38}$$

of inverse image sections has a canonically defined Hilbert space structure with continuous evaluation functionals smoothly dependent on the argument.

It is easy to see that

$$\ker f^* = \{v \in \mathcal{H}_{\mathbb{L}_1} | f^*v = 0\} \tag{3.3.39}$$

is the Hilbert subspace of $\mathcal{H}_{\mathbb{L}_1}$. We define the Hilbert space structure on $f^*\mathcal{H}_{\mathbb{L}_1}$ by the vector spaces identifications

$$f^*\mathcal{H}_{\mathbb{L}_1} \equiv \mathcal{H}_{\mathbb{L}_1} / \ker f^* \equiv (\ker f^*)^\perp \tag{3.3.40}$$

i.e. $f^*\mathcal{H}_{\mathbb{L}_1}$ inherits the Hilbert space structure from the Hilbert subspace $(\ker f^*)^\perp$. In order to prove the property (3.3.35) for $f^*e_\alpha(p) = e_\alpha(f(p))$ we notice that

$$|(f^*v)_\alpha(p)| = |v_\alpha(f(p))| \leqslant M_{\alpha,f(p)}(\|\psi^0\| + \|\psi^\perp\|) \tag{3.3.41}$$

for $p \in f^{-1}(\Omega_\alpha)$. Because of $\psi_\alpha^0(f(p)) = 0$, the left hand side of the inequality (3.3.35) does not depend on $\psi^0 \in \ker f^*$. This results in

$$|(f^*(v)_\alpha(p))| \leqslant M_{\alpha,f(p)} \min_{\psi^0 \in \ker f^*} (\|\psi^0\| + \|\psi^\perp\|) = M_{\alpha,f(p)}\|\psi^\perp\|$$

$$= M_{\alpha,f(p)}\|f^*\psi\|. \tag{3.3.42}$$

The above proves the continuity of the evaluation functionals f^*e_α. The smooth dependence of $f^*e_\alpha(p) = e_\alpha(f(p))$ on p follows from the smoothness of f.

In such a way the category $\mathfrak{H}$ is defined correctly. It is easy to see that the pair $(\mathbb{E} \to \mathbb{CP}(\mathcal{H}), \mathcal{H}_\mathbb{E})$ forms the universal object in this category.

Proposition 3.3.3 *Let* $f : M_2 \to M_1$ *be such that* $f^*\mathbb{L}_1 = \mathbb{L}_2$ *and* $f^*K_1 = K_2$ *then* $f^*\mathcal{H}_{\mathbb{L}_1,K_1} = \mathcal{H}_{\mathbb{L}_2,K_2}$.

Proposition 3.3.2 implies that there is canonically defined covariant functor $\mathcal{F}_{\mathfrak{H},\mathfrak{K}} : \mathfrak{K} \to \mathfrak{H}$:

$$\mathcal{F}_{\mathfrak{H}\mathfrak{K}}(\mathbb{L} \to M, K) = (\mathbb{L} \to M, \mathcal{H}_{\mathbb{L},K}) \tag{3.3.43}$$

from the category $\mathfrak{K}$ of line bundles with distinguished positive Hermitian kernel $K_{\mathbb{L}} \in \Gamma(M \times M, \mathrm{pr}_1^* \overline{\mathbb{L}^*} \otimes \mathrm{pr}_2^* \mathbb{L}^*)$ to the category $\mathfrak{H}$ of line bundles with distinguished Hilbert space $\mathcal{H}_{\mathbb{L}} \subset \Gamma^\infty(M, \overline{\mathbb{L}^*})$ with the condition (3.3.34) on the evaluation functionals.

Now let us discuss the relation between the category $\mathfrak{H}$ and the category of physical systems $\mathfrak{P}$.

Let $(\mathbb{L} \to M, \mathcal{H}_{\mathbb{L}})$ be an object of the category $\mathfrak{H}$. Taking the smooth maps

$$K_\alpha : \Omega_\alpha \to \mathcal{H}_{\mathbb{L}} \setminus \{0\}, \tag{3.3.44}$$

which represent the evaluation functional maps $e_\alpha : \Omega_\alpha \to \mathcal{H}_{\mathbb{L}} \setminus \{0\}$

$$e_\alpha(p) = \langle K_\alpha(p) | \cdot \rangle \tag{3.3.45}$$

on Ω_α, we construct the smooth map $\mathcal{K}^{\mathbb{L}} : M \to \mathbb{CP}(\mathcal{H}_{\mathbb{L}})$ given by

$$\mathcal{K}^{\mathbb{L}}(q) := \mathbb{C}K_\alpha^{\mathbb{L}}(q). \tag{3.3.46}$$

Because of $K_\alpha(q) = g_{\alpha\beta}(q)K_\beta(q)$ the definition (3.3.46) of $\mathcal{K}^{\mathbb{L}}$ is independent on the choice of frame. The smoothness of $\mathcal{K}^{\mathbb{L}}$ is ensured by the one of $e_\alpha : \Omega_\alpha \to \mathcal{H}_{\mathbb{L}}^*$.

Proposition 3.3.4 *The correspondence*

$$\mathcal{F}_{\mathfrak{P}\mathfrak{H}}[(\mathbb{L} \to M, \mathcal{H}_{\mathbb{L}})] := (M, \mathcal{H}_{\mathbb{L}}, \mathcal{K}^{\mathbb{L}} : M \to \mathbb{CP}(\mathcal{H}^*)) \tag{3.3.47}$$

$$\mathcal{F}_{\mathfrak{P}\mathfrak{H}}(f^*) := (f[\varphi_f]),$$

where $(\mathbb{L} \to \Gamma, \mathcal{H}_{\mathbb{L}}) \in \mathcal{O}b(\mathfrak{H})$, $f^* \in \mathrm{Mor}[(\mathbb{L}_1 \to M_1, \mathcal{H}_{\mathbb{L}_1}), (\mathbb{L}_2 \to M_2, \mathcal{H}_{\mathbb{L}_2})]$ *and* $\varphi_f : \mathcal{H}_{\mathbb{L}_2} \to \mathcal{H}_{\mathbb{L}_1}$ *is the monomorphism given by*

$$K_{1\alpha}(f(p)) = K_{1\alpha}^\perp(f(p)) =: \varphi(K_{2\alpha}(p)), \tag{3.3.48}$$

defines a contravariant functor

$$\mathcal{F}_{\mathfrak{P}\mathfrak{H}} : \mathfrak{H} \longrightarrow \mathfrak{P}. \tag{3.3.49}$$

Summing up the statements discussed above one has

Proposition 3.3.5 *The categories $\mathfrak{K}$, $\mathfrak{H}$ and $\mathfrak{P}$ satisfy the following relation*

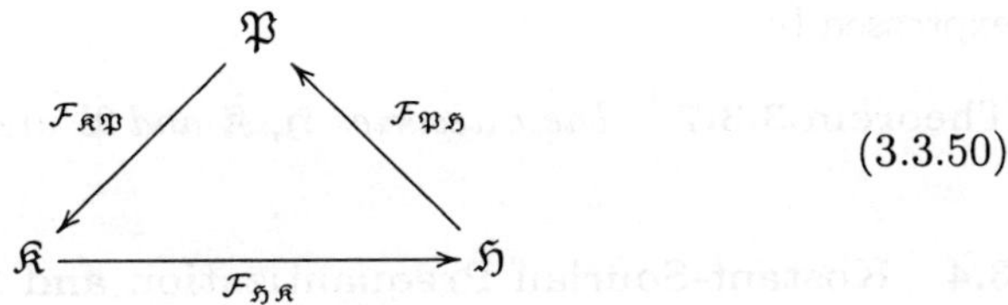

$$(3.3.50)$$

i.e. the functors defined by $\mathcal{F}_{\mathfrak{P}\mathfrak{H}} \circ \mathcal{F}_{\mathfrak{H}\mathfrak{K}} =: \mathcal{F}_{\mathfrak{P}\mathfrak{K}}$, $\mathcal{F}_{\mathfrak{K}\mathfrak{P}} \circ \mathcal{F}_{\mathfrak{P}\mathfrak{H}} =: \mathcal{F}_{\mathfrak{K}\mathfrak{H}}$ and $\mathcal{F}_{\mathfrak{H}\mathfrak{K}} \circ \mathcal{F}_{\mathfrak{K}\mathfrak{P}} =: \mathcal{F}_{\mathfrak{H}\mathfrak{P}}$ are inverse to $\mathcal{F}_{\mathfrak{K}\mathfrak{P}}$, $\mathcal{F}_{\mathfrak{H}\mathfrak{K}}$ and $\mathcal{F}_{\mathfrak{P}\mathfrak{H}}$ respectively. Moreover, the functors $\mathcal{F}_{\mathfrak{K}\mathfrak{H}}$ and $\mathcal{F}_{\mathfrak{K}\mathfrak{P}}$ are given explicitly by

$$\mathcal{F}_{\mathfrak{K}\mathfrak{H}}[(\mathbb{L} \to M, \mathcal{H}_{\mathbb{L}})] = (\mathbb{L} \to M, K = \langle K_\alpha | K_\beta \rangle \operatorname{pr}_1^* \overline{\mathbb{L}}^* \otimes \operatorname{pr}_2^* \mathbb{L}) \quad (3.3.51)$$

$$\mathcal{F}_{\mathfrak{K}\mathfrak{H}}(f^*) = f^*,$$

where $K_\alpha : \Omega_\alpha \to \mathcal{H}_{\mathbb{L}} \setminus \{0\}$ is given by (3.3.45), and

$$\mathcal{F}_{\mathfrak{H}\mathfrak{P}}[(M, \mathcal{H}, \mathcal{K} : M \to \mathbb{CP}(\mathcal{H}))] = (\mathcal{K}^* E \to M, \mathcal{K}^* \mathcal{H}_E) \quad (3.3.52)$$

$$\mathcal{F}_{\mathfrak{H}\mathfrak{P}}(\sigma, \Sigma) = \sigma.$$

The following definition introduces an equivalence among the objects taken into consideration.

Definition 3.3.6

(1) The objects $(\mathbb{L} \to M, K_{\mathbb{L}}), (\mathbb{L}' \to M', K'_{\mathbb{L}'}) \in Ob(\mathfrak{K})$ are equivalent iff $M = M'$ and there exists a bundle isomorphism $\kappa : \mathbb{L} \to \mathbb{L}'$ such that $\kappa^* K'_{\mathbb{L}'} = K_{\mathbb{L}}$.

(2) The objects $(\mathbb{L} \to M, \mathcal{H}_{\mathbb{L}}), (\mathbb{L}' \to M', \mathcal{H}'_{\mathbb{L}'}) \in Ob(\mathfrak{h})$ are equivalent iff $M = M'$ and there is a bundle isomorphism $\kappa : \mathbb{L} \to \mathbb{L}'$ such that $\kappa^* H'_{\mathbb{L}'} = \mathcal{H}_{\mathbb{L}}$.

(3) The objects $(M, \mathcal{H}, \mathcal{K} : M \to \mathbb{CP}(M)), (M', \mathcal{H}', \mathcal{K}' : M' \to \mathbb{CP}(M')) \in Ob(\mathcal{P})$ are equivalent iff $M = M'$ and there is an automorphism $\Sigma : \mathbb{CP}(\mathcal{H}) \to \mathbb{CP}(\mathcal{H}')$ such that $\mathcal{K}' = \Sigma \circ \mathcal{K}$.

These equivalences are preserved by morphisms between all three categories. This allows us to define the categories $\tilde{\mathfrak{H}}$, $\tilde{\mathfrak{K}}$ and $\tilde{\mathfrak{P}}$ whose objects consist of the above described equivalence classes and morphisms that are generated by morphisms of the categories $\mathfrak{H}$, $\mathfrak{K}$ and $\mathfrak{P}$ respectively.

The main result of the considerations given above, which shows that there are three independent ways of presentation of a physical systems, is expressed by

Theorem 3.3.7 *The categories $\tilde{\mathfrak{H}}$, $\tilde{\mathfrak{K}}$ and $\tilde{\mathfrak{P}}$ are isomorphic.*

3.4   Kostant-Souriau Prequantization and Positive Hermitian Kernels

Now, we shall present indispensable for the investigated theory of physical systems, elements of the geometric quantization in the sense of Kostant [9] and Souriau [23]. The theory is based on the notion of a complex line bundle $\mathbb{L} \to M$ with fixed Hermitian metric $H \in C^\infty(M, \overline{\mathbb{L}}^* \otimes \mathbb{L}^*)$ and metrical connection $\nabla : C^\infty(\Omega, \mathbb{L}) \to C^\infty(\Omega, \mathbb{L} \otimes T^*M)$, i.e.

(1)

$$\nabla(fs) = df \otimes s + f\nabla s \tag{3.4.1}$$

(2)

$$dH(s,t) = H(\nabla s, t) + H(s, \nabla t) \tag{3.4.2}$$

for every local smooth sections $s, t \in C^\infty(\Omega, \mathbb{L})$ and $f \in C^\infty(\Omega)$, where Ω is the open subset of M. Let $s_\alpha : \Omega_\alpha \to \mathbb{L}$, $\alpha \in I$ be a local trivialization of $\mathbb{L} \to M$, see (3.3.10). According to the property of (3.4.1) one gives ∇ and H univocally by defining them on the local frames

$$\nabla s_\alpha = k_\alpha \otimes s_\alpha \tag{3.4.3}$$

$$H(s_\alpha, s_\alpha) = H_{\overline{\alpha}\alpha}, \tag{3.4.4}$$

where $k_\alpha \in C^\infty(\Omega_\alpha, T^*M)$ and $0 < H_{\overline{\alpha}\alpha} \in C^\infty(\Omega_\alpha)$ and assuming the transformation rules

$$k_\alpha(m) = k_\beta(m) + g_{\alpha\beta}^{-1}(m)dg_{\alpha\beta}(m) \tag{3.4.5}$$

$$H_{\overline{\alpha}\alpha}(m) = |g_{\alpha\beta}(m)|^2 H_{\overline{\beta}\beta}(m) \tag{3.4.6}$$

for $m \in \Omega_\alpha \cap \Omega_\beta$, where the cocycle $g_{\alpha\beta} : \Omega_\alpha \cap \Omega_\beta \to \mathbb{C} \setminus \{0\}$ is defined by $s_\alpha = g_{\alpha\beta}s_\beta$. Let us remark here that since $\mathbb{L} \to M$ is a complex line bundle, the connection 1-form

$$k_\alpha(x) = k_{\alpha\mu}(x)dx^\mu, \tag{3.4.7}$$

where $(x^1, \ldots, x^n)$ are real coordinates on Ω_α, takes complex values, i.e.

$$k_{\alpha\mu} : \Omega_\alpha \longrightarrow \mathbb{C}. \tag{3.4.8}$$

The consistency condition (3.4.4) locally has the form

$$d \log H_{\overline{\alpha}\beta} = \overline{k}_\alpha + k_\alpha. \tag{3.4.9}$$

Thus from the gauge transformation (3.4.5) one obtains that

$$\operatorname{curv} \nabla := dk_\alpha \qquad \text{on } \Omega_\alpha \tag{3.4.10}$$

is a globally defined $i\mathbb{R}$-valued 2-form, i.e. a curvature form for the Hermitian connection defined on the $\mathbb{C}^*$-principal bundle $\mathbb{L}' \to M$. By definition we shall consider $\mathbb{L}' \to M$ as the subbundle of $\mathbb{L} \to M$ consisting of elements $\xi \in \pi^{-1}(m)$ with $H(m)(\xi, \xi) \neq 0$.

If one assumes

$$g_{\alpha\beta} = e^{2\pi i c_{\alpha\beta}} \tag{3.4.11}$$

then

$$c_{\alpha\beta\gamma} := c_{\alpha\beta} + c_{\beta\gamma} + c_{\gamma\alpha} \tag{3.4.12}$$

is $\mathbb{Z}$-valued cocycle on M related to the covering $\{\Omega_\alpha\}_{\alpha \in I}$, which defines an element $c_1(\mathbb{L}) \in H^2(M, \mathbb{Z})$ called the first Chern class of the bundle $\mathbb{L} \to M$, see for example [9]. Because of

$$2\pi i dc_{\alpha\beta} = k_\alpha - k_\beta \tag{3.4.13}$$

the real-valued form

$$\omega := \frac{1}{2\pi i} \operatorname{curv} \nabla \tag{3.4.14}$$

satisfies

$$[\omega] = c_1(\mathbb{L}) \in H^2(M, \mathbb{Z}), \tag{3.4.15}$$

i.e. it has integer cohomology class. So, (3.4.15) is a necessary condition for the closed form $2\pi i\omega$ to be a curvature form of a Hermitian connection on the complex line bundle. It follows from the Narasimhan and Ramanan result, see [12], that this is also a sufficient condition. We will come back below to this question.

One has the identity

$$[\nabla_x, \nabla_y] - \nabla_{[x,y]} = 2\pi i \omega(x,y) \qquad (3.4.16)$$

which can be proved by direct computation.

Now, let us assume that the curvature 2-form is non-singular. Thus, since $d\omega = 0$, ω is a symplectic form and one can define the Poisson bracket for $f, g \in C^\infty(M, \mathbb{R})$ as usual by

$$\{f, g\} = \omega(X_f, X_g) = -X_f(g), \qquad (3.4.17)$$

where X_f is the Hamiltonian vector field defined by

$$\omega(X_f, \cdot) = -df. \qquad (3.4.18)$$

It was the idea of Souriau (and Kostant) to consider the differential operator $Q_f : C^\infty(M, \mathbb{L}) \to C^\infty(M, \mathbb{L})$ defined by

$$Q_f := \nabla_{X_f} + 2\pi i f \qquad (3.4.19)$$

for $f \in C^\infty(M, \mathbb{R})$. It is easy to see from (3.4.1) and (3.4.16) that

$$Q_{\{f,g\}} = [Q_f, Q_g] \qquad (3.4.20)$$

i.e. the map Q called **Kostant-Souriau prequantization** is a homomorphism of the Poisson-Lie algebra $(C^\infty(M, \mathbb{R}), \{\cdot, \cdot\})$ into the Lie algebra of first-order differential operators acting in the space $C^\infty(M, \mathbb{L})$ of smooth sections of the line bundle $\mathbb{L} \to M$.

At this moment we are far from the quantization of a classical mechanical observable $f \in C^\infty(M, \mathbb{R})$. For this reason it is necessary to construct a Hilbert space $\mathcal{H}_{\mathbb{L}}$ related to $C^\infty(M, \mathbb{L})$ in which the differential operator Q_f can be extended to a self-adjoint operator $\overline{Q_f}$ being the quantum counterpart of f. An effort in this direction was done by using the notion of polarization, see for example [25]. In [17] we explain how one can obtain the polarization from the coherent states map, which is the most fundamental physical object.

After this short review of the Kostant-Souriau geometric prequantization, we shall describe how it is related to our model of a mechanical (physical) system. In order to do this let us fix a line bundle $\mathbb{L} \to M$ with distinguished positive Hermitian kernel $K_{\mathbb{L}}$, which as it was shown, equivalently describes the mechanical system. We define the differential 2-forms

$\omega_{1,2}$ and $\omega_{2,1}$ on the product $M \times M$ by

$$\omega_{12} = id_1 d_2 \log K_{\overline{\alpha}_1 \alpha_2} \tag{3.4.21}$$

$$\omega_{21} = id_2 d_1 \log K_{\overline{\alpha}_2 \alpha_1}, \tag{3.4.22}$$

where the $K_{\overline{\alpha}_1 \alpha_2}$ are coordinates of $K_{\mathbb{L}}$ in the local frames

$$\mathrm{pr}_1^* \, \overline{s}_{\alpha_1}^* \otimes \mathrm{pr}_2^* \, s_{\alpha_2}^* : \Omega_{\alpha_1} \times \Omega_{\alpha_2} \to \mathrm{pr}_1^* \, \overline{\mathbb{L}}^* \otimes \mathrm{pr}_2 \, \mathbb{L}^*. \tag{3.4.23}$$

The operations d_1 and d_2 are differentials with respect to the first and the second component of the product $M \times M$, respectively. The complete differential on $M \times M$ is their sum $d = d_1 + d_2$.

From the transformation rule (3.3.18) and from the hermicity of $K_{\mathbb{L}}$ we get the following properties of ω_{12} and ω_{21}.

Proposition 3.4.1 *The 2-form ω_{12} does not depend on the choice of a trivialization and has the following properties:* $d\omega_{21} = 0$, $\omega_{12} = \overline{\omega_{21}} = -\omega_{21}$.

Let us also consider 1-forms

$$k_{2\alpha_2} := d_2 \log K_{\overline{\alpha}_1 \alpha_2} \tag{3.4.24}$$

$$k_{1\overline{\alpha}_1} := d_1 \log K_{\overline{\alpha}_1 \alpha_2} \tag{3.4.25}$$

which are independent of the indices $\overline{\alpha}_1$ and $\overline{\alpha}_2$, respectively, and satisfy the transformation rules

$$k_{2\alpha_2} = k_{2\beta_2} + d_2 \log g_{\alpha_2 \beta_2} \tag{3.4.26}$$

$$k_{1\overline{\alpha}_1} = k_{1\overline{\beta}_2} + d_1 \log \overline{g_{\alpha_1 \beta_1}}. \tag{3.4.27}$$

Let $\Delta : M \to M \times M$ be the diagonal embedding i.e. $\Delta(m) = (m, m)$ for $m \in M$. We introduce the following notation

$$\Delta^* K = H \qquad \frac{1}{2\pi i} \Delta^* \omega_{12} = \omega \quad \text{and} \quad \Delta^* k_{2\alpha} = k_\alpha. \tag{3.4.28}$$

Now, it is easy to see that the following proposition is valid.

Proposition 3.4.2

(1) H defined by (3.4.28) is a positive Hermitian metric on $\mathbb{L}$.

(2) The 1-form $k_\alpha \in C^\infty(\Omega_\alpha, \mathbb{L} \otimes T^*M)$ $(\overline{k}_\alpha \in C^\infty(\Omega_\alpha, \overline{\mathbb{L}} \otimes T^*M)$ defined by (3.4.28) gives a local representative of a connection ∇ $(\overline{\nabla})$ on the bundle $\mathbb{L}$ $(\overline{\mathbb{L}})$.

(3) $\mathrm{curv}\,\nabla = 2\pi i\omega$ for ω defined by (3.4.28).

(4) The connection ∇ is metric with respect to H.

According to [9] we introduce the following terminology.

Definition 3.4.3 A line bundle $\mathbb{L} \to M$ with distinguished Hermitian metric H and a connection ∇ satisfying the consistency condition (3.4.2) we shall call **pre-quantum bundle** and denote it by $(\mathbb{L} \to M, H, \nabla)$.

The pre-quantum line bundles form a category with the morphisms defined in the standard way. We shall call $\mathcal{L}$ the category of pre-quantum bundles.

From [12; 14] one can obtain

Proposition 3.4.4 *For any pre-quantum bundle $(\mathbb{L} \to M, H, \nabla)$ there exists a smooth map $\mathcal{K} : M \to \mathbb{CP}(\mathcal{H})$ such that*

$$(\mathbb{L} \to M, H, \nabla) = (\mathcal{K}\mathbb{E} \to M, \mathcal{K}^*H_{FS}, \mathcal{K}^*\nabla_{FS}) \tag{3.4.29}$$

i.e. the line bundle $\mathbb{L} \to M$, the Hermitian metric H and the metric connection ∇ can be obtained as the corresponding pullbacks of their counterparts $\mathbb{E} \to \mathbb{CP}(\mathcal{H})$, H_{FS} and ∇_{FS} on the complex projective Hilbert space $\mathbb{CP}(\mathcal{H})$

From Proposition 3.4.4 and Theorem 3.3.7 one concludes that the construction given by formula (3.4.24)-(3.4.28) defines a covariant functor from the category of positive Hermitian kernels $\mathfrak{K}$ to the category of pre-quantum line bundles $\mathcal{L}$.

Taking the above remarks into account this metric structure H, the connection ∇ and curvature form ω related to the positive Hermitian kernel $K_{\mathbb{L}}$ by (3.4.28) are given equivalently by the coherent states map $\mathcal{K} : M \to \mathbb{CP}(\mathcal{H})$ as follows

$$H_{\overline{\alpha}\alpha}(q, q) = K_{\overline{\alpha}\alpha}(q, q) = \langle K_\alpha(q)|K_\alpha(q)\rangle \tag{3.4.30}$$

$$k_\alpha(q) = \frac{\langle K_\alpha(q)|dK_\alpha(q)\rangle}{\langle K_\alpha(q)|K_\alpha(q)\rangle} \tag{3.4.31}$$

$$\omega(q) = \frac{1}{2\pi i}d\left(\frac{\langle K_\alpha|dK_\alpha\rangle}{\langle K_\alpha|K_\alpha\rangle}\right)(q) \tag{3.4.32}$$

for $q \in \Omega_\alpha$.

In order to find the quantum mechanical interpretation of the connection ∇ and its curvature form $2\pi i\omega$ let us take a sequence $q = q_1, \ldots, q_{N-1}, q_N = p$ of points $q_i \in \Omega_{\alpha_i}$, for which we assumed $\Omega_{\alpha_1} = \Omega_\alpha$ and $\Omega_{\alpha_N} = \Omega_\beta$. According to the multiplication property of the transition amplitude, the following expression

$$a_{\alpha\beta}(q, q_2, \ldots, q_{N-1}, p) := a_{\overline{\alpha}_1\alpha_2}(q, q_2) \cdots a_{\overline{\alpha}_{N-1}\beta}(q_{N-1}, p) \qquad (3.4.33)$$

gives the transition amplitude from the state $i([K_\alpha(q)])$ to the state $i([K_\beta(p)])$ under the condition that the system has gone through all the intermediate coherent states $i([K_{\alpha_2}(q_2)]), \ldots, i([K_{\alpha_{N-1}}(q_{N-1})])$. We shall call the sequence

$$i([K_{\alpha_1}(q_1)]), \ldots, i([K_{\alpha_N}(q_N)]) \qquad (3.4.34)$$

of coherent states a process starting at q and ending at p. Consequently, $a_{\alpha\beta}(q, q_2, \ldots, q_{N-1}, p)$ will be called the transition amplitude for that process.

Let us investigate further the process in $i(\mathcal{K}(M))$ parametrized by a piecewise smooth curve $\gamma : [\tau_i, \tau_f] \to M$ such that $\gamma(\tau_k) = q_k$ for $\tau_k \in [\tau_i, \tau_f]$ defined by $\tau_{k+1} - \tau_k = \frac{1}{N-1}(\tau_f - \tau_i)$. Then in the limit $N \to \infty$ this γ-process may be viewed as a process approximately described by a discrete one $(q, q_2, \ldots, q_{N-1}, p)$. The transition amplitude for the process γ is obtained from (3.4.33) by the limit $N \to \infty$

$$a_{\overline{\alpha}\beta}(q, \gamma, p) = \lim_{N \to \infty} \prod_{k=1}^{N-1} a_{\overline{\alpha}_k, \alpha_{k+1}}(\gamma(\tau_k), \gamma(\tau_{k+1})). \qquad (3.4.35)$$

Taking into account the smoothness of $K_\alpha : \Omega_\alpha \to \mathcal{H}$ and piecewise smoothness of γ we define

$$\Delta K_{\alpha_k}(\gamma(\tau_k)) := K_{\alpha_k}(\gamma(\tau_{k+1})) - K_{\alpha_k}(\gamma(\tau_k)), \qquad (3.4.36)$$

where we put in (3.4.36) that $\gamma(\tau_k), \gamma(\tau_{k+1}) \in \Omega_{\alpha_k}$. Then, using (3.4.26) and assuming that $\gamma([\tau_i, \tau_f]) \subset \Omega_{\alpha_k}$ one has

$$a_{\overline{\alpha}\beta}(q, \gamma, p) = \lim_{N \to \infty} \left(\frac{\langle K_\alpha(q)|K_\alpha(q)\rangle}{\langle K_\beta(p)|K_\beta(p)\rangle} \right)^{\frac{1}{2}} \qquad (3.4.37)$$

$$\times \prod_{k=1}^{N-1} \left(1 - \frac{\langle K_{\alpha_k}(\gamma(\tau_k))|\Delta K(\gamma(\tau_l k))\rangle}{\langle K_{\alpha_k}(\gamma(\tau_k))|K_{\alpha_k}(\gamma(\tau_k))\rangle} \right)$$

$$= \left(\frac{\langle K_\alpha(q)|K_\alpha(q)\rangle}{\langle K_\beta(p)|K_\beta(p)\rangle}\right)^{\frac{1}{2}} \lim_{N\to\infty} \exp \sum_{k=1}^{N-1} \frac{\langle K_{\alpha_k}(\gamma(\tau_l))|\Delta K_{\alpha_k}(\gamma(\tau_l))\rangle}{\langle K_{\alpha_k}(\gamma(\tau_l))|K_{\alpha_k}(\gamma(\tau_l))\rangle}$$

$$= \left(\frac{\langle K_\alpha(q)|K_\alpha(q)\rangle}{\langle K_\beta(p)|K_\beta(p)\rangle}\right)^{\frac{1}{2}} \exp \left(\int_{\tau_i}^{\tau_f} k_{\alpha}{\llcorner}\frac{d\gamma}{d\tau}d\tau\right)$$

$$= \exp i \int_{\tau_i}^{\tau_f} Im\frac{\langle K_\alpha|dK_\alpha\rangle}{\langle K_\alpha|K_\alpha\rangle}{\llcorner}\frac{d\gamma}{d\tau}d\tau.$$

After expressing the connection $\nabla = \mathcal{K}^* \nabla_K^{FS}$ in the unitary gauge frame

$$u_\alpha := \frac{1}{H(s_\alpha, s_\alpha)^{\frac{1}{2}}} s_\alpha, \tag{3.4.38}$$

i.e.

$$\nabla u_\alpha = iIm\frac{\langle K_\alpha|dK_\alpha\rangle}{\langle K_\alpha|K_\alpha\rangle} \otimes u_\alpha, \tag{3.4.39}$$

we obtain that the transition for the piecewise process $\gamma([\tau_i, \tau_f])$ starting at q and ending at p is given by the parallel transport

$$a_{\overline{\alpha}\beta}(q,\gamma,p) = \exp i \int_{\gamma([\tau_i,\tau_f])} Im\frac{\langle K|dK\rangle}{\langle K|K\rangle} \tag{3.4.40}$$

from $\mathbb{L}_q$ to $\mathbb{L}_p$ along γ with respect to the connection ∇. In (3.4.40) we applied the notation $\frac{\langle K|dK\rangle}{\langle K|K\rangle} := \frac{\langle K_\alpha|dK_\alpha\rangle}{\langle K_\alpha|K_\alpha\rangle}$ on Ω_α and by the integral $\int_{\gamma([\tau_i,\tau_f])} Im\frac{\langle K|dK\rangle}{\langle K|K\rangle}$ we mean the sum of integrals over the pieces of the curve $\gamma([\tau_i, \tau_f])$ which are contained in Ω_α.

Since the connection ∇ is metric, one has

$$|a_{\overline{\alpha}\beta}(q,\gamma,p)|^2 = 1 \tag{3.4.41}$$

for the transition probability of the considered γ-process. This is a consequence of the continuity of the coherent state map $\iota \circ \mathcal{K} : M \to \mathbb{CP}(\mathcal{H}) \subset H^1(\mathcal{H})$ with respect of the $\|\cdot\|_1$-metric, which causes that $a_{\overline{\alpha}\beta}(q(\tau), q(\tau + \Delta\tau)) \approx 1$ for $\Delta\tau \approx 0$. Therefore, for the classical process, i.e. continuous ones, the interference effects disappear between the infinitely close $q(\tau) \approx q(\tau + \Delta\tau)$ classical pure states. It remains only as a global effect given by the parallel transport (3.4.40) with respect to ∇.

For two piecewise smooth processes starting from q and ending in p one has the following relation

$$a_{\overline{\alpha}\beta}(q,\gamma_2,p) = a_{\overline{\alpha}\beta}(q,\gamma_1,p)\exp\left(2\pi i \int_\sigma \omega\right) \qquad (3.4.42)$$

between the transition amplitudes, where $\partial\sigma = \gamma_1 - \gamma_2$ is the boundary of the surface lying between the curves γ_1 and γ_2. The factor $\exp\left(2\pi i \int_\sigma \omega\right)$ does not depend on the choice of σ. Hence, one concludes that the curvature 2-form ω measures the phase change of the transition amplitude for a cyclic piecewise smooth process.

According to the path-integral approach to the quantum probability amplitude one can define the path integral over the processes starting from q and ending in p by

$$\begin{aligned}
a_{\overline{\alpha}\beta}(q,p) &:= \int \mathcal{D}[\gamma]\exp\left(i\int_\gamma Im\frac{\langle K|dK\rangle}{\langle K|K\rangle}\right) \\
&= \int \prod_{\tau\in[\tau_i,\tau_f]} d_k\gamma(t)\exp\left(i\int_{\tau_i}^{\tau_f} Im\frac{\langle K|dK\rangle}{\langle K|K\rangle}\lrcorner\frac{d\gamma}{d\tau}d\tau\right), \qquad (3.4.43)
\end{aligned}$$

where

$$\begin{aligned}
\int \prod_{\tau\in[\tau_i,\tau_f]} d_k\gamma(t) &:= \lim_{N\to\infty}\int_M \sum_{\delta_2} h_{\delta_2}(\gamma(\tau_2))d\mu_L(\gamma(\tau_2))\cdot\ldots \\
&\times \int_M \sum_{\delta_{N-1}} h_{\delta_{N-1}}(\gamma(\tau_{N-1}))d\mu_L(\gamma(\tau_{N-1})) \qquad (3.4.44)
\end{aligned}$$

and $\mu_L = \overset{n}{\bigwedge}\omega$ is the Liouville measure on (M,ω). We have assumed the completeness condition for the transition amplitudes:

$$a_{\overline{\alpha}\beta}(q,p) = \sum_{\gamma\in I}\int_M a_{\overline{\alpha}\gamma}(q,r)a_{\overline{\gamma}\beta}(r,p)h_\gamma(r)\,d\mu_L(r)\,, \qquad (3.4.45)$$

where $\sum_{\gamma\in I} h_\gamma(r) = 1$ is a partition of unity subordinated to the covering $\cup_{\alpha\in I}\Omega_\alpha = M$.

This point of view on the transition amplitude we shall use to find the Lagrangian description of the system. Having in the mind the energy conservation law we shall admit in (3.4.43) only those trajectories which are confined to the equi–energy surface $h^{-1}(E)$, where $h\in C^\infty(M)$ is the function of total energy of the considered system. Let then $a_{\overline{\alpha}\beta}(q,p;h = E = \text{const})$ denotes the transition amplitude from $\mathcal{K}(q)$ to $\mathcal{K}(p)$ which is the result of the superposition of the equi–energy processes. In order to find $a_{\overline{\alpha}\beta}(q,p; h = E = \text{const})$ one should insert the δ-factor

$$\delta(h(\gamma(\tau_k)) - E)d\mu_L(\gamma(\tau_k)) = \left(\int_{-\infty}^{+\infty} e^{-i(h(\gamma(\tau_k)) - E)\lambda(\tau_k)} d\lambda(\tau_k) \right) d\mu_L(\gamma(\tau_k))$$

$$(3.4.46)$$

into (3.4.43). Thus we obtain

$$a_{\overline{\alpha}\beta}(q,p; h = E = \text{const}) = \int \prod_{\tau \in [\tau_i, \tau_f]} d_{\mathcal{K}}\gamma(\tau)d\lambda(\tau) \cdot \qquad (3.4.47)$$

$$\cdot \exp\left(i \int_{\tau_i}^{\tau_f} \left(Im \frac{\langle K|dK \rangle}{\langle K|K \rangle} \lrcorner \frac{d\gamma}{d\tau} - (h(\gamma(\tau)) - E)\lambda(\tau) \right) d\tau \right).$$

The different parametrizations of the process γ give the same contribution to (3.4.47). The reparametrization invariance may be fixed by introducing $t = \int_{\tau_i}^{\tau} \lambda_0(s)ds$ as a time parametrizing the processes. This way, the integral becomes the integral over the equivalent choices of classical "clocks" and may be dropped out. The resulting amplitude is given by

$$a_{\overline{\alpha}\beta}(q,p; h = E) = e^{-E(t_f - t_i)} \int \prod_{t \in [t_i, t_f]} d_{\mathcal{K}}\gamma(t) \cdot \qquad (3.4.48)$$

$$\cdot \exp\left(i \int_{t_i}^{t_f} \left(Im \frac{\langle K|dK \rangle}{\langle K|K \rangle} \lrcorner \frac{d\gamma}{dt} - h(\gamma(t)) \right) dt \right).$$

Now according to Feynman the Lagrangian L of the system is given by

$$\frac{dL}{dt} = Im \frac{\langle K|dK \rangle}{\langle K|K \rangle} \lrcorner \frac{d\gamma}{dt} - h(\gamma(t)), \qquad (3.4.49)$$

where the summand $Im \frac{\langle K|dK \rangle}{\langle K|K \rangle} \lrcorner \frac{d\gamma}{dt}$ is responsible for the interaction of the system with the effective external field resulting from the way the coherent state map $\mathcal{K} : M \to \mathbb{CP}(\mathcal{H})$ has been realized.

3.5 Relation Between Classical and Quantum Observables

The fundamental problem in the theory of physical systems is to explain how to construct the quantum observables if one has their classical counterparts. Traditionally one calls this procedure the quantization. Let us now explain what we mean by quantization in the framework of our model of a mechanical system. In order to do this let us take two mechanical systems $(M_i, \omega_i, \mathcal{K}_i : M_i \to \mathbb{CP}(\mathcal{H}_i))$, $i = 1, 2$, and consider a symplectomorphism $\sigma : M_1 \to M_2$. By the **quantization** of σ we shall mean the morphism $\Sigma(\sigma)$

$$\Sigma : \mathrm{Sp}C^\infty(M_1, M_2) \ni \sigma \to \Sigma(\sigma) \in \mathrm{Mor}(\mathbb{CP}(\mathcal{H}_1), \mathbb{CP}(\mathcal{H}_2)))$$

defined for such σ for which the diagram (3.2.16) commutes. One has

$$\Sigma(\sigma_1 \circ \sigma_2) = \Sigma(\sigma_2) \circ \Sigma(\sigma_1) \qquad (3.5.1)$$

for $\sigma_1 : M_1 \to M_2$ and $\sigma_2 : M_2 \to M_3$. It is clear that not all elements of $\mathrm{Sp}C^\infty(M_1, M_2)$ are quantizable in this way. If $M_1 = M_2$, $\mathcal{H}_1 = \mathcal{H}_2$ and $\mathcal{K}_1 = \mathcal{K}_2$ the quantizable symplectic diffeomorphisms $\sigma : M \to M$ form a subgroup $\mathrm{SpDiff}_\mathcal{K}(M, \omega)$ of the group $\mathrm{SpDiff}(M, \omega)$ of all symplectic diffeomorphism of M. Since $\Sigma(\sigma) : \mathbb{CP}(\mathcal{H}) \to \mathbb{CP}(\mathcal{H})$ preserve the transition probability it follows from Wigner's theorem, see [24], that there exists a unitary or anti-unitary map $U(\sigma) : \mathcal{H} \to \mathcal{H}$ such that

$$\Sigma(\sigma) = [U(\sigma)] . \qquad (3.5.2)$$

The phase ambiguity in the choice of $U(\sigma)$ in (3.5.2) one removes by passing to the lifting

$$
\begin{array}{ccc}
\mathbb{L}' & \xrightarrow{\;\;\mathcal{K}'\;\;} & \mathbb{E}' \\
\downarrow & & \downarrow \\
M & \xrightarrow{\;\;\mathcal{K}\;\;} & \mathbb{CP}(\mathcal{H})
\end{array}
\qquad (3.5.3)
$$

of the coherent state map $\mathcal{K} : M \to \mathbb{CP}(\mathcal{H})$, where the $\mathbb{C}^*$–principal bundles $\mathbb{L}'$ and $\mathbb{E}'$ are obtained from $\mathbb{L}$ and $\mathbb{E}$ by cutting off zero sections. Fixing the unitary (anti-unitary) representative $U(\sigma)$ one obtains σ' from (3.5.3) and from $\mathbb{E}' \simeq \mathcal{H} \setminus \{0\}$

$$
\begin{array}{ccc}
\mathbb{L}' & \xrightarrow{\;\;\mathcal{K}'\;\;} & \mathbb{E}' \\
\sigma' \downarrow & & \downarrow U(\sigma) \\
\mathbb{L}' & \xrightarrow{\;\;\mathcal{K}'\;\;} & \mathbb{E}'
\end{array}
\qquad (3.5.4)
$$

where the lifting

$$\begin{array}{ccc} \mathbb{L}' & \xrightarrow{\ \sigma'\ } & \mathbb{L}' \\ \downarrow & & \downarrow \\ M & \xrightarrow{\ \sigma\ } & M \end{array} \quad , \tag{3.5.5}$$

is defined by $U(\sigma)$ in an unique way. The map σ' defines a principal bundle automorphism and preserves the positive Hermitian kernel $K_{\mathbb{L}} = \mathcal{K}^* K_{\mathbb{E}}$, i.e.

$$\sigma'(c\xi) = c\sigma'(\xi) \tag{3.5.6}$$

for $c \in \mathbb{C} \setminus \{0\}$ and $\xi \in \mathbb{L}'$ and

$$K_{\mathbb{L}}\left(\sigma'(\xi_1), \sigma'(\xi_2)\right) = K_{\mathbb{L}}(\xi_1, \xi_2) \tag{3.5.7}$$

for $\xi_1, \xi_2 \in \mathbb{L}'$. The inverse statement is also valid.

Proposition 3.5.1 *Let* $\mathbb{L} \to M$ *be a complex line bundle with distinguished positive Hermitian Kernel* $K_{\mathbb{L}}$ *and* $\sigma : M \to M$ *a diffeomorphism that has a lifting* $\sigma' : \mathbb{L}' \to \mathbb{L}'$ *which satisfies (3.5.6) and (3.5.7). Then there are a uniquely defined coherent state map* $\mathcal{K} : M \to \mathbb{CP}(\mathcal{H})$ *and a unitary (anti-unitary) operator* $U(\sigma)$ *with property (3.5.4).*

Definition 3.5.2 The one–parameter subgroup $\sigma(t) \subset \mathrm{Diff}\, M, t \in \mathbb{R}$, we call a **prequantum flow** if and only if it admits lifting $\sigma'(t) \in Diff\mathbb{L}, \ t \in \mathbb{R}$, which preserves the structure of the prequantum bundle $(\mathbb{L} \to \mathbb{M}, \nabla, \mathcal{H})$.

It was shown by Kostant [9] that the Lie algebra $Lie(\mathbb{L}^*, \nabla, \mathcal{H})$ of the vector fields tangent to the prequantum flows is isomorphic to the Poisson algebra $(C^\infty(M, \mathbb{R}), \{\cdot, \cdot\})$ where the Poisson bracket $\{\cdot, \cdot\}$ is defined by (3.4.17). It follows from Proposition 3.4.2 and Proposition 3.4.4 that the prequantum bundle structure is always defined, see (3.4.29) by a coherent state map $\mathcal{K} : M \to \mathbb{CP}(\mathcal{H})$ or, equivalently, by a positive Hermitian kernel $K_{\mathbb{L}} = \mathcal{K}^* K_{\mathbb{E}}$.

Definition 3.5.3 The one–parameter subgroup $\sigma(t) \in \mathrm{SpDiff}\, M, t \in \mathbb{R}$, we call a **quantum flow** if and only if it preserves the structure of the physical system $(M, \mathcal{H}, \mathcal{K} : M \to \mathbb{CP}(\mathcal{H}))$, i.e. there is one parameter subgroup $\Sigma(t)$, $t \in \mathbb{R}$ such that

$$
\begin{array}{ccc}
M & \xrightarrow{\ \mathcal{K}\ } & \mathbb{CP}(\mathcal{H}) \\
\sigma(t)\downarrow & & \downarrow\Sigma(t) \\
M & \xrightarrow[\ \mathcal{K}\ ]{} & \mathbb{CP}(\mathcal{H})
\end{array}
\qquad (3.5.8)
$$

for every $t \in \mathbb{R}$.

Theorem 3.5.4 *The following statements are equivalent:*

(1) The one–parameter subgroup $\sigma(t) \in \operatorname{Diff} M$, $t \in \mathbb{R}$, is a quantum flow of the physical system $(M, \mathcal{H}, \mathcal{K} : M \to \mathbb{CP}(\mathcal{H}))$.

(2) The one–parameter subgroup $\sigma(t) \in \operatorname{Diff} M$, $t \in \mathbb{R}$, has a lifting $\sigma'(t) :$ $\mathbb{L}' \to \mathbb{L}'$, $t \in \mathbb{R}$, which preserves the bundle structure of $\mathbb{L}'$ and the positive Hermitian kernel $K_{\mathbb{L}} = \mathcal{K}^* K_{\mathbb{L}}$.

(3) There are a lifting $\sigma'(t) \in \operatorname{Diff} \mathbb{L}'$, $t \in \mathbb{R}$ and a strong unitary (anti–unitary) one parameter subgroup $U(t) \in \operatorname{Aut}\mathcal{H}$, $t \in \mathbb{R}$, such that

$$
\begin{array}{ccc}
\mathbb{L}' & \xrightarrow{\ \mathcal{K}'\ } & \mathbb{E}' \\
\sigma'(t)\downarrow & & \downarrow U(t) \\
\mathbb{L}' & \xrightarrow[\ \mathcal{K}'\ ]{} & \mathbb{E}'
\end{array}
\qquad (3.5.9)
$$

for every $t \in \mathbb{R}$, where $\mathbb{E}' \cong \mathcal{H} \setminus \{0\}$.

The vector field tangent to the quantum flow $\sigma'(t)$, $t \in \mathbb{R}$, is the lifting of the Hamiltonian field $X_f \in \Gamma^\infty(TM)$ generated by $f \in C^\infty(M, \mathbb{R})$, see [9]. So, the strong unitary one–parameter subgroup $U^f(t)$, $t \in \mathbb{R}$ given by (3.5.9) depends univocally on f. The Stone-von Neumann theorem states that there is a self–adjoint operator F on $\mathcal{H}$ such that

$$
U^f(t) = e^{-itF}. \qquad (3.5.10)
$$

The domain $D(F)$ of F is the linear span $l.s.(\mathcal{K}(M))$ of the set of coherent states. Representing F in $\mathcal{H}_{\mathbb{K},\mathbb{L}} \subset \Gamma^\infty(M, \mathbb{L}^*)$ we obtain

$$
-iF\Psi = \lim_{t \to 0} \frac{(U^f(t) - 1)\Psi}{t} = \lim_{t \to 0} \frac{1}{t}(\sigma'(-t)\Psi - \Psi) = \left(\nabla_{X_f} + 2\pi i f\right)\Psi
$$
$$
(3.5.11)
$$

for $\Psi \in D(F)$. The second equality in (3.5.11) is true since $l.s.(\mathcal{K}(M))$ is $U^f(t)$ invariant , $t \in \mathbb{R}$. Hence, the Kostant–Souriau operator $-iQ_f$, see (3.4.19) is essentially self-adjoint on $l.s.(\mathcal{K}(M))$ and the infinitesimal generator of $U^f(t)$ is its closure.

Let us denote by $C_{\mathcal{K}}^{\infty}(M,\mathbb{R})$ the space of functions which generate the quantum flows on $(M,\mathcal{H},\mathcal{K} : M \to \mathbb{CP}(\mathcal{H}))$. It follows from (2) of Theorem 3.5.4 that it is the Lie subalgebra of Poisson algebra $C^{\infty}(M,\mathbb{R})$. One also has

$$[Q_f, Q_g] = iQ_{\{f,g\}} \tag{3.5.12}$$

what means that $-iQ$ defines a Lie algebra homomorphism, i.e. it is a quantization in Kostant-Souriau sense. We remark that one does not use the notion of polarization, which plays the crucial role in the Kostant–Souriau geometric quantization [25]. In the theory developed here the polarization does not have the crucial meaning. It could be reconstructed from the coherent state map or from the positive Hermitian kernel [17].

In the paper [17] we investigate the quantum algebra $\mathcal{A}_{\mathcal{K}}$ related to the physical system $(M,\mathcal{H},\mathcal{K} : M \to \mathbb{CP}(\mathcal{K}))$. The structure of the algebra $\mathcal{A}_{\mathcal{K}}$ describes on the quantum level this part of the geometry of the classical phase space (M,ω) which is fixed by the choice of polarization on it. In our case it is given by the choice of the coherent state map. In the case when $M = \mathbb{C}^N$ and $\mathcal{K}_G : \mathbb{C}^N \to \mathbb{CP}(\mathcal{H})$ is given by $\mathcal{K}_G(z) = \mathbb{C}K_G(z)$, where

$$K_G(z) = \sum_{k_1,\ldots,k_N=0}^{\infty} \frac{z_1^{k_1}\ldots z_N^{k_N}}{\sqrt{k_1!\ldots k_N!}}|k_1,\ldots,k_N\rangle \tag{3.5.13}$$

and

$$\langle k_1,\ldots,k_N|l_1,\ldots,l_N\rangle = \delta_{k_1 l_1}\ldots\delta_{k_N l_n} \, ,$$

is the Gauss coherent state map, the quantum algebra $\mathcal{A}_{\mathcal{K}_G}$ is the Heisenberg-Weyl algebra. The polarization related to $\mathcal{K}_G$ is given by the complex structure of $\mathbb{C}^N$.

The theory of the algebras $\mathcal{A}_{\mathcal{K}}$ and their relation with $*$–product quantization [4] is investigated in [15; 17].

3.6   Examples

Example 1. MIC–Kepler System

By the MIC–Kepler system [20] one understands a charged massive particle in the field of the Coulomb potential modified by the centrifugal potential and a Dirac monopole magnetic field. In order to define the classical phase space of the MIC–Kepler system let us consider the twistor space $\mathbb{T} \cong \mathbb{C}^4$ and assume that the twistor Hermitian form is given by

$$\langle t, t \rangle := \eta^\dagger \eta - \upsilon^\dagger \upsilon \,, \tag{3.6.1}$$

where $\eta, \upsilon \in \mathbb{C}^2$ are spinor coordinates of the twistor $t = \binom{\eta}{\upsilon} \in \mathbb{T}$. Let Ω_μ be the submanifold of $\mathbb{T}$ defined as follows $\Omega_\mu := \{t \in \mathbb{T} : \langle t, t \rangle = \mu = \text{const}\}$. By $\widetilde{\Omega}$ we denote the quotient manifold $\Omega / U(1)$. The manifold $\widetilde{\Omega}$ has the structure of a complex vector bundle

$$\begin{array}{ccc} \mathbb{C}^2 & \longrightarrow & \widetilde{\Omega} \\ & & \downarrow \\ & & \mathbb{CP}(1) \end{array} \tag{3.6.2}$$

over the Riemann sphere $\mathbb{CP}(1)$. The open sets $\Omega := \{[t] \in \widetilde{\Omega} : \eta_1 \neq 0\}$ and $\Omega' := \{[t] \in \widetilde{\Omega} : \eta_2 \neq 0\}$, where $\eta = \binom{\eta_1}{\eta_2}$ and $\upsilon = \binom{\upsilon_2}{\upsilon_1}$ cover $\widetilde{\Omega}$ and the maps

$$\Omega \ni [t] \xrightarrow{\varphi} (z, u, w) := \left(\frac{\eta_2}{\eta_1}, \eta_1 \overline{\upsilon_1}, \eta_1 \overline{\upsilon_2} \right) \tag{3.6.3}$$

$$\Omega' \ni [t] \xrightarrow{\varphi'} (z', u', w') := \left(\frac{\eta_1}{\eta_2}, \eta_2 \overline{\upsilon_1}, \eta_2 \overline{\upsilon_2} \right) \tag{3.6.4}$$

form a holomorphic atlas on $\widetilde{\Omega}$. We define the coherent state map $\mathcal{K} : \widetilde{\Omega} \to \mathbb{CP}(\mathcal{H})$ by giving its trivialization in the above defined atlas

$$K_\lambda \circ \varphi^{-1}(z, u, w) = \sum_{k_1 + k_2 + \lambda \geqslant l} \frac{z^l u^{k_1} w^{k_2}}{[(k_1 + k_2 + \lambda - l)! l! k_1! k_2!]^{\frac{1}{2}}}$$
$$\times \, |k_1 + k_2 + \lambda - l, l, k_1, k_2 \rangle \tag{3.6.5}$$

$$K_\lambda \circ \varphi'^{-1}(z', u', w') = \sum_{k_1 + k_2 + \lambda \geqslant l} \frac{z'^l u'^{k_1} w'^{k_2}}{[(k_1 + k_2 + \lambda - l)! l! k_1! k_2!]^{\frac{1}{2}}}$$
$$\times \, |k_1 + k_2 + \lambda - l, l, k_1, k_2 \rangle, \tag{3.6.6}$$

where $\{|l_1, l_2, k_1, k_2\rangle\}_{l_1, l_2, k_1, k_2 = 0}^{\infty}$ is a fixed orthonormal basis in the Hilbert space $\mathcal{H}$. The complete description of the MIC–Kepler system in the framework of the theory presented here one can find in [20].

Example 2. The Scalar Conformal Massive Particle

In this case [13; 14] the classical phase space is given by the future tube $\mathbb{M}^{++} := \{w = x + iy : x \in \mathbb{R}^n \text{ and } y \in C_+\}$, where $C_+ = \{y \in \mathbb{R}^n : y^{0^2} - \vec{y}^2 > 0 \text{ and } y^0 > 0\}$. The coherent state map $\mathcal{K}_\lambda : \mathbb{M}^{++} \to \mathbb{CP}(\mathcal{H})$ is defined by $\mathcal{K}_\lambda(w) := \mathbb{C}K_\lambda(w)$ where

$$K_\lambda(w) = 2^{2n-2} \left[det((E - iW)] \right]^{-n} \tag{3.6.7}$$

$$\sum_{m=0}^{\infty} \sum_{2j=0}^{\infty} \sum_{q_1=-j}^{j} \sum_{q_2=-j}^{j} \triangle_{q_1 q_2}^{jm} (Z(W)) |j, m, q_1, q_2\rangle ,$$

$$Z(W) = (E - iW)^{-1}(E + iW) , \qquad W = w^\mu \sigma_\mu \tag{3.6.8}$$

and the σ_μ are the Pauli matrices. The polynomials $\triangle_{q_1 q_2}^{jm}$ are defined by

$$\triangle_{q_1 q_2}^{jm} (Z) := \left(N^{jm} \right)^{-1} \left[\frac{(j + q_1)!(j - q_!)!}{(j + q_2)!(j - q_2)!} \right]^{\frac{1}{2}}$$

$$\times \sum_{max\{0, q_1+q_2\} \leqslant S \leqslant min\{j+q_1, j+q_2\}} \binom{j+q_2}{S} \binom{j-q_2}{S-q_1-q_2}$$

$$\times z_{11}^s z_{12}^{j+q_1-s} z_{21}^{j+q_2-s} z_{22}^{s-q_1-q_2} \tag{3.6.9}$$

where

$$N^{jm} = \left[(n-1)(n-2)^2(n-3) \frac{(n-3)!(n-4)!m!(m+2j+1)!}{(2j+1)!(m+n-2)!(m+2j+n-1)!} \right]^{\frac{1}{2}} \tag{3.6.10}$$

and

$$Z = \begin{pmatrix} z_{11} & z_{12} \\ z_{21} & z_{22} \end{pmatrix} \tag{3.6.11}$$

and $-3 < \frac{\lambda}{h} = n$, h is Planck constant and $\{|j, m, q_1, q_2\rangle\}$ is a orthonormal basis in $\mathcal{H}$. Introducing coordinates $(x^\mu, p^\mu + \lambda \frac{y^\mu}{y^z})$ we obtain

$$\omega_\lambda = \mathcal{K}_\lambda^* \omega_{FS} = dx_\mu \wedge dp^\mu . \tag{3.6.12}$$

So, (p^μ) is relativistic four–momentum. Provided Compton wavelength $\frac{h}{mc}$, $(mc)^2 = p^{0^2} - \vec{p}^2$, and four–velocity $\frac{p^\mu}{mc}$ are small the scalar massive conformal particle is localized in the region of $\mathbb{M}^{++}$ close to the Minkowski space time $(y^\mu \approx 0)$ which is the Shilov boundary of $\mathbb{M}^{++}$. For the detailed description of this model see [13; 14].

Example 3. The Physical Systems Related to Quantum Optical Models

Here the classical phase space is given by the complex disc $\mathbb{D} \subset \mathbb{C}$ of radius $\mathcal{R}(0)$ where $\mathcal{R}$ is a meromorphic function on $\mathbb{C}$ such that $\mathcal{R}(q^n) > 0$ for $0 < q < 1$, $n \in \mathbb{N} \cup \{\infty\}$ and $\mathcal{R}(1) = 0$. The coherent state map is $\mathcal{K}(z) := \mathbb{C}K(z)$, where

$$K_R(z) = \sum_{n=0}^{\infty} \frac{z^n}{\sqrt{\mathcal{R}(q)\ldots\mathcal{R}(q^n)}}|n\rangle \qquad (3.6.13)$$

and $\{|n\rangle\}_{n=0}^{\infty}$ is an orthonormal basis. The annihilation operator A defined by

$$AK(z) = zK(z) \qquad (3.6.14)$$

for $z \in \mathbb{D}$ and its Hermitian conjugation A^* satisfy the relations (see [16])

$$\begin{aligned}
A^*A &= \mathcal{R}(Q) \\
AA^* &= \mathcal{R}(qQ) \\
AQ &= qQA \\
QA^* &= qA^*Q \,,
\end{aligned} \qquad (3.6.15)$$

where Q is defined by $Q|n\rangle = q^n|n\rangle$.

The quantum algebra $\mathcal{A}_\mathcal{R}$ generated by A, A^* and Q is obtained from the free N–degree Heisenberg-Weyl algebra by quantum reduction [18]. It describes the symmetry of the quantum optical system considered.

Acknowledgements

The author would like to thank S.T. Ali, T. Goliński and A. Tereszkiewicz, for discussions and their interest in the paper.

Bibliography

[1] S. Ali, J. Antoine, and J. Gazeau. *Coherent States, Wavelets and their Generalizations*. Springer-Verlag, New York-Berlin-Heidelberg, 2000.

[2] P. Bona. Extended quantum mechanics. *Acta Physica Slovaca*, 50:1–198, 2000.

[3] R. P. Feynman and A. Hibbs. *Quantum Mechanics and Path Integrals*. McGraw-Hill Book Co. New York, 1965.

[4] M. Flato, A. Lichnerowicz, and D. Sternheimer. Deformation of Poisson brackets, Dirac brackets and application. *J. Math. Phys.*, 17, 1976.

[5] G. Gasper and M. Rahman. *Basic Hypergeometric Series*. Cambridge University Press, 1990.

[6] R. Glauber. *Phys. Rev. Lett.*, 10:84, 1983.

[7] M. Horowski, A. Odzijewicz, and A. Tereszkiewicz. Some integrable systems in nonlinear quantum optics. *Journal of Mathematical Physics*, 44(2):480–506, 2002.

[8] R. Klauder and B.-S. Skagerstam. *Coherent States – Applications in Physics and Mathematical Physics*. Singapore: World Scientific, 1985.

[9] B. Kostant. Quantization and unitary representation. *Lect. Notes in Math.*, 170:87–208, 1970.

[10] J. Moyal. Quantum mechanics as a statistical theory. *Proc. Cambridge Math. Soc.*, 45:99–124, 1949.

[11] G. J. Murphy. C^*-*algebras and Operator Theory*. Academic Press, 1990.

[12] M. Narasimhan and M. Ramanan. Existence of universal connections. *Am. J. Math.*, 83:563–572, 1961.

[13] A. Odzijewicz. On reproducing kernels and quantization of states. *Commun. Math. Phys.*, 114:577–597, 1988.

[14] A. Odzijewicz. Coherent states and geometric quantization. *Commun. Math. Phys.*, 150:385–413, 1992.

[15] A. Odzijewicz. *Covariant and Contravariant Berezin Symbols of Bounded Operators. Quantization and Infinite-Dimensional Systems*, pages 99–108. New York-London: Plenum Press, 1994.

[16] A. Odzijewicz. Quantum algebras and q-special functions related to coherent states maps of the disc. *Commun. Math. Phys.*, 192:183–215, 1998.

[17] A. Odzijewicz. Non-commutative Kähler-like structures in quantization. *to appear*, 2003.

[18] A. Odzijewicz, M. Horowski, and A. Tereszkiewicz. Integrable mulit-boson systems and orthogonal polynomials. *J. Phys. A: Math. Gen.*, 34:4353–4376, 2001.

[19] A. Odzijewicz and T. S. Ratiu. Banach Lie–Poisson spaces and reduction. *Comm. Math. Phys.*, 243:1–54, 2003.

[20] A. Odzijewicz and M. Świętochowski. Coherent states map for MIC-Kepler system. *J. Math. Phys.*, 38:5010–5030, 1997.

[21] M. Perelomov. *Commun. Math. Phys.*, 26:222, 1972.

[22] E. Schrödinger. *Naturwissenschaften*, 14:664–666, 1926.

[23] J. Souriau. *Structure des Systemes dynamiques*. Paris: Dunod, 1970.

[24] V. Varadarajan. *Geometry of Quantum Theory*. Springer-Verlag New York-Berlin-Heidelberg, 1985.

[25] N. J. Woodhouse. *Geometric Quantization*. Clarendon Pr, 1992.

Chapter 4

The Group of Volume Preserving Diffeomorphisms and the Lie Algebra of Unimodular Vector Fields: Survey of Some Classical and Not-so-classical Results

Claude Roger[1]

Abstract: We survey various results about the Lie algebra of unimodular vector fields and the corresponding group, such as computations of cohomology, extensions, and rigidity properties. This subject is closely related to recent problems in mathematical physics, e.g quantization of branes.

Sto lat konferencji dla
"Metod geometrycznych w fizyce" w Białowieży !

4.1 Introduction

We shall discuss in this article several results concerning the Lie algebra of unimodular vector fields and its associated "Lie" group, the group of volume preserving diffeomorphisms. We shall consider various cohomological aspects, derivations, extensions, deformations; we recall the proof of the rigidity of this Lie algebra, then describe some coadjoint orbits of the group, and discuss their geometric quantization, following Ismagilov. Analogous objects in supergeometry are as well introduced. Physical motivations for this kind of work are well known since the works of V. I. Arnold in the late sixties (cf. [A-K]): geometrical models of hydrodynamics are constructed

[1]Institut Girard Desargues - UMR 5028, Université Claude Bernard (Lyon 1), F 69622 Villeurban Cedex. France

from the group of volume preserving diffeomorphisms and its various generalizations. But very recently, new interest appeared in the physical literature for unimodular vector fields, with problems coming from branes ([A-L-M-Y],[Ho],[M-S],[S]) linked with some kind of deformation quantization, analogous to the now well known theory in the Hamiltonian case (cf. [Fe], [Ko]); we hope this quick survey of new and old results — but not so well known as they ought to be — will be useful for that purpose.

Basic definitions and forewords

Let V be an orientable n dimensional compact manifold and let ω in $\Omega^n(V)$ be a volume form. Unless otherwise stated, we shall suppose $n \geq 3$. The Lie algebra of vector fields on V will be denoted by $\mathrm{Vect}(V)$. For $X \in \mathrm{Vect}(V)$ and any differential form α, let us denote by $i(X)\alpha$ and $L_X\alpha$ the inner product and the Lie derivative respectively. The *divergence* $\mathrm{Div}(X)$ of X is then the function on V defined as $L_X\omega = \mathrm{d}\big(i(X)\omega\big) = \mathrm{Div}(X)\omega$. We shall say that a vector field $X \in \mathrm{Vect}(V)$ is *unimodular* if $\mathrm{Div}(X) = 0$ and *exact unimodular* if there exists a $(n-2)$ form α such that $i(X)\omega = \mathrm{d}\,\alpha$. A direct and easy computation shows that the space of unimodular (resp. exact unimodular) vector fields is stable under Lie bracket, so they are Lie subalgebras of $\mathrm{Vect}(V)$, denoted by $\overline{\mathrm{SVect}}(V)$ and $\mathrm{SVect}(V)$ respectively. It is now very natural to introduce the group corresponding to $\overline{\mathrm{SVect}}(V)$, the group $\mathcal{SD}(V)$ of volume preserving diffeomorphisms (i.e. $f^*(\omega) = \omega$). For technical reasons, we shall only consider such diffeomorphisms which are isotopic to the identity through volume preserving diffeomorphisms (in other words,the connected component of the identity in the whole group). This group is a closed subgroup of the group of all diffeomorphisms; it admits a structure of Fréchet–Lie group: one has a manifold structure on $\mathcal{SD}(V)$, modeled on a Fréchet space, such that the structure maps of the group are differentiable, with all the traditional difficulties of differentiability in Fréchet spaces (see [H] or the more detailed [K-M] for a "clean" theory of groups of diffeomorphisms). But to which extent is $\mathcal{SD}(V)$ the Lie group integrating the Lie algebra of unimodular vector fields? Let $\varphi_{t,\ t\in[0,1]}$ be a one parameter group of diffeomorphisms and $X(x) = \dfrac{\mathrm{d}}{\mathrm{d}\,t}\varphi_t(x)|_{t=0}$ the vector field which is its infinitesimal generator; then φ_t belongs to $\mathcal{SD}(V)$ if and only if X is in $\overline{\mathrm{SVect}}(V)$. So $\overline{\mathrm{SVect}}(V)$ can be considered as the Lie algebra of $\mathcal{SD}(V)$ in that weak sense; the main difference with the finite dimensional situation is that one parameter groups don't generate the whole group (cf. [H]). Below, we shall consider the more delicate problem of finding the subgroup of $\mathcal{SD}(V)$ corresponding to $\mathrm{SVect}(V)$.

Before ending this introductory part, we should say some words about the classification of Lie algebras of vector fields in order to make the framework of our problem more precise. Since the very beginning of differential geometry, one of the most important problems has been the classification of all possible structures on a manifold defined by an atlas; this lead to the theory of Lie pseudogroups and G-structures. To a Lie pseudogroup corresponds a Lie algebra of vector fields, and these Lie algebras, simple, primitive, transitive, have been classified by E. Cartan around 1930 (for a precise proof see [S-S]); the most well known are the Lie algebra of all vector fields, the Lie algebra of hamiltonian vector fields (in even dimension), the Lie algebra of contact vector fields (in odd dimension), and the Lie algebra of unimodular vector fields. Indeed, all simple primitive transitive Lie algebras can be deduced from those four (sometimes called "Cartan algebras").

4.2 Some Results on the Cohomology of the Lie Algebra of Unimodular Vector Fields

Systematic exploration of algebraic properties of Lie algebras of vector fields began around 1968 with the work of Gelfand and Fuks about the cohomology of $\mathrm{Vect}(S^1)$ and $\mathrm{Vect}(V)$ (cf. [F]). Meanwhile, the other Lie algebras of the Cartan family were studied from the point of view of extensions, derivations and deformations by A. Lichnerowicz and his collaborators.

We shall begin with the sketch of some results of the important paper [L]. One has a naturally defined exact sequence:

$$0 \longrightarrow \mathrm{SVect}(V) \longrightarrow \overline{\mathrm{SVect}}(V) \longrightarrow H_{\mathrm{d}R}^{n-1}(V) \longrightarrow 0.$$

$$(4.1)$$

The second mapping associates to each $X \in \overline{\mathrm{SVect}}(V)$ the cohomology class of $i(X)\omega$ in $H_{\mathrm{d}R}^{n-1}(V)$. This is an exact sequence of Lie algebras if $H_{\mathrm{d}R}^{n-1}(V)$ is equipped with the trivial bracket. One has the following results.

Proposition 4.2.1 ([L])

(1) One has $\left[\overline{\mathrm{SVect}}(V), \overline{\mathrm{SVect}}(V)\right] = \mathrm{SVect}(V)$.
In other words $H_1(\overline{\mathrm{SVect}}(V)) = H_{\mathrm{d}R}^{n-1}(V)$.

> *(2) All derivations of $\overline{\mathrm{SVect}}(V)$ are inner.*
> *(so $H^1(\overline{\mathrm{SVect}}(V), \overline{\mathrm{SVect}}(V)) = 0$).*
> *(3) All derivations of $\mathrm{SVect}(V)$ are given by brackets with elements of $\overline{\mathrm{SVect}}(V)$.*
> *(In cohomological terms $H^1(\mathrm{SVect}(V), \mathrm{SVect}(V)) = H_{\mathrm{d}R}^{n-1}(V)$.)*

This results remain valid when $n = 2$; in that case $\overline{\mathrm{SVect}}(V) = \mathrm{Sp}(V)$ the Lie algebra of symplectic vector fields and $\mathrm{SVect}(V) = \mathrm{Ham}(V)$, the Lie algebra of hamiltonian vector fields. One must stress the remarkable fact that no special hypothesis such as continuity, smoothness or locality are needed for this proposition; the cohomologies are purely algebraically defined.

We shall now say some words about the Lie algebra cohomology of $\overline{\mathrm{SVect}}(V)$ with trivial coefficients. Here we need some continuity hypothesis on the cochains; we shall limit ourselves to Lie algebra cochains which are continuous with respect to the natural C^∞ topology on $\overline{\mathrm{SVect}}(V)$. So we must adapt the tools of Gelfand–Fuks cohomology, as detailed in the book [F]. Roughly speaking, two ingredients enter the cohomology: the geometric part, coming from the topology of the underlying manifold and producing De Rham cohomology classes and the formal part, consisting of cohomology classes of the Lie algebra of formal vector fields, corresponding to our Lie algebra of vector fields. Let's recall the construction of the Lie algebra $\mathrm{Vect}(n)$ of all formal vector fields: it's nothing but the Lie algebra of all derivations of the ring of formal power series in n indeterminates. So any $X \in \mathrm{Vect}(n)$ reads as $X = X^i \dfrac{\partial}{\partial x_i}$ where $i = 1, \ldots, n$ and X^i is a formal power series. The Lie algebra of formal vector fields corresponding to $\mathrm{SVect}(V)$ is then $\mathrm{SVect}(n)$; $X \in \mathrm{SVect}(n)$ iff $\displaystyle\sum_{i=1}^{n} \frac{\partial X^i}{\partial x_i} = 0$. The cohomology of $\mathrm{Vect}(n)$ has been completely calculated by Gelfand and Fuks [F], but for $\mathrm{SVect}(n)$, one has only:

Proposition 4.2.2   (Guillemin and Shnider [G-S])

> $H^k\big(\mathrm{SVect}(n)\big) = 0, for \quad k = 1, \ldots, n-1;$
> $H^n\big(\mathrm{SVect}(n)\big) = \mathbb{R}, \text{ the generator being given by a "formal volume form" } \varpi(X_1, \ldots, X_n) = \mathrm{Det}\big(X_1(0), \ldots, X_n(0)\big).$

Their argument turns out to be fruitful in many other situations: one considers the maximal abelian subalgebra of $\mathrm{SVect}(n)$, i.e. the subalgebra of constant vector fields, and use some algebraic reduction argument to obtain this vanishing result. But computation of higher degree cohomology groups rely on complicated invariant computations, and seems to be an open problem. Now, these computations can be used directly for the case of Lie algebra of vector fields, generalizing a little bit the techniques of Gelfand and Fuks for $\text{Vect}(V)$.

Proposition 4.2.3   (cf. [R])
One has $H^k\big(\overline{\text{SVect}}(V), \mathbb{R}\big) = H_{n-k}(V, \mathbb{R})$ for $k = 0, 1, \ldots, n$.

This isomorphism is easy to describe explicitly; one associates to every $(n-k)$ cycle Δ the cochain $\overline{\Delta}$ defined as follows

$$\overline{\Delta}(X_1, \ldots, X_k) = \int_\Delta i(X_1)i(X_2)\cdots i(X_k)\omega \,.$$

The proof makes use of a direct construction following the lines of Fuks' book: one has a map of complexes

$$C^*\big(\text{SVect}(n)\big) \otimes C_*(V) \longrightarrow C^*\big(\overline{\text{SVect}}(V)\big)$$

inducing a spectral sequence just as described in ([F] pp. 121).

One encounters here a specific difficulty: the space $\overline{\text{SVect}}(V)$ is not the space of sections of a vector bundle, unlike $\text{Vect}(V)$. In sheaf theoretic terms, the sheaf $\underset{\sim}{\overline{\text{SVect}}}$ on V defined by $\underset{\sim}{\overline{\text{SVect}}}(U) = \overline{\text{SVect}}(U)$ for any open set $U \subset V$ is not a fine sheaf, with vanishing cohomology; one has a natural fine resolution by considering the truncated De Rham complex. In other words:

$$0 \longrightarrow \underset{\sim}{\overline{\text{SVect}}} \longrightarrow \underset{\sim}{\Omega}^{\,n-1} \longrightarrow \underset{\sim}{\Omega}^{\,n} \longrightarrow 0$$

is an exact sequence of sheaves, $\underset{\sim}{\Omega}^{\,p}$ indicates the pth De Rham sheaf (simply remark that for acyclic U, $\overline{\text{SVect}}(U) = \text{SVect}(U)$). See [L-R] for more details on this point of view; one can recover from this resolution the exact sequence (1) above and an isomorphism between $H^1\big(V, \underset{\sim}{\overline{\text{SVect}}}\big)$ and $\mathbb{R}$.

As a corollary of this proposition, one has

$$H^2\big(\overline{\text{SVect}}(V), \mathbb{R}\big) = H_{n-2}(V, \mathbb{R})$$

and it is not hard to check that one has as well

$$H^2\big(\text{SVect}(V), \mathbb{R}\big) = H_{n-2}(V, \mathbb{R})\,.$$

This cohomology group classifies all *central extensions* of $\mathrm{SVect}(V)$. Recall that an exact sequence of Lie algebras

$$0 \longrightarrow E \longrightarrow \widehat{\mathfrak{g}} \longrightarrow \mathfrak{g} \longrightarrow 0$$

is a central extension of $\mathfrak{g}$ by E if E is central in $\mathfrak{g}$. Each cohomology class in $H^2(\mathfrak{g}, \mathbb{R})$ defines a central extension

$$0 \longrightarrow \mathbb{R} \longrightarrow \widehat{\mathfrak{g}} \longrightarrow \mathfrak{g} \longrightarrow 0.$$

In other terms one has a *versal* central extension

$$0 \longrightarrow H_2(\mathfrak{g}) \longrightarrow \widehat{\mathfrak{g}} \longrightarrow \mathfrak{g} \longrightarrow 0.$$

from which any non trivial central extension is induced.

If one has $H_1(\mathfrak{g}) = 0$, this extension is *universal* (see [H-S] for details on those results). Here, one obtains a versal central extension:

$$0 \longrightarrow H_{DR}^{n-2}(V) \longrightarrow \widehat{\mathrm{SVect}}(V) \longrightarrow \mathrm{SVect}(V) \longrightarrow 0$$

$$(4.2)$$

(since $H_{n-2}(V, \mathbb{R})^* = H_{DR}^{n-2}(V)$).

From the proposition above, the cocycle of this extension is given by $\langle C(X, Y), \Delta \rangle = \int_{\Delta} i(X)i(Y)\omega$ where Δ is any $(n-2)$ cycle, and $\langle \, , \, \rangle$ stands for the pairing between homology and cohomology. This central extension appears implicitly in [L] and its universality is proven in [R].

Generalizing a little bit the approach of [L], one can give a more natural and geometric construction of this extension. Let $\Omega_p(V)$ be the space of contravariant antisymmetric tensor fields on V; such a field $A \in \Omega_p(V)$ will be written in a local coordinates system $A|_U = A^{i_1 \cdots i_p} \partial_{i_1} \wedge \ldots \wedge \partial_{i_p}$ where as usual ∂_i stands for $\dfrac{\partial}{\partial x_i}$.

One has on the direct product $\Omega_*(V) = \bigoplus_{p=0}^{n} \Omega_p(V)$ a structure of an associative graded commutative algebra defined by exterior product of tensor fields. Then one has, for each $p = 0, \ldots, n$, a natural isomorphism between $\Omega_p(V)$ and $\Omega^{n-p}(V)$, given by $A \longrightarrow i(A)\omega$. The inverse of this isomorphism will be denoted by $\sharp$; we can use it to transfer the De Rham differential on the contravariant side and one obtains the codifferential $\delta = \sharp \circ \mathrm{d} \circ \sharp^{-1}$. One has a commutative diagram for each $1 \leq p \leq n$

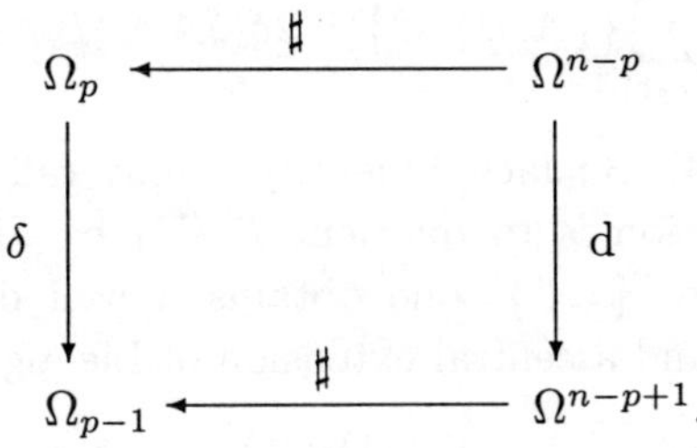

One has $\delta \circ \delta = 0$, and for local coordinates $(x_1, \ldots, x_n)$ such that $\omega = \mathrm{d}\,x_1 \wedge \ldots \wedge \mathrm{d}\,x_n$, one has $(\delta A)^{i_1 \cdots i_{p-1}} = \partial_i A^{i\,i_1 \cdots i_{p-1}}$, so for $p = 1$, $\delta A = \mathrm{Div}(A)$. The obstruction for δ being a derivation is called *Schouten* bracket; one has

$$[A, B] = \delta(A \wedge B) - \delta A \wedge B - (-1)^{|A|} A \wedge \delta B \tag{4.3}$$

where $|A|$ denotes as usual the degree of A.

This bracket defines on $\Omega_*(V)[1]$ (the graduation has been shifted by one degree lower) a graded Lie algebra structure. Moreover, the exterior product has a good property with respect to $[\,,\,]$: one computes easily

$$[A \wedge B, C] = A \wedge [B, C] + (-1)^{|B|(|C|-1)}[A, C] \wedge B \,.$$

This kind of algebraic structure, strongly analogous to a Poisson algebra but with a shift in graduation is called a *Gerstenhaber algebra*. Here the graded Lie algebra bracket is defined from the codifferential δ by formula (4.3); this structure is called *Batalin–Vilkovisky algebra* (shortly, B.V.). One can then summarize the above construction as follows: one associates to any manifold with volume form (V, ω) a B.V. algebra $(\Omega_*(V), \wedge, \delta)$. The relation of this construction with unimodular vector fields is now easy to find: if $X \in \mathrm{SVect}(V)$ and $\alpha \in \Omega^{n-2}(V)$ such that $i(X)\omega = \mathrm{d}\,\alpha$, set $\sharp \alpha = A \in \Omega_2(V)$, then $X = \delta A$; the converse is immediately checked to be true. So there exists a surjective map

$$\delta \colon \Omega_2(V) \longrightarrow \mathrm{SVect}(V) \longrightarrow 0 \,.$$

One can define a bracket $\{\,,\,\}$ on $\Omega_2(V)$:

$$\Omega_2(V) \times \Omega_2(V) \longrightarrow \Omega_2(V)$$
$$(A \quad , \quad B) \quad \longrightarrow \{A, B\} = \delta A \wedge \delta B$$

It satisfies $\delta(\{A, B\}) = [\delta A, \delta B]$, but unfortunately, it is not a Lie bracket. One computes easily

$$\sum_{\text{cycl}} \{\{A, B\}, C\} = 2\delta(\delta A \wedge \delta B \wedge \delta C) \tag{4.4}$$

(cf. [L] formula 15.4). So Jacobi identity is only valid modulo divergence of 3-tensors; the solution is to quotient $\Omega_2(V)$ by $\delta(\Omega_3(V))$ which is central with respect to $\{\ ,\ \}$, one obtains a well defined Lie bracket on $\Omega_2(V)/\delta(\Omega_3(V))$, and a central extension of Lie algebras:

$$0 \longrightarrow \frac{\operatorname{Ker}\delta}{\delta(\Omega_3(V))} \longrightarrow \frac{\Omega_2(V)}{\delta(\Omega_3(V))} \xrightarrow{\ \bar{\delta}\ } \operatorname{SVect}(V) \longrightarrow 0 \tag{4.5}$$

It's easy to check that (4.5) and (4.2) are naturally isomorphic.

Another way of considering this extension would simply be to take $(\Omega_*(V), \delta)$ but truncated at $* \geq 2$, and consider it as an equivalent data to $\operatorname{SVect}(V)$. This point of view could be fruitful for the quantization problem.

4.3   More Results about Cohomology of Lie Algebra of Unimodular Vector Fields: The Rigidity Theorem

We shall not develop here the general theory of formal deformations of Lie algebra structures (see e.g. [Fe], [L-R]). Simply recall that the second adjoint cohomology group $H^2(\mathfrak{g}, \mathfrak{g})$ of a Lie algebra $\mathfrak{g}$ classifies its infinitesimal deformations, and the group $H^3(\mathfrak{g}, \mathfrak{g})$ contains the obstructions to inductive prolongations of these deformations. For $\mathfrak{g} = \operatorname{S\overline{Vect}}(V)$ one can consider the local cohomology $H^*_{\text{loc}}(\operatorname{S\overline{Vect}}(V), \operatorname{S\overline{Vect}}(V))$: we try to compute the cohomology of the complex of local cochains, i.e. those which satisfy

$$\operatorname{Supp} C(X_1, \ldots, X_p) \subset \bigcap_{i=1}^{p} \operatorname{Supp} X_i.$$

According to a well known theorem of Peetre, these cochains are given locally by multidifferential operators. Of course, this local cohomology governs *local deformations*, the only natural ones if one thinks of the notion of locality in field theory. The result is as follows:

Proposition 4.3.1 $H^2_{loc}(\operatorname{S\overline{Vect}}(V); \operatorname{S\overline{Vect}}(V)) = 0$ *if $n = \dim V > 3$.*

For $n = 3$, $H^2_{\text{loc}}(\operatorname{S\overline{Vect}}(V); \operatorname{S\overline{Vect}}(V))$ is one dimensional and the obstruction to further prolongation is always non trivial.

This implies:

Theorem 4.3.2 *The Lie algebra* $\overline{\mathrm{SVect}}(V)$ *is rigid (it admits no non trivial formal deformations).*

The necessary cohomological computations encounter the same difficulty as above, coming from the fact that the associated sheaf $\underline{\mathrm{S}\widetilde{\mathrm{Vect}}}$ is not fine, so one must use a fine resolution, such as above:

$$0 \longrightarrow \underline{\mathrm{S}\widetilde{\mathrm{Vect}}} \longrightarrow \underline{\Omega}^{\,n-1} \longrightarrow \underline{\Omega}^{\,n} \longrightarrow 0.$$

One can now give a nice geometrical description of the generator of $H^2_{\mathrm{loc}}\big(\overline{\mathrm{SVect}}(V); \overline{\mathrm{SVect}}(V)\big)$ for $n = 3$; we think it could be useful for further more sophisticated constructions, in spite of the fact that the deformation fails.

There exists a class Φ in $H^2_{\mathrm{loc}}\big(\mathrm{Vect}(V), \Omega^2(V)\big)$ which appeared for the first time in the work of I. M. Gelfand (Proceedings of ICM, Nice, 1970). Let $\mathcal{P}(V) \xrightarrow{\pi} V$ be the frame bundle on V and ϖ be a connection on it; one can associate to vector fields X and Y on V their canonical lifts $\widetilde{X}$ and $\widetilde{Y}$ on $\mathcal{P}(V)$. One then has on $\mathcal{P}(V)$ a tensor valued 2-form $L_{\widetilde{X}}\varpi \wedge L_{\widetilde{Y}}\varpi$ and an associated scalar 2-form $\mathrm{Tr}(L_{\widetilde{X}}\varpi \wedge L_{\widetilde{Y}}\varpi)$; it's easy to check that it is a basic 2-form, so there exists $\varphi(X, Y) \in \Omega^2(V)$ such that $\pi^*\big(\varphi(X, Y)\big) = \mathrm{Tr}(L_{\widetilde{X}}\varpi \wedge L_{\widetilde{Y}}\varpi)$. One checks that $X, Y \to \varphi(X, Y)$ defines a 2-cocycle on $\mathrm{Vect}(V)$ whose cohomology class is called Φ. Its restriction defines naturally a class in $H^2_{\mathrm{loc}}\big(\overline{\mathrm{SVect}}(V), \Omega^2(V)\big)$, and one induces the required class in $H^2_{\mathrm{loc}}\big(\overline{\mathrm{SVect}}(V), \overline{\mathrm{SVect}}(V)\big)$ through the exact sequence of sheaves

$$0 \longrightarrow \underline{\mathrm{S}\widetilde{\mathrm{Vect}}} \longrightarrow \underline{\Omega}^{\,2} \longrightarrow \underline{\Omega}^{\,3} \longrightarrow 0$$

(cf. [L-R] Appendix 3). One can also recall the key role of this class Φ in deformation theory for the hamiltonian case: if one has a Poisson tensor Λ, the Vey class in the local cohomology of the Poisson algebra is defined by $S(f, g) = \langle \Phi(H_f, H_g), \Lambda \rangle$ for every functions f and g, H_f and H_g being the associated Hamiltonian fields defined by $H_f = i(\mathrm{d}\,f)\Lambda$ and $H_g = i(\mathrm{d}\,g)\Lambda$. This Vey class is the leading term of the formal deformation of the Poisson bracket.

To conclude this part, one must say that our rigidity result destroys completely the naïve hope for a deformation theory for unimodular fields with $n = 3$ generalizing directly the $\star$–product theory for hamiltonian fields with $n = 2$. The need for a deformation quantization theory in the unimodular case is now evident in various physical situations (see e.g. [MS], [S], [A-L-M-Y], [Ho]); more sophisticated constructions will be necessary for that purpose. We hope to discuss them in a forthcoming paper.

4.4 Unimodular Vector Fields in Supergeometry

We shall consider generalizations of Lie algebra of unimodular vector fields in a supergeometric framework (see [M] for basic definitions and results in supergeometry). One must first choose a supermanifold $\mathcal{V}$ such that V is the underlying differentiable manifold. Let $\mathcal{V} = \prod T^*V$ be the superization of V whose dimension is $(n|n)$: one adds odd variables as the cotangent bundle on V, so the space $\mathcal{O}(\mathcal{V})$ of functions on $\mathcal{V}$ is isomorphic to $\Omega_*(V)$. The Schouten bracket on $\Omega_*(V)$ then appears as an odd Poisson bracket for the odd symplectic form canonically defined on $\prod T^*V$: this odd symplectic form comes from the usual symplectic form on T^*V after a change of parity.

Let us now consider once more the codifferential $\delta\colon \Omega_*(V) \to \Omega_*(V)$ and set $S\overline{\Omega}_*(V) = \mathrm{Ker}\,\delta$, $S\Omega_*(V) = \mathrm{Im}\,\delta$.

Formula (4.3) implies $\delta([A, B])+[\delta A, B]+(-1)^{|A|}[A, \delta B] = 0$, so $S\overline{\Omega}_*(V)$ is a subalgebra of $\Omega_*(V)$ and $S\Omega_*(V)$ is an ideal of $S\overline{\Omega}_*(V)$ for Schouten bracket. Moreover if $A, B \in S\overline{\Omega}_*(V)$ then $[A, B] = \delta(A \wedge B) \in S\Omega_*(V)$; and one checks easily that the quotient superalgebra $S\overline{\Omega}_*(V)/S\Omega_*(V)$ is isomorphic to $H^*_{DR}(V)$ with the trivial bracket.

It is possible to generalize Lichnerowicz's arguments [L] to the supercase, and one obtains:

Proposition 4.4.1

> (1) $S\Omega_*(V)$ is the derived superalgebra of $S\overline{\Omega}_*(V)$.
> (2) All (graded) derivations of $S\overline{\Omega}_*(V)$ are inner.
> (3) All (graded) derivations of $S\Omega_*(V)$ are given by Schouten bracket with elements of $S\overline{\Omega}_*(V)$.

In cohomological terms, one has

$$H^1_{\mathrm{gr}}\big(S\overline{\Omega}_*(V), \mathbb{R}\big) = H^*_{DR}(V), \qquad H^1_{\mathrm{gr}}\big(S\Omega_*(V), S\Omega_*(V)\big) = H^*_{DR}(V)$$
$$H^1_{\mathrm{gr}}\big(S\overline{\Omega}_*(V), S\overline{\Omega}_*(V)\big) = 0.$$

We have not been able to generalize to this supercase the computations of Gelfand–Fuks cohomology, but it is possible to extend the construction of the central extension. One defines a graded antisymmetric bracket by

$\{A, B\} = \delta A \wedge \delta B$, for any A and B in $\Omega_*(V)$, one still has relation (4) and $\delta(\{A, B\}) = [\delta A, \delta B]$. So there exists on exact sequence of graded spaces

$$0 \longrightarrow S\overline{\Omega}_*(V) \longrightarrow \Omega_*(V) \xrightarrow{\ \delta\ } S\Omega_*(V) \longrightarrow 0$$

which is not an exact sequence of graded Lie algebras, since $\{\ ,\}$ doesn't satisfy Jacobi identity. But one can divide out by $\mathrm{Im}(\delta) = S\Omega_*(V)$ and one obtains a central extension of graded Lie algebras

$$0 \longrightarrow H_{DR}^{n-1-*}(V) \longrightarrow \Omega_{*+1}(V)/S\Omega_{*+1}(V) \xrightarrow{\ \delta\ } S\Omega_*(V) \longrightarrow 0$$

$$(4.6)$$

(beware of the shift of graduation in the cohomology space).

In degree one, it gives

$$0 \longrightarrow H_{DR}^2(V) \longrightarrow \Omega_2(V)/\delta(\Omega_3(V)) \longrightarrow \mathrm{SVect}(V) \longrightarrow 0$$

and we recover extension (4.2).

So our constructions really give a super version of constructions valid for $\mathrm{SVect}(V)$; but we choose here a specific "superization" of V: a more systematic approach would be to consider a general supermanifold $\mathcal{V}$ whose underlying C^∞ manifold is V, take the superversion of a volume form (Berezinian) and the corresponding Lie superalgebra of vector fields which leave this structure invariant; but this kind of object is in general too huge to be tractable.

Before going to more particular cases, we shall make the

4.4.2 Conjecture. *The extension* (5) *is the versal central extension of* $S\Omega_*(V)$.

We shall now describe the low dimensional cases in more details.

For $n = 2$ this is the hamiltonian case. One has

$$\Omega_2(V) = \Omega_0(V) = C^\infty(V), \quad \Omega_1(V) = \mathrm{Vect}(V),$$

the 2-tensors being identified with functions through $f \to f\Lambda$ where Λ denotes the volume tensor. The bracket then gives $\{f\Lambda, g\Lambda\} = \{f, g\}\Lambda$ where $\{f, g\}$ is the "usual" Poisson bracket of functions f and g.

One has $S\overline{\Omega}_2(V) = \mathbb{R}$, $S\Omega_2(V) = \{0\}$, $S\overline{\Omega}_1(V) = \mathrm{Sp}(V)$ the symplectic vector fields, $S\Omega_1(V) = \mathrm{Ham}(V)$ the hamiltonian vector fields, $S\overline{\Omega}_0(V) = C^\infty(V)$ and $S\Omega_0(V) = \overline{C^\infty}(V)$ the space of functions with vanishing integral (remind V is compact connected).

So one obtains here the familiar extensions:

$$0 \longrightarrow H^1_{DR}(V) \longrightarrow \mathrm{Vect}(V)/\mathrm{Ham}(V) \longrightarrow \overline{C^\infty}(V) \longrightarrow 0$$

$$0 \longrightarrow H^0_{DR}(V) \longrightarrow C^\infty(V) \longrightarrow \mathrm{Ham}(V) \longrightarrow 0$$

$$\| \qquad\qquad\qquad\qquad \|$$

$$\mathbb{R} \qquad\qquad\qquad \text{Poisson algebra.}$$

For $n = 3$: This is the natural generalization of $n = 2$ hamiltonian case, and also the most interesting case in view of its applications to hydrodynamics (see [A-K]).

One has again $S\overline{\Omega}_0(V) = C^\infty(V)$ and $S\Omega_0(V) = \overline{C^\infty}(V)$

$$S\overline{\Omega}_1(V) = \overline{\mathrm{SVect}}(V), \quad S\Omega_1(V) = \mathrm{SVect}(V), \quad S\Omega_3(V) = 0$$
$$S\overline{\Omega}_3(V) = \mathbb{R}, \quad S\overline{\Omega}_2(V) = \{A \in \Omega_2(V) \mid \mathrm{d}(i(A)\omega) = 0\}$$
$$S\Omega_2(V) = \{\mathrm{grad}(f) \mid f \in C^\infty(V)\} \quad \text{where} \quad \mathrm{grad}(f) = i(\mathrm{d}f)\Lambda$$

denotes the generalized gradient (Λ being as usual the volume 3-tensor).

One can split $S\Omega_*(V)$ into its odd and even part $S\Omega_*(V) = \mathfrak{g}_{(0)} \oplus \mathfrak{g}_{(1)}$ with $\mathfrak{g}_{(0)} = S\Omega_1(V) = \mathrm{SVect}(V)$ and $\mathfrak{g}_{(1)} = \overline{C^\infty}(V) \oplus S\Omega_2(V)$. We shall identify elements of $\mathfrak{g}_{(1)}$ with couples $(f, \varphi) \in \overline{C^\infty}(V)^2$ with φ identified with $\mathrm{grad}(\varphi)$. The bracket $\mathfrak{g}_{(0)} \times \mathfrak{g}_{(1)} \to \mathfrak{g}_{(1)}$ is then simply given by the natural action of $\mathrm{SVect}(V)$ on functions.

The only non trivial part of the symmetric bracket

$$\mathfrak{g}_{(1)} \times \mathfrak{g}_{(1)} \longrightarrow \mathfrak{g}_{(0)}$$

is then $[f, \varphi] = i(\mathrm{d}f \wedge \mathrm{d}\varphi)\Lambda$ (yes, it *is* a symmetric bracket !).

For the central extension, we recover the even part (2)

$$0 \longrightarrow H^1_{DR}(V) \longrightarrow \Omega_2(V)/S\Omega_2(V) \longrightarrow \mathrm{SVect}(V) \longrightarrow 0$$

the odd part being:

$$0 \to H^0_{DR}(V) \oplus H^2_{DR}(V) \to \frac{\Omega_1(V)}{S\Omega_1(V)} \oplus \Omega_3(V) \to \overline{C^\infty}(V) \oplus S\Omega_2(V) \to 0.$$

It is an easy exercise to identify the various components of the cocycle of this extension of superalgebras (don't forget the symmetric part).

Remarks 4.4.3 (1) We do not know any place in the literature in which the theory of central extensions of Lie superalgebras is developed, but it is straightforward to generalize the classical theory to the super case.

(2) The case $n = 3$ has still another particular feature, the existence of a kind of Killing form. The *Arnold invariant* is easily defined: for X and Y in SVect(V), let α and β be two 1-forms such that $d\alpha = i(X)\omega$ and $d\beta = i(Y)\omega$; the Arnold invariant $I(X,Y)$ is equal to $\int_V \alpha \wedge d\beta$. One checks easily that $(X,Y) \longrightarrow I(X,Y)$ is a symmetric, bilinear, invariant, non degenerate form.

Before leaving supergeometry , we shall propose an open problem: check all "superizations" of SVect(V), analogous to the superalgebra $S\Omega_*(V)$ just discussed, and prove their rigidity(or find their deformations!).

4.5 About the Group of Volume Preserving Diffeomorphisms

We already introduced this group $S\mathcal{D}(V)$ in the beginning of the article, now we shall try to generalize, (or better to say "integrate") the previous constructions to $S\mathcal{D}(V)$. For example, what is the subgroup of $S\mathcal{D}(V)$ which corresponds to SVect(V) ? We shall first "integrate" the exact sequence of Lie algebras (1), following the methods of Ismagilov [I].

Let $\widehat{V} \xrightarrow{\pi} V$ be the universal covering of V and $\widehat{\omega} = \pi^*(\omega)$ the corresponding volume form on $\widehat{V}$; let $\Gamma = \pi_1(V)$ the fundamental group. We shall consider the group $\widehat{S\mathcal{D}(V)}$ of diffeomorphisms of $\widehat{V}$, such that $f^*(\widehat{\omega}) = \widehat{\omega}$ and commuting with Γ; then $\widehat{\Gamma} = \Gamma \cap \widehat{S\mathcal{D}(V)}$ is central in both groups, so one obtains a central extension of groups:

$$1 \longrightarrow \widehat{\Gamma} \longrightarrow \widehat{S\mathcal{D}(V)} \xrightarrow{p} S\mathcal{D}(V) \longrightarrow 1.$$

It is an interesting exercise to work out this construction in more details for V a torus with canonical volume form.

We shall now construct the so called *"Calabi invariant"* (cf. [C]). Let $\alpha \in Z^1(V)$ a closed 1-form and $f \in \widehat{S\mathcal{D}(V)}$, $\widetilde{f} = p(f)$. For $x \in V$, take $\xi \in \pi^{-1}(x)$ and ℓ a path in $\widehat{V}$ from ξ to $f(\xi)$; consider the path $p(\ell)$ in V and set $C^\alpha(f)(x) = \int_{p(\ell)} \alpha$. Well-definedness is checked immediately, as well as the cocycle relation

$$C^\alpha(fg)(x) = C^\alpha(f)(x) + C^\alpha(g)\big(\widetilde{f}(x)\big).$$

Moreover, if $\alpha = \mathrm{d}\,\varphi$ then $C^\alpha(f)(x) = \varphi\big(\widetilde{f}(x)\big) - \varphi(x)$ so it is a coboundary. So we constructed a mapping $H^1_{DR}(V) \longrightarrow H^1\big(\widehat{\mathcal{SD}(V)}, C^\infty(V)\big)$, the last group being of course group cohomology of $\widehat{\mathcal{SD}(V)}$ with coefficients in the module $C^\infty(V)$.

One can deduce another mapping $H^1_{DR}(V) \longrightarrow \mathrm{Hom}\big(\widehat{\mathcal{SD}(V)}, \mathbb{R}\big)$ defined by $\alpha \longrightarrow \big[f \to \int_V C^\alpha(f)\omega\big]$; its restriction to $\widehat{\Gamma}$ is nothing but $[\gamma] \longrightarrow \int_\gamma \alpha$, i.e. the classical identification $H^1_{DR}(V) \longrightarrow \mathrm{Hom}\big(\pi_1(V), \mathbb{R}\big)$. Upon dualizing one obtains $\widehat{\mathcal{SD}(V)} \longrightarrow H_1(V, \mathbb{R})$ which restricts to the abelianization: $\widehat{\Gamma} \subset \pi_1(V) \overset{h}{\longrightarrow} H_1(V, \mathbb{R})$. Finally, we divide out by $\widehat{\Gamma}$ to obtain the "Calabi invariant", a homomorphism

$$\mathcal{SD}(V) \longrightarrow H_1(V, \mathbb{R})/h(\widehat{\Gamma}) \qquad (h(\widehat{\Gamma}) \text{ is a lattice}).$$

One obtains an exact sequence of groups:

$$1 \longrightarrow \mathcal{SD}_0(V) \longrightarrow \mathcal{SD}(V) \longrightarrow H_1(V, \mathbb{R})/h(\widehat{\Gamma}) \longrightarrow 1$$

$$(4.7)$$

which integrates the sequence (4.1): more precisely if one considers the exact sequence of tangent spaces at the unit element, one recovers sequence (4.1) (don't forget $H_1(V, \mathbb{R}) = H^{n-1}_{DR}(V)$).

There exists another description of $\mathcal{SD}_0(V)$; for $U \subset V$ a coordinate chart diffeomorphic to $\mathbb{R}^n$ one defines $\mathcal{SD}_U(V)$ as the group of elements of $\mathcal{SD}(V)$ whose support is contained in U. It is immediate to check that $\mathcal{SD}_U(V) \subset \mathcal{SD}_0(V)$ but one can prove that the $\mathcal{SD}_U(V)$ for various charts U generate $\mathcal{SD}_0(V)$ (see [I] pp. 99 sqq).

The natural question is now to construct and interpret the central extension for $\mathcal{SD}(V)$; we will then need some other stuff, such as the coadjoint orbits.

4.6   About Coadjoint Orbits of $\mathcal{SD}(V)$ and its Central Extensions

Every Lie group acts naturally on its Lie algebra by the adjoint action and on the dual of its Lie algebra by the coadjoint action. This action became popular following the work of A. A. Kirillov on the orbit method (cf. [K] and bibliography inside); one tries to classify the coadjoint orbits and the symplectic structures on them and then construct representations of the group from this data. It works mainly for finite dimensional groups, but there exist also some infinite dimensional examples (see for example [G]). Of course, when the dimension is infinite, one must make choices for elements of the dual to consider: taking the full algebraic dual space obviously leads to pathologies. One way is to restrict to *regular* duals for Lie algebras of vector fields, i. e. linear forms (sometimes called " momenta") which are defined by smooth densities. We shall note $\mathfrak{g}^*_{\mathrm{reg}}$ the regular dual of $\mathfrak{g}$. For example, one has

$$\mathrm{Vect}(V)^*_{\mathrm{reg}} = \Omega^1(V) \underset{C^\infty(V)}{\otimes} \Omega^n(V),$$

an element $\alpha \otimes \widetilde{\omega} \in \Omega^1(V) \otimes \Omega^n(V)$ acts as follows

$$\langle \alpha \otimes \widetilde{\omega}, X \rangle = \int_V \alpha(X)\widetilde{\omega}.$$

Things are simpler for $\overline{\mathrm{SVect}}(V)$ and $\mathrm{SVect}(V)$, since the volume form is fixed. One can identify $\mathrm{SVect}^*(V)_{\mathrm{reg}}$ and $\overline{\mathrm{SVect}}^*(V)_{\mathrm{reg}}$ with $\Omega^1(V)/Z^1(V)$ and $\Omega^1(V)/B^1(V)$ respectively; a one-form α acts on a vector X giving $\int_V \alpha(X)\omega$. The coadjoint action is then given by the natural action of diffeomorphisms on forms.

For our applications to the group, we shall consider different momenta, given by integration currents rather than smooth densities; a $(n-2)$ dimensional compact submanifold $\Sigma \subset V$ acts on $\mathrm{SVect}(V)$ as follows: $\langle \Sigma, X \rangle = \int_\Sigma \eta$ where $\mathrm{d}\eta = i(X)\omega$. We consider also, more globally, spaces of submanifolds:let $\Sigma \subset V$ be a fixed compact $(n-2)$ dimensional submanifold of V, and let $P(\Sigma, V)$ be the space of embeddings of Σ into V, isotopic to identity $\Sigma \xrightarrow{i} \Sigma \subset V$. The group of diffeomorphisms $\mathrm{Diff}(\Sigma)$ acts on the right on $P(\Sigma, V)$ by composition, and so one obtains a fibration:

$$\mathrm{Diff}(\Sigma) \longrightarrow P(\Sigma, V)$$
$$\downarrow$$
$$\mathcal{M}(\Sigma)$$

where $\mathcal{M}(\Sigma)$ is the space of submanifolds of V isotopic to Σ. The theorem of prolongations of isotopies (cf. Hirsch [Hi] Chap 8) implies easily it is a fibration. Moreover, if one considers the ad-hoc manifold structure on $\mathcal{M}(\Sigma)$ and $P(\Sigma, V)$, it is a locally trivial fibre bundle (cf. Hamilton [H]).

We can now identify the tangent space: let $T(\Sigma)$ and $\nu(\Sigma)$ be the tangent and normal bundles to Σ respectively, so one has an exact sequence of bundles on Σ

$$0 \longrightarrow T(\Sigma) \longrightarrow i^*(T(V)) \longrightarrow \nu(\Sigma) \longrightarrow 0 \ ;$$

then $\quad T_i P(\Sigma, V) = \Gamma_\Sigma\big(i^*(T(V))\big), \quad T_{\mathrm{Id}}\big(\mathrm{Diff}(\Sigma)\big) = \Gamma_\Sigma(T(\Sigma)) = \mathrm{Vect}(\Sigma)$

$\qquad T_i\big(\mathcal{M}(\Sigma)\big) = \Gamma_\Sigma(\nu(\Sigma))$

(here Γ_Σ symbolizes as usual the space of sections over Σ). All of these spaces are " tame" Fréchet spaces (cf. [H]). We shall define a symplectic form on the manifold $\mathcal{M}(\Sigma)$. Let ξ and η be two tangent vectors to $\mathcal{M}(\Sigma)$ in i; for every $x \in \Sigma$, let $\overline{\xi}(x)$ and $\overline{\eta}(x)$ in $T_x V$ which are lifts of $\xi(x)$ and $\eta(x)$ in $\nu_x(\Sigma)$, one obtains vector fields $\overline{\xi}$ and $\overline{\eta}$ on Σ since we can do all that smoothly. Define Ω by $\Omega(\xi, \eta) = \displaystyle\int_\Sigma i(\overline{\xi})i(\overline{\eta})\omega$; it doesn't depend on the particular choices of lifts $\overline{\xi}$ and $\overline{\eta}$. So we have constructed an antisymmetric non degenerate bilinear form

$$\Omega \colon T_i\big(\mathcal{M}(\Sigma)\big) \times T_i\big(\mathcal{M}(\Sigma)\big) \longrightarrow \mathbb{R}\,.$$

We can then easily define the form in any point of $\mathcal{M}(\Sigma)$. Finally, one has

Theorem 4.6.1   (Ismagilov [I]) *The form Ω makes $\mathcal{M}(\Sigma)$ a symplectic manifold which can be embedded in the dual of* $\mathrm{SVect}(V)$ *as a coadjoint orbit of the group* $\mathcal{SD}_0(V)$. *The embedding is defined by* $\langle \widetilde{\Sigma}, X \rangle = \displaystyle\int_{\widetilde{\Sigma}} \eta$ *where* $\mathrm{d}\,\eta = i(X)\omega$.

Remarks 4.6.2 (1) One should say more correctly that $\mathcal{M}(\Sigma)$ is a *weak* symplectic manifold, as usual in non hilbertian infinite dimensional differential geometry: Ω induces a monomorphism of $T\mathcal{M}(\Sigma)$ in its dual, not an isomorphism.

(2) The action of $\mathcal{SD}_0(V)$ on $\mathcal{M}(\Sigma)$ is simply given by isotopies of the generalized knot $(\Sigma \subset V)$ (use Milnor's theorem of prolongation of isotopies once more).

(3) The only delicate part of the proof is to show that Ω is closed: it makes use of a covering of $\mathcal{M}(\Sigma)$ by open sets on which ω is exact (see [I]).

(4) The case $n = 3$ is easier. One simply has to consider the space of knots in a compact 3-dimensional manifold, and its properties are richer than in the general case. See the book of J. L. Brylinski [B] for details.

The next step is to construct a geometric prequantization of those manifolds $\mathcal{M}(\Sigma)$. It is possible to construct a prequantum bundle

$$S^1 \longrightarrow (E, \theta) \xrightarrow{\pi} \mathcal{M}(\Sigma),$$

which is as in finite dimension a circle bundle with a connection form θ such that $d\theta = \pi^*(\Omega)$. Following an idea of J. M. Souriau ([So] Chap 5) one then considers the group of automorphisms of this bundle which respect the connection form ("quantomorphisms"): this group is a central extension of the group of symplectic diffeomorphisms of the base, so restricting to the group $\mathcal{SD}_0(V)$ one obtains a central extension:

$$1 \longrightarrow S^1 \longrightarrow A_\Sigma(E, \theta) \longrightarrow \mathcal{SD}_0(V) \longrightarrow 1.$$

So one has obtained a central extension for each $\Sigma \subset V$; in fact, it depends only on the homology class $[\Sigma] \in H_{n-2}(V, \mathbb{R})$. Then, we can sum all these extensions over all possible $[\Sigma] \in H_{n-2}(V, \mathbb{R})$, and one obtains a central extension:

$$1 \longrightarrow H^{n-2}(V, \mathbb{R})/L \longrightarrow A(V) \longrightarrow \mathcal{SD}_0(V) \longrightarrow 1$$

$$\tag{4.8}$$

L being a lattice isomorphic to $H^{n-2}(V, \mathbb{Z})$ modulo torsion.

This extension integrates extension (4.2): to show that, one must use local group cocycles since no global cocycle exist because of non triviality of extension (4.8) as a torus bundle: this difficulty is classical in group cohomology; nevertheless, these local cocycles are enough to obtain the Lie algebra extension (cf. [I] for a proof) after derivation.

One can give one more interpretation of this central extension $A(V)$. We already met the groups $\mathcal{SD}_U(V) \subset \mathcal{SD}_0(V)$ of diffeomorphisms whose support is contained in a coordinate chart U; remark now that if $U_1 \subset U_2$, then $\mathcal{SD}_{U_1}(V) \subset \mathcal{SD}_{U_2}(V)$, so these groups form an inductive system. It is now easy to see that the extension (4.8) trivializes over those subgroups, so one obtains maps, compatible with inclusion: $\mathcal{SD}_U(V) \longrightarrow A(V)$. One has

Proposition 4.6.3   ([I] p. 118) $A(V)$ *is isomorphic to the inductive limit* $\varinjlim_U \mathcal{SD}_U(V)$.

The proof of the analogous result for Lie algebras is an easy exercise. We didn't discuss some aspects of volume preserving diffeomorphisms, because it would have made this paper much too long; after discussing the dual and coadjoint orbits we could have explained the applications to fluid mechanics, and above all the interpretation of Euler equation as a hamiltonian system on the dual. The interested reader is referred to the book of Arnold and Khesin [A-K]. Similarly we gave only a very short sketch of the Arnold invariant, without explaining its beautiful geometric interpretations, such as generalized linking number; see the work of Gambaudo and Ghys [G-G] for example. Our main motivation is to work out efficient tools to attack the deformation quantization problem; we plan to discuss this subject in more details in another article.

Before leaving the subject, we shall say a few words about the link with the somewhat mysterious theory of branes, mentioned in the introduction. It is a far reaching generalization of string theory: instead of curves, the basic objects are branes(cf mem-branes), i.e. n-dimensional (for $n > 1$) in spacetime. Several point of view have then been developed;the brane can represent boundary constraints for open strings,the ends of the strings must stay on some fixed branes. One can also consider branes as dynamical objects,directly generalizing strings,which move inside spacetime; the latter point of view is reminiscent of various problems in riemannian geometry. In both cases the branes come equipped with an intrinsic volume form, and the group of volume preserving diffeomorphisms turns out to be a part of the symmetry group of the dynamical lagrangian, just as Virasoro group for conformal field theory. See [Ho][M-S][S] for more details about the role of the group of volume preserving diffeomorphisms in brane theory, and the big book of Michio Kaku[Ka] for a general presentation of branes in the context of string theory.

Bibliography

[A-K] Arnold, Vladimir I.; Khesin, Boris A. Topological methods in hydrodynamics. Applied Mathematical Sciences, 125. Springer-Verlag, New York, 1998. xvi+374 pp.

[A-L-M-Y] Awata, Hidetoshi; Li, Miao; Minic, Djordje; Yoneya, Tamiaki On the Quantization of Nambu Brackets. hep-th/9906248

[B] Brylinski, Jean-Luc Loop spaces, characteristic classes and geometric quantization. Progress in Mathematics, 107. Birkhäuser Boston Inc., Boston, MA, 1993. xvi+300 pp.

[C] Calabi, Eugenio On the group of automorphisms of a symplectic manifold. Problems in analysis (Lectures at the Sympos. in honor of Salomon Bochner, Princeton Univ., Princeton, N.J., 1969), pp. 1–26. Princeton Univ. Press, Princeton, N.J., 1970.

[Fe] Fedosov, Boris. Deformation quantization and index theory. Mathematical Topics, 9. Akademie Verlag, Berlin, 1996. 325 pp.

[F] Fuks, D. B. Когомологии бесконечномерных алгебр Ли (Russian) [Cohomology of infinite-dimensional Lie algebras.] "Nauka", Moscow, 1984. 272 pp. English version, Consultants Bureau, New York, 1986. 339 pp.

[G] Golenistcheva-Kutuzova, M. I. Generic orbits of the diffeomorphism group of a two-manifold in the space $\mathfrak{G}^*_{reg}$ of regular momenta. Proceedings of the Winter School on Geometry and Physics (Srní, 1990). Rend. Circ. Mat. Palermo (2) Suppl. No. 26 (1991), 171–178.

[G-G] Gambaudo, Jean-Marc; Ghys, Etienne. Signature asymptotique d'un champ de vecteurs en dimension 3. Duke Math. J. 106 (2001), no. 1, 41–79.

[G-S] Guillemin, Victor; Shnider, Steven. Some stable results on the cohomology of the classical infinite-dimensional Lie algebras. Trans. Amer. Math. Soc. 179 (1973), 275–280.

[H] Hamilton, Richard S. The inverse function theorem of Nash and Moser. Bull. Amer. Math. Soc. (N.S.) 7 (1982), no. 1, 65–222.

[Hi] Hirsch, Morris W. Differential topology. Graduate Texts in Mathematics, 33. Springer-Verlag, New York, 1994. x+222 pp.

[Ho] Hoppe, Jens. On M-algebras, the quantisation of Nambu-mechanics, and volume preserving diffeomorphisms. Papers honouring the 60th birthday of Klaus Hepp and of Walter Hunziker, Part II (Zürich, 1995). Helv. Phys. Acta 70 (1997), no. 1-2, 302–317.

[H-S] Hilton, P. J.; Stammbach, U. A course in homological algebra. Second edition. Graduate Texts in Mathematics, 4. Springer-Verlag, New York, 1997 .

[I] Ismagilov, R. S. Representations of infinite-dimensional groups. Translated from the Russian manuscript by D. Deart. Translations of Mathematical Monographs, 152. American Mathematical Society, Providence, RI, 1996. x+197 pp.

[Ka] Kaku, Michio. Strings, conformal fields, and M-theory. Second edition. Graduate Texts in Contemporary Physics. Springer-Verlag, New York, 2000. xvi+531 pp.

[K] Kirillov, A. A. Merits and demerits of the orbit method. Bull. Amer. Math. Soc. (N.S.) 36 (1999), no. 4, 433–488.

[Ko] Kontsevich, Maxim. Deformation quantization of Poisson manifolds, I. q-alg/9709040

[K-M] Kriegl, Andreas; Michor, Peter W. The convenient setting of global analysis. Mathematical Surveys and Monographs, 53. American Mathematical Society, Providence, RI, 1997. x+618 pp.

[L-R] Lecomte, Pierre B. A.; Roger, Claude. Rigidité de l'algèbre de Lie des champs de vecteurs unimodulaires. J. Differential Geom. 44 (1996), no. 3, 529–549.

[L] Lichnerowicz, André. Algèbre de Lie des automorphismes infinitésimaux d'une structure unimodulaire. Ann. Inst. Fourier (Grenoble) 24 (1974), no. 3, xiv, 219–266.

[M] Manin, Yuri I. Gauge field theory and complex geometry. Translated from Russian by N. Koblitz and J. R. King. Grundlehren der Mathematischen Wissenschaften [Fundamental Principles of Mathematical Sciences], 289. Springer-Verlag, Berlin, 1988. x+297 pp.

[M-S] Matsuo, Y.; Shibusa, Y. Volume Preserving Diffeomorphism and Noncommutative Branes. hep-th/0010040

[R] Roger, Claude. 1- Extensions centrales d'algèbres et de groupes de Lie de dimension infinie, algèbre de Virasoro et généralisations. Mathematics as language and art (Białowieża, 1993). Rep. Math. Phys. 35 (1995), no. 2-3, 225–266.
2-Unimodular vector fields and deformation quantization. Deformation quantization,Strasbourg, 2001. IRMA Lectures in mathematics and theoretical physics 1. De Gruyter, Berlin New-York, 2002, pp 137-148.

[S] Schomerus, Volker. D-branes and Deformation Quantization. hep-th/9903205

[S-S] Singer, I. M.; Sternberg, Shlomo. The infinite groups of Lie and Cartan. I. The transitive groups. J. Analyse Math. 15 (1965), 1–114.

[So] Souriau, J.-M. Structure des systèmes dynamiques. Dunod, Paris, 1970. xxxii+414 pp.

Chapter 5

Moduli Space of Germs of Symplectic Connections of Ricci Type

Michel Cahen[1]

5.1 Introduction

On any smooth, finite dimensional, paracompact manifold M, there exists a smooth riemannian metric g. The space of riemannian metrics on M, $\mathcal{E}(M)$, is infinite dimensional. One may impose restrictions to the metric; for example by means of a variational principle. If the functional is chosen to be

$$\int_M \rho_g \, d\mu_g$$

where ρ_g is the scalar curvature of g and $d\mu_g$ is the standard measure associated to g, the critical points are the so called Einstein metrics. Riemannian geometers have studied the existence of Einstein metrics on a given manifold M; in the case there is existence they have looked at the moduli space of Einstein metrics on M, i.e. the space of Einstein metrics modulo the action of the diffeomorphism group of M.

On any smooth, finite dimensional, paracompact manifold M, there does not exist a smooth symplectic form ω. The manifold must be even dimensional, orientable; but these 2 conditions are far from sufficient as exemplified by the spheres S^{2n} ($n \geq 2$) which do not admit a symplectic structure. We shall thus consider a symplectic manifold (M, ω). A symplectic connection ∇ is a linear connection which is torsion free and for which

[1]Université Libre de Bruxelles, CP218, boulevard du Triomphe, B1050 Bruxelles, Belgique, mcahen@ulb.ac.be

ω is parallel. The space of symplectic connections on (M, ω), $\mathcal{E}(M, \omega)$ is infinite dimensional. One may impose restrictions to the connection; for example by means of a variational principle. Let the functional be chosen to be

$$\int_M r^2 \omega^n,$$

where $\dim M = 2n$ and r denotes the Ricci tensor of the connection ∇ (i.e. $r(X, Y) = \mathrm{tr}\ [Z \to R(X, Z)Y]$, where X, Y, Z are vector fields on M and $R(X, Z)$ is the curvature endomorphism associated to X and Z for the connection ∇). Finally r^2 is the scalar defined as follows. Let ρ be the endomorphism

$$\omega(X, \rho Y) \underset{\mathrm{def}}{=} r(X, Y).$$

Then

$$r^2 \underset{\mathrm{def}}{=} \mathrm{tr}\ \rho^2.$$

Remark that $\mathrm{tr}\ \rho = 0$ as ρ_x belongs to the symplectic algebra of $(T_x M, \omega_x)$.

The Euler Lagrange equations of this functional are :

$$\underset{X,Y,Z}{\bigoplus}\ (\nabla_X r)(Y, Z) = 0,$$

where $\bigoplus$ denotes the sum over cyclic permutations of the indicated quantities.

A connection ∇ satisfying these field equations is said to be **preferred**. By analogy with the riemannian situation we can formulate :

Problem 1   *Can one describe the moduli space of preferred connections on (M, ω), i.e. the space of preferred connections on (M, ω) modulo the action of the symplectic diffeomorphism group.*

The following has been proven in [1].

Theorem 5.1   *Let (M, ω) be a compact symplectic surface; let ∇ be a complete preferred symplectic connection. Then*

> *(i) if $M = S^2$, ∇ is the Levi Civita connection associated to a metric of constant positive curvature*
> *(ii) if $M = T^2$, the connection ∇ is flat*
> *(iii) if M is a surface of genus $g \geq 2$, ∇ is the Levi Civita connection associated to a metric of constant negative curvature.*

In dimension $2n \geq 4$ very little is known. Fortunately a subclass of preferred connections may be described with some detail. Let me first define the subclass.

Let (M, ω) be a symplectic manifold and ∇ a symplectic connection. At a point $x \in M$, the curvature tensor R_x of ∇ is a tensor of type $\binom{0}{4}$ having the following symmetries

$$R_x(X, Y, Z, T) \underset{\text{def}}{=} \omega(R(X, Y)Z, T)$$

- (i) $R_x(X, Y, Z, T) = -R_x(Y, X, Z, T)$
- (ii) $R_x(X, Y, Z, T) = R_x(X, Y, T, Z)$
- (iii) $\underset{X,Y,Z}{\oplus} R_x(X, Y, Z, T) = 0.$

From (i) and (ii), $R_x \in \Lambda^2 T_x^* M \otimes \odot^2 T_x^* M$, where $\odot^k V$ is the symmetrized k-tensor product of the vector space V. Recall Koszul's exact sequences :

$$0 \leftrightarrows \odot^4 V \leftrightarrows V \otimes \odot^3 V \overset{s}{\underset{a}{\leftrightarrows}} \Lambda^2 V \otimes \odot^2 V \overset{s}{\underset{a}{\leftrightarrows}} \Lambda^3 V \otimes V \leftrightarrows \Lambda^4 V \leftrightarrows 0$$

where

$$a(u_1 \wedge \ldots \wedge u_p \otimes v_1 \ldots v_q) \underset{\text{def}}{=} \sum_{i=1}^{q} u_1 \wedge \ldots \wedge u_p \wedge v_i \otimes v_1 \ldots \hat{v}_i \ldots v_q$$

$$s(u_1 \wedge \ldots \wedge u_p \otimes v_1 \ldots v_q) \underset{\text{def}}{=} \sum_{j=1}^{p} u_1 \wedge \ldots \wedge \hat{u}_j \wedge \ldots \wedge u_p \otimes u_j v_1 \ldots v_q (-1)^{p-j}.$$

Then

$$a^2 = s^2 = 0$$

$$(as + sa)|_{\Lambda^p V \otimes \odot^q V} = (p + q) id|_{\Lambda^p V \otimes \odot^q V}.$$

Since

$$(aR_x)(X, Y, Z, T) = \underset{X,Y,Z}{\oplus} R_x(X, Y, Z, T) = 0$$

we see from (iii) that the space $\mathcal{R}_x$ of curvature tensors at x is

$$\mathcal{R}_x = \ker a \subset \Lambda^2 T_x^* M \otimes \odot^2 T_x^* M.$$

The group $Sp(T_xM, \omega_x)$ acts on $\mathcal{R}_x$. Under this action the space $\mathcal{R}_x$ decomposes in 2 stable subspaces :

$$\mathcal{R}_x = \mathcal{E}_x \oplus \mathcal{W}_x.$$

The action of $Sp(T_xM, \omega_x)$ on each of the subspaces is irreducible [2]. These subspaces may be described as follows.

Let $t \in \odot^2 T_x^* M$; the map $j : \odot^2 T_x^* M \to \mathcal{R}_x : t \to as(\omega_x \otimes t)$ is injective and $Sp(T_xM, \omega_x)$ equivariant. The image $j \odot^2 T_x^* M$ is the stable subspace $\mathcal{E}_x$.

The symplectic form ω_x induces a non degenerate scalar product on $\Lambda^2 T_x^* M \otimes \odot^2 T_x^* M$; its restriction to $\mathcal{E}_x$ is also non degenerate. Hence :

$$\mathcal{R}_x = \mathcal{E}_x \oplus \mathcal{E}_x^{\perp} \cap \mathcal{R}_x := \mathcal{E}_x \oplus \mathcal{W}_x.$$

If r_x denotes as above the Ricci tensor associated to R_x, one checks that the Ricci tensor associated to $j(r_x)$ is $-2(n+1)r_x$. Hence the decomposition of the curvature tensor R_x into its $\mathcal{E}_x$ component (denoted E_x) and its $\mathcal{W}_x$ component (denoted W_x) reads:

$$R_x = E_x + W_x.$$

$$E_x(X, Y, Z, T) = -\frac{1}{2(n+1)}\big[2\omega_x(X, Y)r_x(Z, T) + \omega_x(X, Z)r_x(Y, T)$$

$$+\omega_x(X, T)r_x(Y, Z) - \omega_x(Y, Z)r_x(X, T) - \omega_x(Y, T)r_x(X, Z)\big].$$

A connection ∇ is said to be **of Ricci type** if, at each point x, $W_x = 0$ (such connections were called reducible by Vaisman [2]).

Observe that in dimension 2 $(n = 1)$, the space $\mathcal{W}$ vanishes identically. Thus, in a certain sense, the condition for a connection to be of Ricci type, generalizes in higher dimension the surface situation.

Lemma 5.1 *Let (M, ω) be a symplectic manifold of dimension $2n$ $(n \geq 2)$ and let ∇ be a symplectic connection of Ricci type. Then ∇ is a preferred connection.*

This leads to

Problem 2 *Can one describe the moduli space of symplectic connections of Ricci type on the symplectic manifold (M, ω).*

This paper is a contribution to the solution of problem 2. It describes work done in collaboration with Simone Gutt and Lorenz Schwachhöfer.

5.2   Some Properties of Ricci Type Connections

Let (M, ∇) be a real analytic manifold endowed with a torsion free analytic linear connection ∇. Let $x_0 \in M$ and let U_0 be a convex normal neighbourhood of x_0; let $\exp^{-1} : U_0 \to V_0 (\subset T_{x_0} M)$ be the corresponding logarithmic chart and denote by $\{ y^\alpha ; \alpha \leq d = \dim M \}$ the coordinates associated to a choice of basis in $T_{x_0} M$. The Christoffel symbols $\Gamma^\gamma_{\alpha\beta} (\nabla_{\partial_\alpha} \partial_\beta \underset{\mathrm{def}}{=} \Gamma^\gamma_{\alpha\beta} \partial_\gamma)$ at a point $u \in U_0$ may be expressed as a converging series

$$\Gamma^\gamma_{\alpha\beta}(u) = \sum_{|A| \geq 1} C^\gamma_{A\alpha\beta} y^A(u)$$

where A is a multiindex $(A = (l_1, \ldots, l_d))$, $|A| = l_1 + \ldots + l_d$ and $y^A = y_1^{l_1} \ldots y_d^{l_d}$. The coefficients are universal polynomials in the curvature tensor and its covariant derivatives (up to order $(|A| - 1)$ for C_A) evaluated at the point x_0 [3]. In particular this says that on a connected manifold M, the connection ∇ is uniquely determined by the value of the curvature tensor and its covariant derivatives at a point x_0 in M.

Lemma 5.2   *[4] Let (M, ω) be a symplectic manifold of dim $2n$ $(n \geq 2)$; let ∇ be a Ricci type symplectic connection. Then*

(i)   the curvature endomorphism is given by

$$R(X, Y) = -\frac{1}{2(n+1)} [-2\omega(X, Y)\rho - \rho Y \otimes \underline{X} + \rho X \otimes \underline{Y}$$
$$- X \otimes \underline{\rho Y} + Y \otimes \underline{\rho X}]$$

where $\underline{X}$ denotes the 1-form $i(X)\omega$ (for X a vector field on M) and where ρ is the endomorphism associated to the Ricci tensor :

$$r(U, V) \underset{\mathrm{def}}{=} \omega(U, \rho V);$$

(ii)   there exists a vector field u such that

$$\nabla_X \rho = -\frac{1}{2n+1} [X \otimes \underline{u} + u \otimes \underline{X}];$$

(iii)   there exists a function f such that

$$\nabla_X u = -\frac{2n+1}{2(n+1)} \rho^2 X + fX;$$

(iv) there exists a real number K such that

$$tr\rho^2 + \frac{4(n+1)}{2n+1}f = K;$$

(v) the hamiltonian vector field associated to f, X_f, reads

$$X_f = -\frac{1}{n+1}\rho u$$

and its covariant derivative reads

$$\nabla_Y X_f = -\frac{1}{(n+1)(2n+1)}u\omega(u,Y) + \frac{2n+1}{2(n+1)^2}\rho^3 Y - \frac{1}{n+1}f\rho Y.$$

By (i) one sees that the curvature tensor is determined by ρ; by (ii) one sees that the 1st covariant derivative of the curvature is determined by u; by (iii) one sees that the second covariant derivative of the curvature is determined by ρ and f; by (iv) one sees that f is determined by ρ and K; by (v) one sees that the 3rd covariant derivative of the curvature is determined by u, ρ, K and similarly for the 4th covariant derivative of the curvature. Clearly this $\{\rho, u, K\}$ dependence extends to all orders. Hence

Corollary 5.1   *Let (M, ω) be a symplectic manifold of dimension $2n$ ($n \geq 2$) and let ∇ be a Ricci type connection. Let $x_0 \in M$; then the curvature R_{x_0} and its covariant derivatives $(\nabla^k R)_{x_0}$ (for all k) are determined by the values at x_0 of (ρ_{x_0}, u_{x_0}, K).*

Corollary 5.2   *Let (M, ω, ∇) (resp. (M', ω', ∇')) be 2 symplectic manifolds of the same dimension $2n$ ($n \geq 2$) endowed with a symplectic connection of Ricci type., Assume they are both analytic and that there exists a linear map $f : T_{x_0}M \to T_{x_0'}M'$ such that (i) $f^*\omega'_{x_0'} = \omega_{x_0}$ (ii) $fu_{x_0} = u'_{x_0'}$ (iii) $f \circ \rho_{x_0} \circ f^{-1} = \rho'_{x_0'}$. Assume further that $K = K'$. Then there exists a normal neighborhood of x_0 (resp. x_0') U_{x_0} (resp. $U'_{x_0'}$) and a symplectic affine diffeomorphism $\varphi : (U_{x_0}, \omega, \nabla) \to (U'_{x_0'}, \omega', \nabla')$ such that $\varphi(x_0) = x_0'$ and $\varphi_{*x_0} = f$.*

Theorem 5.2   *Let (M, ω, ∇) (resp. (M', ω', ∇')) be a real analytic symplectic manifold endowed with a Ricci type symplectic connection. Assume M (resp. M') is simply connected and ∇ (resp. ∇') is geodesically complete. Assume there exists o (resp. o') in M (resp. M') and a linear symplectic isomorphism $f : T_oM \to T_{o'}M'$ such that $fu_o = u'_{o'}$ and $f\rho_o f^{-1} = \rho'_{o'}$. Assume finally that $K = K'$. Then there exists a global symplectic affine diffeomorphism $\varphi : M \to M'$ such that $\varphi(o) = o'$ and $\varphi_{*o} = f$.*

Proof. This is a direct consequence of corollary 2 and of the theorem of extensions of affine transformations [5].

The triple (ρ, u, K) which determines the geometry of (M, ω, ∇) can be used in a geometrical construction. Consider the symplectic vector space $(\mathbb{R}^{2n}, \dot{\Omega})$ and the symplectic vector space $(\mathbb{R}^{2n+2}, \Omega)$, having chosen basis where

$$\dot{\Omega} = \begin{pmatrix} 0 & I_n \\ -I_n & 0 \end{pmatrix} \qquad \Omega = \begin{pmatrix} \epsilon & 0 \\ 0 & \dot{\Omega} \end{pmatrix} \qquad \varepsilon = \begin{pmatrix} 0 & 1 \\ -1 & 0 \end{pmatrix},$$

and consider the natural inclusion of the symplectic group $Sp(n, \mathbb{R}) := Sp(\mathbb{R}^{2n}, \dot{\Omega})$ in the group $Sp(n+1, \mathbb{R}) := Sp(\mathbb{R}^{2n+2}, \Omega)$:

$$i(A) = \begin{pmatrix} I_2 & 0 \\ 0 & A \end{pmatrix} \qquad A \in Sp(n, \mathbb{R}).$$

The group $Sp(n, \mathbb{R})$ acts on the Lie algebra $sp(n+1, \mathbb{R})$ by adjoint action. Let (M, ω, ∇) be a symplectic manifold of dimension $2n$ with a Ricci type connection; let $B(M) \xrightarrow{\pi} M$ be the principal bundle of symplectic frames over M. Denote by $\tilde{u} : B(M) \to \mathbb{R}^{2n}$ the $Sp(n, \mathbb{R})$ equivariant function given by $\tilde{u}(\xi) = \xi^{-1}u(x)$ where $\pi(\xi) = x$. Similarly denote by $\tilde{\rho} : B(M) \to sp(n, \mathbb{R})$ the $Sp(n, \mathbb{R})$ equivariant function given by $\tilde{\rho}(\xi) = \xi^{-1}\rho(x)\xi$; we view $sp(n, \mathbb{R}) \subset \text{End } \mathbb{R}^{2n}$; the symmetry of the Ricci tensor implies that $\tilde{\rho}(\xi)$ belongs to the symplectic algebra.

Define the $Sp(n, \mathbb{R})$ equivariant map $\tilde{A} : B(M) \to sp(n+1, \mathbb{R})$

$$\tilde{A}(\xi) = \begin{pmatrix} 0 & \dfrac{(\pi^* f)(\xi)}{2(n+1)(2n+1)} & \dfrac{-\underline{\tilde{u}}(\xi)}{2(n+1)(2n+1)} \\ 1 & 0 & 0 \\ 0 & \dfrac{-\tilde{u}(\xi)}{2(n+1)(2n+1)} & \dfrac{-\tilde{\rho}(\xi)}{2(n+1)} \end{pmatrix}$$

where $\underline{\tilde{u}}(\xi) = i(u(\xi))\dot{\Omega}$. The function $\tilde{A}$ is associated to a section $\underline{A}$ of the Lie algebra bundle $E = B(M) \underset{Sp(n, \mathbb{R})}{\times} sp(n+1, \mathbb{R})$.

Lemma 5.3 *There exist a 1-form B on M, with values in E such that for any vector field X on M :*

$$\widetilde{\nabla_X \underline{A}}(\xi) = [\widetilde{B}(X)(\xi), \tilde{A}(\xi)].$$

This 1-form has for expression :

$$\widetilde{B(X)}(\xi) = \begin{pmatrix} 0 & \dfrac{\dot{\Omega}(\tilde{u}(\xi), \tilde{X}(\xi))}{2(n+1)(2n+1)} & \dfrac{\tilde{\rho}(\xi)\tilde{X}(\xi)}{2(n+1)} \\ 0 & 0 & \tilde{X}(\xi) \\ -\tilde{X}(\xi) & \dfrac{\tilde{\rho}(\xi)\tilde{X}(\xi)}{2(n+1)} & 0 \end{pmatrix}$$

where , as above, $\widetilde{B(X)}$ is the equivariant function associated to the section $B(X)$ of E and $\tilde{X}$ the equivariant function associated to X.

Remark. B is not uniquely defined; one may add to B any 1-form with values in the commutant of $\underline{A}$.

Proof. From the definition of $\underline{A}$ and lemma 2 one gets the above formula as one possible solution.

Define another covariant derivative, ∇', of sections of E by :

$$\nabla'_X = \nabla_X - [B(X), \cdot].$$

The section $\underline{A}$ is parallel relative to ∇'. Furthermore

Lemma 5.4   *The curvature of ∇' is the 2-form, with values in E*

$$R' = -2\omega\underline{A}.$$

Proof. If X, Y are vector fields on M and D is a smooth section of E :

$$(\nabla'_X\nabla'_Y - \nabla'_Y\nabla'_X - \nabla'_{[X,Y]})D = (\nabla_X\nabla_Y - \nabla_Y\nabla_X - \nabla_{[X,Y]})D$$

$$- [(\nabla_X B)(Y) - (\nabla_Y B)(X) - [B(X), B(Y)], D].$$

The first term gives :

$$(\nabla_X\nabla_Y - \widetilde{\nabla_Y\nabla_X} - \nabla_{[X,Y]})D|_\xi = [R_\xi(\bar{X}, \bar{Y}), \tilde{D}(\xi)]$$

where R is the curvature of ∇ and $\bar{X}$ (resp. $\bar{Y}$) is the horizontal lift of X (resp. Y) at ξ. Furthermore :

$$R_\xi(\bar{X}, \bar{Y}) = -\frac{1}{2(n+1)}[-2\omega_x(X,Y)\tilde{\rho} - \underline{\rho\tilde{Y}} \otimes \underline{\tilde{X}} + \underline{\rho\tilde{X}} \otimes \underline{Y} - \tilde{X} \otimes \underline{\rho\tilde{Y}} + \tilde{Y} \otimes \underline{\rho\tilde{X}}].$$

Using lemma 2 we get for the second term :

$$(\nabla_X B)(Y) - (\widetilde{\nabla_Y B)(X}) - [B(X), B(Y)] =$$

$$\begin{pmatrix} 0 & \dfrac{\pi^* f(\xi)\omega(X,Y)}{(n+1)(2n+1)} & \dfrac{-\tilde{\underline{u}}(\xi)\omega(X,Y)}{(n+1)(2n+1)} \\[2ex] 2\omega_x(X,Y) & 0 & 0 \\[2ex] 0 & \dfrac{-\tilde{u}(\xi)\omega(X,Y)}{(n+1)(2n+1)} & \dfrac{1}{2(n+1)}\left[\begin{array}{c} \tilde{X}\otimes\widetilde{\rho Y} - \tilde{Y}\otimes\widetilde{\rho X} \\ +\widetilde{\rho X}\otimes\underline{\tilde{Y}} - \widetilde{\rho Y}\otimes\underline{\tilde{X}} \end{array} \right] \end{pmatrix}$$

grouping the terms we get for the curvature of ∇':

$$R'(X,Y) = -2\omega(X,Y)\underline{A}.$$

5.3 Examples of Manifolds with Ricci Type Connections

We now describe examples of symplectic manifolds admitting a connection of Ricci type and we prove that these examples are sufficient to describe locally all possible situations.

Let $0 \neq A \in sp(n+1,\mathbb{R})$ and denote by Σ_A, the closed hypersurface $\Sigma_A \subset \mathbb{R}^{2n+2}$ with equation :

$$\Omega(x, Ax) = 1.$$

As before Ω is the standard symplectic form on $\mathbb{R}^{2n+2}$; in order for Σ_A to be non empty we replace, if necessary, A, by $-A$. Let $\dot{\nabla}$ be the standard, flat symplectic affine connection on $\mathbb{R}^{2n+2}$. Then if X, Y are vector fields tangent to Σ_A :

$$\nabla_X Y = \dot{\nabla}_X Y - \Omega(AX,Y)x$$

defines a torsion free linear connection along Σ_A. The vector field Ax is an affine vector field for this connection; it is clearly complete. Denote by H the 1-parametric group of diffeomorphisms of Σ_A generated by this vector field.

If $x \in \Sigma_A$, $T_x\Sigma_A = \rangle Ax \langle^\perp$ where $\rangle a_1, \cdots a_n \langle$ is the subspace generated by $a_1 \cdots a_n$; let $\mathcal{H}_x (\subset T_x\Sigma_A) = \rangle x, Ax \langle^\perp$; then

$$T_x\mathbb{R}^{2n+2} = (\mathcal{H}_x \oplus \mathbb{R}Ax) \oplus \mathbb{R}x.$$

A vector belonging to $\mathcal{H}_x$ will be called horizontal.

Since the vector field Ax is nowhere 0 on Σ_A, for any $x_0 \in \Sigma_A$, there exists a neighborhood $U_{x_0}(\subset \Sigma_A)$, a ball D of radius r_0, centered at the origin, an interval $I(\subset H)$ symmetric with respect to the neutral element of H and a diffeomorphism $\chi : D \times I \to U_{x_0}$ such that (i) $\chi(0,1) = x_0$ (ii)

$\chi(y,h) = h \cdot \chi(y,1)$ (where $\cdot$ denotes the action of H on Σ_A). We shall denote $\pi : U_{x_0} \to D$; $\pi = p_1 \circ \chi^{-1}$. If we view Σ_A as a constraint manifold in $\mathbb{R}^{2n+2}$, D is a local version of the

Marsden-Weinstein reduction of Σ_A. A symplectic form on D, $\omega^{(1)}$ is defined by

$$\omega_y^{(1)}(X,Y) = \Omega_x(\bar{X},\bar{Y}) \qquad y = \pi(x)$$

where $\bar{X}$ (resp. $\bar{Y}$) denotes the horizontal lift of X (resp. Y). A symplectic connection $\nabla^{(r)}$ on D is defined by

$$\overline{\nabla_X^{(r)} Y}(x) = \nabla_{\bar{X}} \bar{Y}(x) + \Omega(\bar{X},\bar{Y})Ax.$$

Proposition 5.1   *[5] The manifold $(D,\omega^{(1)})$ is a symplectic manifold and $\nabla^{(r)}$ is a symplectic connection of Ricci type. Furthermore*

$$\overline{\rho^{(r)} X}(x) = -2(n+1)\overline{A\bar{X}}$$

$$\bar{u}(x) = -2(n+1)(2n+1)\overline{A^2 x}$$

$$(\pi^* f)(x) = 2(n+1)(2n+1)\Omega(A^2 x, Ax)$$

$$K = 4(n+1)^2 \mathrm{tr} A^2.$$

Theorem 5.3   *Let (M,ω,∇) be a real analytic symplectic manifold endowed with a Ricci type connection. Let $x_0 \in M$ and let ξ_0 be a symplectic frame at x_0. Let*

$$A_0 \underset{\mathrm{def}}{=} \tilde{A}(\xi_0) = \begin{pmatrix} 0 & \dfrac{(p^* f)(\xi_0)}{2(n+1)(2n+1)} & -\dfrac{-\tilde{u}(\xi_0)}{2(n+1)(2n+1)} \\[2ex] 1 & 0 & 0 \\[2ex] 0 & -\dfrac{\tilde{u}(\xi_0)}{2(n+1)(2n+1)} & -\dfrac{\tilde{\rho}(\xi_0)}{2(n+1)} \end{pmatrix}$$

where $p : B(M) \to M$ is the canonical projection. Let $\Sigma_{A_0} \subset \mathbb{R}^{2n+2} = \{x | \Omega(x, A_0 x) = 1\}$; denote by (e_0, e_0', e_j) $(1 \le j \le 2n)$ the standard basis of $\mathbb{R}^{2n+2}$ (as used in §2). The point $e_0 \in \Sigma_{A_0}$, $T_{e_0}\Sigma_{A_0} =\rangle e_0', e_j \langle$; let U_{e_0} be the neighborhood of e_0 in Σ_{A_0} mentioned above and let $\pi : U_{e_0} \to D_0(\subset \mathbb{R}^{2n})$ be the projection of U_{e_0} onto the reduced manifold D_0. Then there exists a normal neighborhood $\tilde{U}_{x_0}$ of x_0 in M, a neighborhood V_0 of 0 in D_0 and a symplectic affine diffeomorphism $\varphi : \tilde{U}_{x_0} \to V_0$ such that $\varphi(x_0) = 0$.

Proof. This is a direct consequence of the fact that in the symplectic basis $\eta_0 = \{\pi_{*e_0} e_j; 1 \le j \le 2n\}$ of $T_0 D_0$ the matrix $\tilde{A}(\eta_0)$ associated to $(D_0, \omega^{(1)}, \nabla^r)$ is precisely A_0.

This says that locally any symplectic manifold (M, ω) admitting a Ricci type connection ∇, is the reduced space of a quadratic surface in flat space. Now, if $(A, x) \in sp(n+1, \mathbb{R}) \times \mathbb{R}^{2n}$ is such that $\Omega(x, Ax) \ne 0$, there exists a unique multiple of A such that $\Omega(x, kAx) = 1$. The pair (kA, x) gives a local model. Now if $S \in Sp(n+1, \mathbb{R})$, the pair $(kSAS^{-1}, Sx)$ gives an isomorphic local model. Conversely if (A_1, x_1) (resp. (A_2, x_2)) give rise to isomorphic local models it can be proven [6] that there exists an element $S \in Sp(n+1, \mathbb{R})$ and 2 real numbers k, l such that

$$x_2 = lSx_1, \quad A_2 = kSA_1S^{-1} \quad kl^2 = 1.$$

Hence :

Theorem 5.4 *Let $\mathcal{C}$ be the set of equivalence classes of germs of real analytic symplectic manifolds admitting a Ricci type analytic connection. Let $\mathcal{D}$ be the open set of $sp(n+1, \mathbb{R}) \times \mathbb{R}^{2n}$ composed of pairs (A, x) $(A \in sp(n+1, \mathbb{R}), x \in \mathbb{R}^{2n})$ such that $\Omega(x, Ax) \ne 0$. Let $\mathcal{D}_1$ be the set of orbits of $\mathbb{R} \times Sp(n+1, \mathbb{R})$ in $\mathcal{D}$; the action being given by $(l, S) \cdot (A, x) = (\frac{1}{l^2} SAS^{-1}, lx)$. Then there is a natural bijection between $\mathcal{C}$ and $\mathcal{D}_1$.*

5.4   Complex Projective Space

If $0 \ne A$ is an element of $sp(n+1, \mathbb{R})$ and $A^2 = \lambda I$ for a certain $\lambda \in \mathbb{R}$ then the natural map $\Sigma_A \to M^{(r)}$(=the reduced manifold) endows Σ_A with a structure of circle or line bundle over $M^{(r)}$. Furthermore $(M^{(r)}, \omega^{(r)}, \nabla^{(r)})$ is a symplectic symmetric space. In fact all symmetric spaces, whose canonical connection is of Ricci type are of this type. The only compact simply connected one is $\mathbb{P}_n(\mathbb{C})$ [7].

Let $(\mathbb{P}_n(\mathbb{C}), \dot{\omega})$ be the complex projective space with its standard symplectic structure; let $\dot{\nabla}$ be the Levi Civita connection associated to the Fubini-Study metric; $\dot{\nabla}$ is a symplectic connection of Ricci type. Assume ∇ is another symplectic connection which is close to $\dot{\nabla}$; this means that if $L = \nabla - \dot{\nabla}$ (L is a 1 form with values in the endomorphisms), then

$$\max_x ||L_x|| < \epsilon \qquad \max_x ||(\nabla L)_x|| < \epsilon \qquad \max_x ||(\nabla^2 L)_x|| < \epsilon$$

$$\max_x \|(\nabla^3 L)_x\| < \epsilon$$

for a certain positive ϵ to be made precise below; $\| \ \|$ is the norm induced on tensors at each point by the Fubini-Study metric.

For $\mathbb{P}_n(\mathbb{C})$ the constraint manifold Σ_A ($\subset \mathbb{C}^{n+1} = \mathbb{R}^{2n+2}$) is the standard sphere S^{2n+1} and the projection $\pi : \Sigma_A \to \mathbb{P}_n(\mathbb{C})$ is the Hopf fibration. We choose a basis of $\mathbb{C}^{n+1}$ in which the standard symplectic form has matrix

$$\Omega = \begin{pmatrix} 0 & 1 & 0 \\ -1 & 0 & 0 \\ 0 & 0 & \dot\Omega \end{pmatrix} \qquad \dot\Omega = \begin{pmatrix} 0 & I_n \\ -I_n & 0 \end{pmatrix}$$

$$A = \begin{pmatrix} 0 & -1 & 0 \\ 1 & 0 & 0 \\ 0 & 0 & j_0 \end{pmatrix} \qquad j_0 = \dot\Omega^{-1}.$$

Then

$$\rho^{(r)}(\pi(e_0)) = -2(n+1)j_0$$

$$\bar u(\pi(e_0)) = 0$$

$$f(\pi(e_0)) = -2(n+1)(2n+1)$$

$$K = -8n(n+1)^2.$$

The "close to $\dot\nabla$" assumption implies that the Ricci endomorphism associated to ∇, ρ, is invertible. Furthermore the symmetric bilinear form $\omega(\cdot, \rho\cdot)$ is positive definite. This implies that ρ has no real eigenvalues; furthermore if $(a+ib)$ is a complex eigenvalue and X_1+iX_2 a corresponding eigenvector :

$$0 = \omega(\rho X_1, X_2) + \omega(X_1, \rho X_2) = 2a\omega(X_1, X_2)$$

$$0 < \omega(X_1, \rho X_1) = -b\omega(X_1, X_2)$$

which implies $a = 0$ and $\omega(X_1, X_2) \neq 0$. Thus all eigenvalues of ρ are pure imaginary. Finally as the 2 plane spanned by the real and imaginary part of an eigenvector is symplectic, one sees , by induction, that ρ is semi simple. The function f is strictly negative, it can not be constant as this would imply (lemma 2) that $u = 0$; which implies ∇ is locally symmetric. But this implies $\nabla - \dot\nabla = 0$. It can be proven [7] that the reduction model, in the case of $\mathbb{P}_n(\mathbb{C})$ is a global one.

Consider then a point $x_0 \in \mathbb{P}_n(\mathbb{C})$ which is a critical point of f. At x_0,

$u(x_0) = 0$ (lemma 2). There exists a symplectic frame ξ_0 at x_0 such that $\tilde{A}(\xi_0)$ has the form

$$\tilde{A}(\xi_0) = \begin{pmatrix} 0 & -a_0 & 0 \\ 1 & 0 & 0 \\ 0 & 0 & D \end{pmatrix} \qquad D = \begin{pmatrix} 0 & -\delta_n \\ +\delta_n & 0 \end{pmatrix}$$

$$\delta_n = \operatorname{diag}(a_1, \ldots, a_n)$$

and $a_0, a_1, \ldots, a_n$ are positive real numbers. The vector field generating the 1 parametric group $\exp t\tilde{A}(\xi_0)$ is

$$\tilde{A}(\xi_0)x = -a_0 x^{0'} e_0 + x^0 e_0' - a_j x_{n+j} e_j + a_j x_j e_{n+j}.$$

The corresponding flow is :

$$(x^0 + ia_0^{1/2} x^{0'})(t) = (x^0 + ia_0^{1/2} x^{0'})(0)e^{ia_0^{1/2}t}$$

$$(x^j + ix^{n+j})(t) = (x^j + ix^{n+j})(0)e^{ia_j^{1/2}t}.$$

If there exists $\alpha \neq \beta \in \{0, \ldots, n\}$ such that $a_\alpha^{1/2}/a_\beta^{1/2}$ is not rational the quotient manifold would not be Hausdorff; hence is certainly not $\mathbb{P}_n(\mathbb{C})$. To see this it is sufficient to consider the points of $\Sigma_{\tilde{A}(\xi_0)}$ such that $z^\lambda = 0$ $\forall \lambda \neq \alpha, \beta$; thus

$$|z^\alpha|^2 + |z^\beta|^2 = 1$$

$$z^0 = x^0 + ia^{1/2} x^{0'} \qquad z^v = x^v + ix^{n+v}.$$

The sphere S^3 is partitioned in 2 tori and 2 circles. On each of the tori the flow is a "irrational" flow and thus the quotient is not Hausdorff.
If all ratio $a_\alpha^{1/2}/a_\beta^{1/2}$ are rational but there is a pair for which this ratio is $\neq 1$, the quotient is an orbifold, hence not $\mathbb{P}_n(\mathbb{C})$. Consider again the sphere S^3

$$|z^\alpha|^2 + |z^\beta|^2 = 1.$$

On this sphere the flow reads

$$z^\alpha \to z^\alpha e^{ipt} \qquad z^\beta \to z^\beta e^{iqt}$$

with p, q relatively prime integers. The action is not a free one. Let $\nu : S^3 \to \mathbb{P}_1(\mathbb{C}) : (z^\alpha, z^\beta) \to [(z^\alpha)^q, (z^\beta)^p]$ where $[(a, b)]$ is the equivalence class $[a, b] = [\lambda a, \lambda b]$, $\lambda \in \mathbb{C}^*$. This map clearly factorizes through the orbit space. Conversely if $\nu(z^\alpha, z^\beta) = \nu(z'^\alpha, z'^\beta)$, there exists $\lambda \in \mathbb{C}^*$ such that

$$(z'^\alpha)^q = \lambda(z^\alpha)^q \qquad (z'^\beta)^p = \lambda(z^\beta)^p.$$

Clearly $|\lambda| = 1$ and one then checks that the points (z^α, z^β) and (z'^α, z'^β) belong to the same orbit. Thus the map $\tilde{\nu}$: orbit space $\to \mathbb{P}_1(\mathbb{C})$ is bijective.

Hence at the topological level the orbit space is homeomorphic to $\mathbb{P}_1(\mathbb{C})$. But ν_* is not a submersion everywhere. Hence the orbit space is an orbifold and hence not $\mathbb{P}_1(\mathbb{C})$. On $\mathbb{P}_n(\mathbb{C})$ we must have all eigenvalues equal

$$a_0 = a_1 = \ldots = a_n.$$

In this situation $A^2 = -kId$ $(k > 0)$ and the connection is the standard symmetric one. Thus we have

Theorem 5.5 *Let $(\mathbb{P}_n(\mathbb{C}), \omega_0)$ be complex projective space with its standard symplectic form. Let ∇ be a symplectic connection which is close to the Fubini study connection $\dot{\nabla}$ in a C^3 sense. Assume ∇ is Ricci type. Then $\nabla = \dot{\nabla}$, i.e. in the moduli space of Ricci type connections on $\mathbb{P}_n(\mathbb{C})$, the connection $\dot{\nabla}$ corresponds to an isolated point.*

Bibliography

[1] F. Bourgeois and M. Cahen, A variational principle for symplectic connections. *J. Geom. Phys.* **30** (1999) 233–265.

[2] M. De Visher, Mémoire de licence, Bruxelles, 1999. See also I. Vaisman, Symplectic Curvature Tensors *Monats. Math.* **100** (1985) 299–327.

[3] S. Kobayashi and K. Nomizu, *Foundations of differential geometry. Vol II.* John Wiley & Sons, New York–London, 1963.

[4] M. Cahen, S. Gutt, J. Horowitz and J. Rawnsley, Homogeneous symplectic manifolds with Ricci-type curvature, *J. Geom. Phys.* **38** (2001) 140–151.

[5] P. Baguis, M. Cahen, A construction of symplectic connections through reduction. L.M.P. 57 (2001), pp. 149-160.

[6] M. Cahen, L. Schwachhöfer : in preparation

[7] M. Cahen, S. Gutt and J. Rawnsley, Symmetric symplectic spaces with Ricci-type curvature, in Conférence Moshe Flato 1999, vol 2, G. Dito et D. Sternheimer (eds), Math. Phys. Studies 22 (2000) 81–91.

Chapter 6

Banach Lie-Poisson Spaces

Anatol Odzijewicz[1] and Tudor S. Ratiu[2]

Abstract: This paper gives a brief review of some of the results of the authors regarding Banach Lie-Poisson spaces. The category of these spaces is described and the properties of the morphisms is discussed. Its relationship to W^*-algebras and Banach Lie algebras is also presented. Several examples are given.

6.1 Introduction

The present paper is a review of the theory of Banach Lie-Poisson spaces developed by the authors over the past several years. We felt that a short introduction to the subject would be beneficial to those interested in entering this interesting area of research. Some parts of this theory are quite technical and these are deliberately ignored in the present review, precisely to improve the readability. The proofs as well as the relationship to quantum mechanics can be found in the original articles.

The paper is organized as follows. Section 6.2 presents the category $\mathfrak{B}$ of Banach Lie-Poisson spaces. The material in this section comes exclusively from [6] where all the proofs, as well as additional information, can be found. Section 6.3 is devoted to infinite dimensional examples of Banach Lie-Poisson spaces. The first class of examples are the preduals

[1]Institute of Physics, University of Bialystok, Lipowa 41, PL-15424 Bialystok, Poland, `aodzijew@labfiz.uwb.edu.pl`

[2]Centre Bernoulli and Section de Mathématiques, École Polytechnique Fédérale de Lausanne, CH–1015 Lausanne, Switzerland, `tudor.ratiu@epfl.ch`

of W^*-algebras motivated by a result of [1] on trace class operators. The proofs for this class of examples are to be found in [6]. The second class of examples have to do with extensions of Banach Lie-Poisson spaces. The method of extensions constructs new Banach Lie-Poisson spaces out of old. The proofs for the sketch of this theory presented here, as well as several other examples, can be found in [8]. The third class of examples is devoted to the Banach Lie-Poisson spaces of lower triangular p-diagonal trace class operators, generalizing that of bidiagonal operators in [7]. Section 6.4 reviews the theory of coadjoint orbits for Banach Lie groups whose Banach Lie algebra admits a predual invariant under the coadjoint action. The proofs of the statements in this section can be found in [6].

6.2   Banach Lie-Poisson spaces

The goal of this section is to describe the category of Banach Lie-Poisson spaces. For this, it is necessary to introduce the notion of a **Banach Poisson manifold** which is a pair $(P, \{\cdot, \cdot\})$ consisting of a smooth Banach manifold and a bilinear operation $\{\cdot, \cdot\}$ on the space of smooth functions on any open subset U of P satisfying the following conditions:

(i) $(C^\infty(P), \{\cdot, \cdot\})$ is a Lie algebra;

(ii) $\{\cdot, \cdot\}$ satisfies the Leibniz identity on each factor;

(iii) the vector bundle map $\sharp : T^*P \to T^{**}P$ covering the identity satisfies
$\sharp(T^*P) \subset TP.$

The vector bundle map in (iii) is given in the following way: for any real valued smooth function $f : U \to \mathbb{R}$, where U is an open subset of P, define $\sharp(df)$ by $\langle \sharp(df), dg \rangle = \{g, f\}$ for any smooth function $g : U \to \mathbb{R}$, where $\langle \cdot, \cdot \rangle : T^{**}P \times T^*P \to \mathbb{R}$ is the natural duality pairing. The symbol d denotes the exterior differential on P and the definition of $\sharp$ is correct since, by property (ii), the Poisson bracket $\{g, f\}$ depends only on the point values of the derivatives of the functions f and g.

Condition $\sharp(T^*P) \subset TP$ is automatically satisfied in certain cases: if P is a smooth manifold modeled on a reflexive Banach space, that is, $\flat^{**} = \flat$, or if P is a strong symplectic manifold with symplectic form ω. Thus, for finite dimensional manifolds, the definition given above coincides with the usual one. If (P, ω) is a strong symplectic manifold, one verifies condition (iii) in the following manner. First, recall that **strong** means that for each $p \in P$ the map $v_p \in T_pP \mapsto \omega(p)(v_p, \cdot) \in T_p^*P$ is a bijective continuous linear map. Thus, for any smooth function $f : P \to \mathbb{R}$ there exists a vector field X_f, called the **Hamiltonian vector field**, such that $df = \omega(X_f, \cdot)$. Second, define the Poisson bracket by $\{f, g\} = \omega(X_f, X_g) = \langle df, X_g \rangle$, which shows that $\sharp df = X_f$ and hence $\sharp(T^*P) \subset TP$.

On the other hand, a weak symplectic manifold is not a Banach Poisson manifold in the sense defined above. Recall that **weak** means that the map $v_p \in T_pP \mapsto \omega(p)(v_p, \cdot) \in T_p^*P$ is an injective continuous linear map that is, in general, not surjective. Therefore, one cannot construct the map that associates to every differential df of a smooth function $f : P \to \mathbb{R}$ the Hamiltonian vector field X_f. Since the definition of the Poisson bracket should be $\{f, g\} = \omega(X_f, X_g)$, one cannot define this operation on functions and hence weak symplectic manifold structures do not define, in general, Banach Poisson manifolds. For example, if $\mathfrak{b}$ is a Banach space, the canonical symplectic structure on $\mathfrak{b} \times \mathfrak{b}^*$, given by

$$\omega((b_1, \beta_1), (b_2, \beta_2)) := \langle \beta_1, b_1 \rangle - \langle \beta_1, b_2 \rangle, \tag{6.1}$$

for $b_1, b_2 \in \mathfrak{b}$, $\beta_1, \beta_2 \in \mathfrak{b}^*$ and where $\langle \cdot, \cdot \rangle : \mathfrak{b}^* \times \mathfrak{b} \to \mathbb{R}$ (or $\mathbb{C}$) is the duality pairing, is only weak, in general. There are various ways to deal with the non existence of globally defined Hamiltonian vector fields on weak symplectic manifolds. One of them is to restrict the space of functions on which one is working, as is often done in field theory. Another is to deal with densely defined vector fields and invoke the theory of (nonlinear) semigroups; see [3] for this approach.

Returning to condition (iii) in the definition of a Banach Poisson manifold, note that its purpose is to be able to define for any locally defined function $h \in C^\infty(U)$, U open in P, the **Hamiltonian vector field** by $X_h := \sharp(dh)$, or equivalently,

$$X_h[f] := \langle X_h, df \rangle = \{f, h\}$$

where f is an arbitrary smooth locally defined function on P. **Hamilton's equations** $\dot{p}(t) = X_h(p(t))$ are then equivalent to the Poisson bracket formulation $\dot{f} = \{f, h\}$ for any smooth locally defined function f.

In this paper we shall study exclusively linear Poisson brackets. For this, we need to recall that a **Banach Lie algebra** $(\mathfrak{g}, [\cdot, \cdot])$ is a Banach space that is also a Lie algebra such that the Lie bracket is a bilinear continuous map $\mathfrak{g} \times \mathfrak{g} \to \mathfrak{g}$. Thus the adjoint and coadjoint maps $\mathrm{ad}_x : \mathfrak{g} \to \mathfrak{g}$, $\mathrm{ad}_x y := [x, y]$, and $\mathrm{ad}_x^* : \mathfrak{g}^* \to \mathfrak{g}^*$ are also continuous for each $x \in \mathfrak{g}$. A **Banach Lie-Poisson space** $(\mathfrak{b}, \{\cdot, \cdot\})$ is defined to be a Poisson manifold such that $\mathfrak{b}$ is a Banach space and the dual $\mathfrak{b}^* \subset C^\infty(\mathfrak{b})$ is a Banach Lie algebra under the Poisson bracket operation. Denote by $[\cdot,\cdot]$ the restriction of the Poisson bracket $\{\cdot,\cdot\}$ from $C^\infty(\mathfrak{b})$ to the Lie subalgebra $\mathfrak{b}^*$. The following theorem is fundamental.

Theorem 6.1   ([6]) *The Banach space $\mathfrak{b}$ is a Banach Lie-Poisson space $(\mathfrak{b}, \{\cdot,\cdot\})$ if and only if its dual $\mathfrak{b}^*$ is a Banach Lie algebra $(\mathfrak{b}^*, [\cdot,\cdot])$ satisfying $\mathrm{ad}_x^* \, \mathfrak{b} \subset \mathfrak{b} \subset \mathfrak{b}^{**}$ for all $x \in \mathfrak{b}^*$. Moreover, the Poisson bracket of $f, g \in C^\infty(\mathfrak{b})$ is given by*

$$\{f, g\}(b) = \langle [Df(b), Dg(b)], b \rangle, \tag{6.2}$$

where $b \in \mathfrak{b}$ and D denotes the Fréchet derivative. If h is a smooth function on $\mathfrak{b}$, the associated Hamiltonian vector field is given by

$$X_h(b) = -\,\mathrm{ad}^*_{Dh(b)} \, b. \tag{6.3}$$

For example, if $\mathfrak{b}$ is a reflexive Banach Lie algebra, then its dual $\mathfrak{b}^*$ is a Banach Lie-Poisson space. In particular, the dual of any finite dimensional Lie algebra is a Lie-Poisson space.

The Banach Lie Poisson spaces form the objects of a category $\mathfrak{B}$ whose **morphisms** are defined to be linear continuous maps $\phi : \mathfrak{b}_1 \to \mathfrak{b}_2$ between Banach Lie-Poisson spaces $\mathfrak{b}_1$ and $\mathfrak{b}_2$ that preserve the Poisson brackets, that is,

$$\{f \circ \phi, g \circ \phi\}_1 = \{f, g\}_2 \circ \phi$$

for any $f, g \in C^\infty(\mathfrak{b}_2)$. Such a morphism ϕ is also called a **linear Poisson map**.

Let $\mathfrak{L}$ denote the category of Banach Lie algebras and continuous Lie algebra homomorphisms. Denote by $\mathfrak{L}_0$ the following subcategory of $\mathfrak{L}$. An object of $\mathfrak{L}_0$ is a Banach Lie algebra $\mathfrak{g}$ admitting a predual $\mathfrak{g}_*$, that is, $(\mathfrak{g}_*)^* = \mathfrak{g}$, and satisfying $\mathrm{ad}^*_\mathfrak{g} \, \mathfrak{g}_* \subset \mathfrak{g}_*$ where ad^* is the coadjoint representation of $\mathfrak{g}$ on $\mathfrak{g}^*$. A morphism in the category $\mathfrak{L}_0$ is a Banach Lie algebra homomorphism $\psi : \mathfrak{g}_1 \to \mathfrak{g}_2$ such that the dual map $\psi^* : \mathfrak{g}_2^* \to \mathfrak{g}_1^*$ preserves at least one choice of the corresponding preduals, that is, $\psi^* : (\mathfrak{g}_2)_* \to (\mathfrak{g}_1)_*$, where $(\mathfrak{g}_i)_*$ is one possible predual of $\mathfrak{g}_i$ for $i = 1, 2$. Let $\mathfrak{L}_{0u}$ be the subcategory of $\mathfrak{L}_0$ whose objects have a *unique* predual. The prescriptions $\mathfrak{F}(\mathfrak{b}) = \mathfrak{b}^*$ and $\mathfrak{F}(\phi) = \phi^*$ define a contravariant functor $\mathfrak{F} : \mathfrak{B} \to \mathfrak{L}_0$. On the subcategory $\mathfrak{F}^{-1}(\mathfrak{L}_{0u}) \subset \mathfrak{B}$ this functor is invertible. The inverse of $\mathfrak{F}$ is given by $\mathfrak{F}^{-1}(\mathfrak{g}) = \mathfrak{g}_*$ and $\mathfrak{F}^{-1}(\psi) = \psi^*|_{(\mathfrak{g}_2)_*}$, where $\psi : \mathfrak{g}_1 \to \mathfrak{g}_2$.

The internal structure of morphisms in the category $\mathfrak{B}$ of Banach Lie-Poisson spaces can be described explicitly. Let $\phi : \mathfrak{b}_1 \to \mathfrak{b}_2$ be a linear Poisson map between Banach Lie-Poisson spaces and assume that $\operatorname{im}\phi$ is closed in $\mathfrak{b}_2$. Then the Banach space $\mathfrak{b}_1/\ker\phi$ is predual to $\mathfrak{b}_2^*/\ker\phi^*$, that is, $(\mathfrak{b}_1/\ker\phi)^* \cong \mathfrak{b}_2^*/\ker\phi^*$. In addition, it can be shown that $\mathfrak{b}_2^*/\ker\phi^*$ is a Banach Lie algebra satisfying the condition $\operatorname{ad}_{[x]}^*(\mathfrak{b}_1/\ker\phi) \subset \mathfrak{b}_1/\ker\phi$ for all $[x] \in \mathfrak{b}_2^*/\ker\phi^*$ and that $\mathfrak{b}_1/\ker\phi$ is a Banach Lie-Poisson space. The quotient map $\pi : \mathfrak{b}_1 \to \mathfrak{b}_1/\ker\phi$ is a surjective linear Poisson map. The map $\iota : \mathfrak{b}_1/\ker\phi \to \mathfrak{b}_2$ defined by $\iota([b]) := \phi(b)$, where $b \in \mathfrak{b}_1$ and $[b] \in \mathfrak{b}_1/\ker\phi$ is an injective linear Poisson map. This shows that the map ϕ can be decomposed as a composition $\phi = \iota \circ \pi$ of a surjective and an injective linear Poisson map. The study of linear Poisson maps is thus reduced to that of the surjective and injective ones.

It turns out that the surjective linear Poison maps define a Poisson bracket on the target. More precisely, assume that $(\mathfrak{b}_1, \{\cdot,\cdot\})$ is a Banach Lie-Poisson space and let $\pi : \mathfrak{b}_1 \to \mathfrak{b}_2$ be a continuous linear surjective map onto the Banach space $\mathfrak{b}_2$. Then $\mathfrak{b}_2$ carries a unique Banach Lie-Poisson structure such that π is a linear Poisson map if and only if $\operatorname{im}\pi^* \subset \mathfrak{b}_1^*$ is closed under the Lie bracket $[\cdot,\cdot]_1$ of $\mathfrak{b}_1^*$. This Poisson structure on $\mathfrak{b}_2$ is said to be **coinduced** by π. The map $\pi^* : \mathfrak{b}_2^* \to \mathfrak{b}_1^*$ is a Banach Lie algebra morphism whose dual $\pi^{**} : \mathfrak{b}_1^{**} \to \mathfrak{b}_2^{**}$ maps $\mathfrak{b}_1$ into $\mathfrak{b}_2$.

Injective maps also define Poisson structures, but this time on the source space. Namely, if $\mathfrak{b}_1$ is a Banach space, $(\mathfrak{b}_2, \{\cdot,\cdot\}_2)$ is a Banach Lie-Poisson space, and $\iota : \mathfrak{b}_1 \to \mathfrak{b}_2$ is an injective continuous linear map with closed range, then $\mathfrak{b}_1$ carries a unique Banach Lie-Poisson structure such that ι is a linear Poisson map if and only if $\ker\iota^*$ is an ideal in the Banach Lie algebra $\mathfrak{b}_2^*$. This Poisson structure on $\mathfrak{b}_1$ is said to be **induced** by ι. The map $\iota^* : \mathfrak{b}_2^* \to \mathfrak{b}_1^*$ is a Banach Lie algebra morphism whose dual $\iota^{**} : \mathfrak{b}_1^{**} \to \mathfrak{b}_2^{**}$ maps $\mathfrak{b}_1$ into $\mathfrak{b}_2$. A convenient sufficient condition to verify the criterion just described is the following: if $\operatorname{ad}^{2}{}_{\mathfrak{b}_2^*}^{*}\iota(\mathfrak{b}_1) = \iota(\mathfrak{b}_1)$, then $\ker\iota^*$ is an ideal in the Banach Lie algebra $\mathfrak{b}_2^*$ and thus the map $\iota : \mathfrak{b}_1 \to \mathfrak{b}_2$ induces a Banach Lie Poisson structure on $\mathfrak{b}_1$.

Due to the isomorphism between the subcategory $\mathfrak{F}^{-1}(\mathfrak{L}_{0u}) \subset \mathfrak{B}$ and the category $\mathfrak{L}_{0u}$ of Banach Lie algebras admitting a unique predual, properties of objects in one category can be characterized by properties in the other. This is made more precise in the following manner. Let $\mathfrak{b} \in \operatorname{Ob}(\mathfrak{B})$ be a Banach Lie-Poisson space and let $\mathfrak{g} \in \operatorname{Ob}(\mathfrak{L}_0)$ be a Banach Lie algebra such that $\mathfrak{b}^* = \mathfrak{g}$. Then:

(i) There exists a bijective correspondence between the coinduced Banach Lie-Poisson structures from $\mathfrak{b}$ and the Banach Lie subalgebras of $\mathfrak{g}$. If the surjective continuous linear map $\pi : \mathfrak{b} \to \mathfrak{c}$ coinduces a Banach Lie-Poisson structure on $\mathfrak{c}$, the Banach Lie subalgebra of $\mathfrak{g}$ given by this correspondence is $\pi^*(\mathfrak{c}^*)$.

Conversely, if $\mathfrak{k} \subset \mathfrak{g}$ is a Banach Lie subalgebra then the Banach Lie-Poisson space given by this correspondence is $\mathfrak{b}/\mathfrak{k}^\circ$, where $\mathfrak{k}^\circ := \{b \in \mathfrak{b} \mid \langle b, k \rangle = 0 \text{ for all } k \in \mathfrak{k}\}$ is the annihilator of $\mathfrak{k}$ in $\mathfrak{b}$, and $\pi : \mathfrak{b} \to \mathfrak{b}/\mathfrak{k}^\circ$ is the quotient projection.

(ii) There exists a bijective correspondence between the induced Banach Lie-Poisson structures in $\mathfrak{b}$ (i.e., the Banach Lie-Poisson subspaces of $\mathfrak{b}$) and the Banach Lie ideals of $\mathfrak{g}$. If the injection $\iota : \mathfrak{c} \to \mathfrak{b}$ with closed range induces a Banach Lie-Poisson structure on $\mathfrak{c}$, then the ideal in $\mathfrak{g}$ given by this correspondence is $\ker \iota^*$.

Conversely, if $\mathfrak{i} \subset \mathfrak{g}$ is a Banach Lie ideal, then the Banach Lie-Poisson subspace of $\mathfrak{b}$ given by this correspondence is $\mathfrak{i}^\circ$, where $\mathfrak{i}^\circ$ is the annihilator of $\mathfrak{i}$ in $\mathfrak{b}$ and $\iota : \mathfrak{i}^\circ \to \mathfrak{b}$ is the inclusion.

6.3   Examples of Banach Lie-Poisson Spaces

This entire section is dedicated to the description of some examples of infinite dimensional Banach Lie-Poisson spaces.

Example 1: Preduals of W^*-algebras. A W^*-algebra is a C^*-algebra $\mathfrak{m}$ which possesses a predual Banach space $\mathfrak{m}_*$, i.e. $\mathfrak{m} = (\mathfrak{m}_*)^*$; this predual is unique ([9]). Since $\mathfrak{m}^* = (\mathfrak{m}_*)^{**}$, the predual Banach space $\mathfrak{m}_*$ canonically embeds into the Banach space $\mathfrak{m}^*$ dual to $\mathfrak{m}$. Thus we shall always think of $\mathfrak{m}_*$ as a Banach subspace of $\mathfrak{m}^*$. The existence of $\mathfrak{m}_*$ allows the introduction of the $\sigma(\mathfrak{m}, \mathfrak{m}_*)$-topology on the W^*-algebra $\mathfrak{m}$; for simplicity we shall call it the σ-topology in the sequel. Recall that a net $\{x_\alpha\}_{\alpha \in A} \subset \mathfrak{m}$ converges to $x \in \mathfrak{m}$ in the σ-topology if $\lim_{\alpha \in A} \langle x_\alpha, b \rangle = \langle x, b \rangle$ for all $b \in \mathfrak{m}_*$. The predual space $\mathfrak{m}_*$ is known to equal the subspace of $\mathfrak{m}^*$ of all σ-continuous linear functionals on $\mathfrak{m}$ ([9]).

Theorem 6.2   ([6]) *Let $\mathfrak{m}$ be a W^*-algebra and $\mathfrak{m}_*$ be the predual of $\mathfrak{m}$. Then $\mathfrak{m}_*$ is a Banach Lie-Poisson space with the Poisson bracket of $f, g \in C^\infty(\mathfrak{m}_*)$ given by (6.2). The Hamiltonian vector field X_f defined by the smooth function $f \in C^\infty(\mathfrak{m}_*)$ is given by (6.3).*

Thus, if $\mathfrak{a}$ is a C^*-algebra, then its dual $\mathfrak{a}^*$ is a Banach Lie-Poisson space because $\mathfrak{a}^{**}$ is isomorphic to the universal enveloping von Neumann algebra of the C^*-algebra $\mathfrak{a}$ ([9; 10]). Thus, if $\mathfrak{m}$ is a W^*-algebra then both its predual $\mathfrak{m}_*$ and its dual $\mathfrak{m}^*$ are Banach Lie-Poisson spaces. Since any W^*-algebra is a Banach Lie algebra relative to the commutator bracket and possesses a unique predual, Theorem 6.2 shows that the category $\mathfrak{W}$ of W^*-algebras is a subcategory of $\mathfrak{L}_{0u}$.

In order to illustrate Theorem 6.2 let us analyze some concrete examples. Let $\mathcal{H}$ be a complex Hilbert space. By $L^1(\mathcal{H})$, $L^2(\mathcal{H})$, and $L^\infty(\mathcal{H})$ we shall denote the involutive Banach algebras of the trace class operators, the Hilbert-Schmidt operators, and the bounded operators on $\mathcal{H}$ respectively. Recall that $L^1(\mathcal{H})$ and $L^2(\mathcal{H})$ are self adjoint algebraic ideals in $L^\infty(\mathcal{H})$. Let $\mathcal{K}(\mathcal{H}) \subset L^\infty(\mathcal{H})$ denote the ideal of all compact operators on $\mathcal{H}$. Then

$$L^1(\mathcal{H}) \subset L^2(\mathcal{H}) \subset \mathcal{K}(\mathcal{H}) \subset L^\infty(\mathcal{H}) \tag{6.4}$$

and the following remarkable dualities hold (see e.g. [5]):

$$\mathcal{K}(\mathcal{H})^* \cong L^1(\mathcal{H}), \quad L^2(\mathcal{H})^* \cong L^2(\mathcal{H}), \quad \text{and} \quad L^1(\mathcal{H})^* \cong L^\infty(\mathcal{H}). \tag{6.5}$$

These are implemented by the strongly non-degenerate pairing

$$\langle x, \rho \rangle = \mathrm{tr}(x\rho) \tag{6.6}$$

where $x \in L^1(\mathcal{H})$, $\rho \in \mathcal{K}(\mathcal{H})$ for the first isomorphism, $\rho, x \in L^2(\mathcal{H})$ for the second isomorphism and $x \in L^\infty(\mathcal{H})$, $\rho \in L^1(\mathcal{H})$ for the third isomorphism. Therefore the spaces $\mathcal{K}(\mathcal{H})$, $L^2(\mathcal{H})$, and $L^1(\mathcal{H})$ are all Banach Lie-Poisson spaces relative to the bracket

$$\{f, g\}(\rho) = \mathrm{tr}([Df(\rho), Dg(\rho)]\rho),$$

where $\rho \in \mathcal{K}(\mathcal{H}), L^2(\mathcal{H})$, and $L^1(\mathcal{H})$ respectively. The Hamiltonian vector field is given by $X_f(\rho) = [Df(\rho), \rho]$. The original observation that $L^1(\mathcal{H})$ is a Banach Lie-Poisson space is due to Bona [1].

Example 2: Extensions of Banach Lie-Poisson spaces. The method of extensions described below enables one to construct new Banach Lie-Poisson spaces out of known ones. We begin by briefly recalling the notion of extension of Lie algebras and its properties.

A sequence of Banach Lie algebras

$$0 \longrightarrow \mathfrak{n} \xrightarrow{\iota} \mathfrak{g} \xrightarrow{\pi} \mathfrak{h} \longrightarrow 0 \tag{6.7}$$

is *exact* if it is exact as a sequence in the category of Banach spaces, that is if ι has closed range and $\ker \pi = \mathrm{im}\, \iota$, and all maps are Banach

Lie algebra homomorphisms. In the categories $\mathfrak{L}_0$ (respectively $\mathfrak{L}_{0u}$) the sequence (6.7) is **exact** if it is exact in the category $\mathfrak{L}$ and the duals of the maps in the sequence preserve at least one choice of (respectively the uniquely associated) predual spaces, that is $\iota(\mathfrak{g}_*) \subset \mathfrak{n}_*$ and $\pi_*(\mathfrak{h}_*) \subset \mathfrak{g}_*$, where $\mathfrak{n}_*, \mathfrak{g}_*, \mathfrak{h}_*$ are preduals of $\mathfrak{n}, \mathfrak{g}, \mathfrak{h}$ respectively. The Lie algebra $\mathfrak{g}$ is called an **extension of $\mathfrak{h}$ by** $\mathfrak{n}$.

For Banach Lie-Poisson spaces the definition is similar. A sequence of Banach Lie-Poisson spaces

$$0 \longrightarrow \mathfrak{a} \overset{j}{\longrightarrow} \mathfrak{b} \overset{p}{\longrightarrow} \mathfrak{c} \longrightarrow 0 \tag{6.8}$$

is **exact** if it is exact as a sequence in the category of Banach spaces and all maps are linear Poisson maps. The Banach Lie-Poisson space $\mathfrak{b}$ is said to be an **extension of $\mathfrak{c}$ by** $\mathfrak{a}$. Extensions of Banach Lie algebras are intimately connected with extensions of Banach Lie-Poisson spaces as the following theorem shows.

Theorem 6.3   ([8]) *The Banach spaces $\mathfrak{a}$, $\mathfrak{b}$, $\mathfrak{c}$ form an exact sequence (6.8) of Banach Lie-Poisson spaces if and only if their duals $\mathfrak{n} := \mathfrak{c}^*$, $\mathfrak{g} := \mathfrak{b}^*$, $\mathfrak{h} := \mathfrak{a}^*$ form an exact sequence of Banach Lie algebras (6.7) in the category $\mathfrak{L}_0$, where $\iota := p^*$ and $\pi := j^*$. In particular, if $\mathfrak{g}$ is the direct sum $\mathfrak{g} \cong \mathfrak{n} \oplus \mathfrak{h}$ of Banach Lie algebras with ι and π the inclusion of the first component and π the projection on the second component, then $\mathfrak{b}$ can be chosen as the direct sum $\mathfrak{a} \oplus \mathfrak{c}$ of the Banach Lie-Poisson spaces $\mathfrak{a}$ and $\mathfrak{c}$ with j the inclusion on the first component and p the projection on the second component.*

Assume now that we are given a continuous bilinear skew symmetric map $\omega : \mathfrak{h} \times \mathfrak{h} \to \mathfrak{n}$ and a continuous linear map $\varphi : \mathfrak{h} \to \operatorname{aut}(\mathfrak{n})$. Then

$$[(\zeta, \eta), (\zeta', \eta')] = ([\zeta, \zeta'] + \varphi(\eta)(\zeta') - \varphi(\eta')(\zeta) + \omega(\eta, \eta'), [\eta, \eta']) \tag{6.9}$$

for $\zeta \in \mathfrak{n}$ and $\eta \in \mathfrak{h}$ endows the Banach space direct sum $\mathfrak{g} := \mathfrak{n} \oplus \mathfrak{h}$ with a Banach Lie algebra structure if and only if

$$\omega([\eta, \eta'], \eta'') + \omega([\eta', \eta''], \eta) + \omega([\eta'', \eta'], \eta')$$
$$- \varphi(\eta)(\omega(\eta', \eta'')) - \varphi(\eta')(\omega(\eta'', \eta)) - \varphi(\eta'')(\omega(\eta, \eta')) = 0 \tag{6.10}$$

and

$$\operatorname{ad}_{\omega(\eta, \eta')} + \varphi([\eta, \eta']) - [\varphi(\eta), \varphi(\eta')] = 0 \tag{6.11}$$

for any $\eta, \eta', \eta'' \in \mathfrak{h}$. Consequently, the Banach Lie algebra $\mathfrak{n} \oplus \mathfrak{h}$ is an extension of the Banach Lie algebra $\mathfrak{n}$ by the Banach Lie algebra $\mathfrak{h}$. So one gets the following result.

Theorem 6.4 ([8]) *Given are two Lie-Poisson Banach spaces $\mathfrak{a}$ and $\mathfrak{c}$ whose duals are the Banach Lie algebras $\mathfrak{h} := \mathfrak{a}^*$ and $\mathfrak{n} := \mathfrak{c}^*$ respectively, a continuous bilinear skew symmetric map $\omega : \mathfrak{h} \times \mathfrak{h} \to \mathfrak{n}$, and a continuous linear map $\varphi : \mathfrak{h} \to \mathrm{aut}(\mathfrak{n})$ satisfying (6.10) and (6.11). The Banach space $\mathfrak{c} \oplus \mathfrak{a}$ is an extension of $\mathfrak{c}$ by $\mathfrak{a}$ if and only if*

$$\varphi(\eta)^*(\mathfrak{c}) \subset \mathfrak{c}, \qquad (\varphi(\cdot)\zeta)^*(\mathfrak{c}) \subset \mathfrak{a}, \qquad \omega(\eta, \cdot)^*(\mathfrak{c}) \subset \mathfrak{a} \qquad (6.12)$$

for all $\eta \in \mathfrak{h} = \mathfrak{a}^*$ and $\zeta \in \mathfrak{n} = \mathfrak{c}^*$.

With the notations and hypotheses of Theorem 6.4, the Lie-Poisson bracket of $f, g \in C^\infty(\mathfrak{c} \oplus \mathfrak{a})$ is given by

$$\{f, g\}(c, a) = \left\langle a, \left[\frac{\delta f}{\delta a}, \frac{\delta g}{\delta a}\right] \right\rangle$$
$$+ \left\langle c, \left[\frac{\delta f}{\delta c}, \frac{\delta g}{\delta c}\right] - \varphi\left(\frac{\delta g}{\delta a}\right)\frac{\delta f}{\delta c} + \varphi\left(\frac{\delta f}{\delta a}\right)\frac{\delta g}{\delta c} + \omega\left(\frac{\delta f}{\delta a}, \frac{\delta g}{\delta a}\right) \right\rangle \quad (6.13)$$

for $c \in \mathfrak{c}$ and $a \in \mathfrak{a}$, where $\delta f/\delta c \in \mathfrak{c}^*$ and $\delta f/\delta a \in \mathfrak{a}^*$ denote the partial functional derivatives of f defined by

$$D_{\mathfrak{c}} f(c, a)(c') = \left\langle c', \frac{\delta f}{\delta c} \right\rangle \qquad \text{and} \qquad D_{\mathfrak{a}} f(c, a)(a') = \left\langle a', \frac{\delta f}{\delta a} \right\rangle$$

for all $c' \in \mathfrak{c}$ and all $a' \in \mathfrak{a}$; $D_{\mathfrak{c}} f(c, a)$ and $D_{\mathfrak{a}} f(c, a)$ denote the partial Fréchet derivatives of f at $(c, a) \in \mathfrak{c} \oplus \mathfrak{a}$ respectively.

Let us specialize this situation for the case $\mathfrak{h} = L^\infty(\mathcal{H})$, $\mathfrak{h}_* = L^1(\mathcal{H})$, $\mathfrak{n} = \mathfrak{n}_* = \mathcal{H}$ with the trivial Banach Lie algebra structure, $\omega = 0$, and φ is the identity map of $L^\infty(\mathcal{H})$. Formula (6.13) for $f, g \in C^\infty(\mathcal{H} \oplus L^1(\mathcal{H}))$ becomes

$$\{f, g\}(v, \rho) = \mathrm{tr}\left(\rho\left[\frac{\delta f}{\delta \rho}, \frac{\delta g}{\delta \rho}\right]\right) + \left\langle v \left| \frac{\delta f}{\delta \rho}\frac{\delta g}{\delta v} - \frac{\delta g}{\delta \rho}\frac{\delta f}{\delta v} \right.\right\rangle,$$

where $\langle \cdot \mid \cdot \rangle$ denotes the Hermitian inner product on $\mathcal{H}$, $\rho \in L^1(\mathcal{H})$ and $v \in \mathcal{H}$; note that $\delta f/\delta \rho \in L^\infty(\mathcal{H})$ and $\delta f/\delta v \in \mathcal{H}$. Hamilton's equation $\dot{f} = \{f, h\}$ for the Hamiltonian $h \in C^\infty(L^1(\mathcal{H}))$ can be equivalently written as the system of equations

$$|\dot{v}\rangle = -\left(\frac{\delta h}{\delta \rho}\right)^* |v\rangle \qquad \dot{\rho} = \left[\frac{\delta h}{\delta \rho}, \rho\right] + \left|\frac{\delta h}{\delta \rho}\right\rangle\langle v|.$$

Example 3: The p-diagonal trace class operators. Let us consider the following general example. Assume that $\mathfrak{g}$ is a Banach Lie algebra such that $\mathfrak{g} = \mathfrak{h} \oplus \mathcal{I}$ is a Banach space splitting with $\mathcal{I}$ a Banach Lie ideal in $\mathfrak{g}$. Thus, $\mathfrak{g}/\mathcal{I} \cong \mathfrak{h}$ is a Banach Lie algebra. Assume also that $\mathfrak{h}$ has a predual $\mathfrak{h}_*$, that $\mathcal{I}$ also has a predual $\mathcal{I}_*$, and that the respective coadjoint actions preserve these preduals. Then both $\mathfrak{h}_*$ and $\mathcal{I}_*$ are Banach Lie-Poisson spaces. In addition, $\mathfrak{g}_* := \mathfrak{h}_* \oplus \mathcal{I}_*$ is a predual space to $\mathfrak{g}$ and since the coadjoint action of $\mathfrak{g}$ preserves $\mathfrak{g}_*$ by construction, it follows that $\mathfrak{g}$ is also a Banach Lie-Poisson space. In addition, the inclusion of $\mathfrak{h}_*$ into $\mathfrak{g}_*$ is a Poisson map.

As an application of these ideas, let $\mathcal{H}$ be a real separable Hilbert space and denote by $L_+^\infty(\mathcal{H})$ the Banach Lie algebra of bounded upper triangular (relative to this chosen basis) linear operators on $\mathcal{H}$. Fix some $p \in \mathbb{N}$ and define $L_{+,p}^\infty(\mathcal{H})$ to be the Banach subspace of $L_+^\infty(\mathcal{H})$ consisting of all bounded operators having only p upper diagonals. Let $\mathcal{I}_{+,p}$ be the Banach Lie ideal of all elements in $L_+^\infty(\mathcal{H})$ all of whose first p upper diagonals vanish. This is precisely the set-up of the previous considerations by observing that the predual of $L_{+,p}^\infty(\mathcal{H})$ is $L_{-,p}^1(\mathcal{H})$, the space of lower triangular trace class operators with only p non zero diagonals, and that the predual of $\mathcal{I}_{+,p}^\infty$ is $\mathcal{I}_{-,p}^1$, the Banach ideal of $L_-^1(\mathcal{H})$ consisting of all lower triangular trace class operators whose first lower p diagonals vanish. Thus, $L_-^1(\mathcal{H})$, $\mathcal{I}_{-,p}^1$, $L_{-,p}^1(\mathcal{H})$, and $\mathcal{I}_{-,p}^1$ are Banach Lie-Poisson spaces and the inclusion of $L_{-,p}^1(\mathcal{H})$ into $L_-^1(\mathcal{H})$ is a Poisson map. The case $p = 2$ is worked out in great detail in [7].

6.4 Coadjoint Orbits

In this section we shall discuss the coadjoint orbits in a Banach Lie-Poisson space, if its dual admits an underlying Lie group. For this we need to quickly review some concepts that in infinite dimensions take on a different form than the one familiar from finite dimensional manifold theory.

A smooth map $f : M \to N$ between *finite* dimensional manifolds is called an immersion, if for every $m \in M$ the derivative $T_m f : T_m M \to T_{f(m)} N$ is injective. In infinite dimensions there are various notions generalizing this concept. A smooth map $f : M \to N$ between Banach manifolds is called a

(i) *immersion* if for every $m \in M$ the tangent map $T_m f : T_m M \to T_{f(m)} N$ is injective with closed split range;

(ii) *quasi immersion* if for every $m \in M$ the tangent map $T_m f : T_m M \to$

$T_{f(m)}N$ is injective with closed range;

(iii) **weak immersion** if for every $m \in M$ the tangent map $T_m f : T_m M \to T_{f(m)}N$ is injective.

An immersion between Banach manifolds has the same properties as an immersion between finite dimensional manifolds. For example, it is characterized by the property that locally it is given by a map of the form $u \mapsto (u, 0)$, where the model space of the chart on N necessarily splits. This is the concept widely used in the literature. Unfortunately, in the study of Banach Poisson manifolds, not even the weaker concept of quasi immersion is satisfactory and one is forced to work with genuine weak immersions.

If $(P, \{\cdot, \cdot\}_P)$ is a Banach Poisson manifold, the vector subspace $S_p := \{X_f(p) \mid f \in C^\infty(P)\}$ of $T_p P$ is called the **characteristic subspace** at p. Note that S_p is, in general, not a closed subspace of the Banach space $T_p P$. The union $S := \cup_{p \in P} S_p \subset TP$ is called the **characteristic distribution** of the Poisson structure on P. Note that even if S_p were closed and split in $T_p P$ for every $p \in P$, S would not necessarily be a subbundle of TP. However, the characteristic distribution S is always **smooth**, in the sense that for every $v_p \in S_p \subset T_p P$ there is a locally defined smooth vector field (namely some X_f) whose value at p is v_p. Assume that the characteristic distribution is integrable. For finite dimensional manifolds this is automatic by the Stefan-Sussmann theorem (see, e.g. [4], Appendix 3, Theorem 3.9) which, to our knowledge, is not available in infinite dimensions.

Let $\mathcal{L}$ be a leaf of the characteristic distribution, that is,

- $\mathcal{L}$ is a connected smooth Banach manifold,
- the inclusion $\iota : \mathcal{L} \hookrightarrow P$ is a weak injective immersion,
- $T_q \iota(T_q \mathcal{L}) = S_q$ for each $q \in \mathcal{L}$,
- if the inclusion $\iota' : \mathcal{L}' \hookrightarrow P$ is another weak injective immersion satisfying the three conditions above and $\mathcal{L} \subset \mathcal{L}'$, then necessarily $\mathcal{L}' = \mathcal{L}$, that is, $\mathcal{L}$ is maximal.

If we assume, in addition, that on the leaf $\mathcal{L}$

- there is a weak symplectic form $\omega_{\mathcal{L}}$ consistent with the Poisson structure on P,

then $\mathcal{L}$ will be called a **symplectic leaf.**

In order to explain what this consistency means, we need to make a remark regarding the bundle map $\sharp : T^* P \to TP$ associated to the Poisson structure. Recall that if $f : U \subset P \to \mathbb{R}$ is a smooth function and U

is an open subset of P, then $\sharp(df) = X_f$. Thus, for each $p \in P$, the linear continuous map $\sharp_p : T_p^* P \to T_p P$ induces a bijective continuous map $[\sharp_p] : T_p^* P / \ker \sharp_p \to S_p$. By definition, $\omega_{\mathcal{L}}$ is **consistent** with the Poisson structure on P if

$$\omega_{\mathcal{L}}(q)(u_q, v_q) = \varpi(\iota(q)) \left(([\sharp_{\iota(q)}]^{-1} \circ T_q \iota)(u_q), ([\sharp_{\iota(q)}]^{-1} \circ T_q \iota)(v_q) \right), \quad (6.14)$$

where ϖ is the Poisson tensor on P (that is, $\varpi(df, dg) = \{f, g\}$ for any locally defined smooth functions f and g on P). This shows that the weak symplectic form $\omega_{\mathcal{L}}$ consistent with the Poisson structure on P is unique.

For finite dimensional Poisson manifolds, it is known that all leaves are symplectic (see [11]) and so the last assumption above is not necessary. In the infinite dimensional case this question is open, even in the case of a Banach Lie group G whose Lie algebra $\mathfrak{g}$ has a predual $\mathfrak{g}_*$ invariant under the coadjoint action. In this case, $\mathfrak{g}_*$ is a Banach Lie-Poisson space and one can characterize a large class of points in $\mathfrak{g}_*$ whose coadjoint orbits are all weak symplectic manifolds. Their connected components are therefore symplectic leaves. These points in the predual are given by the following theorems proved in [6].

Theorem 6.5   *Let G be a (real or complex) Banach Lie group with Lie algebra $\mathfrak{g}$. Assume that:*

(i) *$\mathfrak{g}$ admits a predual $\mathfrak{g}_*$, that is, $\mathfrak{g}_*$ is a Banach space whose dual is $\mathfrak{g}$;*

(ii) *the coadjoint action of G on the dual $\mathfrak{g}^*$ leaves the predual $\mathfrak{g}_*$ invariant, that is, $\mathrm{Ad}_g^*(\mathfrak{g}_*) \subset \mathfrak{g}_*$, for any $g \in G$;*

(iii) *for a fixed $\rho \in \mathfrak{g}_*$ the coadjoint isotropy subgroup $G_\rho := \{g \in G \mid \mathrm{Ad}_g^* \rho = \rho\}$, which is closed in G, is a Lie subgroup of G (in the sense that it is a submanifold of G and not just injectively immersed).*

Then the Lie algebra of G_ρ equals $\mathfrak{g}_\rho := \{\xi \in \mathfrak{g} \mid \mathrm{ad}_\xi^* \rho = 0\}$ and the quotient topological space $G/G_\rho := \{gG_\rho \mid g \in G\}$ admits a unique smooth (real or complex) Banach manifold structure making the canonical projection $\pi : G \to G/G_\rho$ a surjective submersion. The manifold G/G_ρ is weakly symplectic relative to the two form ω_ρ given by

$$\omega_\rho([g])(T_g\pi(T_e L_g \xi), T_g\pi(T_e L_g \eta)) := \langle \rho, [\xi, \eta] \rangle, \quad (6.15)$$

where $\xi, \eta \in \mathfrak{g}$, $g \in G$, $[g] := \pi(g) = gG_\rho$, and $\langle \cdot, \cdot \rangle : \mathfrak{g}_* \times \mathfrak{g} \to \mathbb{R}$ (or $\mathbb{C}$) is the canonical pairing between $\mathfrak{g}_*$ and $\mathfrak{g}$. Alternatively, this form can be expressed as

$$\omega_\rho([g])(T_g\pi(T_e R_g \xi), T_g\pi(T_e R_g \eta)) := \left\langle \mathrm{Ad}_{g^{-1}}^* \rho, [\xi, \eta] \right\rangle. \quad (6.16)$$

The two form ω_ρ is invariant under the left action of G on G/G_ρ given by $g \cdot [h] := [gh]$, for $g, h \in G$.

Theorem 6.6 *Let the Banach Lie group G and the element $\rho \in \mathfrak{g}_*$ satisfy the hypotheses of Theorem 6.5. Then the map*

$$\iota : [g] \in G/G_\rho \mapsto \mathrm{Ad}^*_{g^{-1}} \rho \in \mathfrak{g}_* \qquad (6.17)$$

is an injective weak immersion of the quotient manifold G/G_ρ into the predual space $\mathfrak{g}_*$. Endow the coadjoint orbit $\mathcal{O} := \{\mathrm{Ad}^*_g \rho \mid g \in G\}$ with the smooth manifold structure making ι into a diffeomorphism. The push forward $\iota_*(\omega_\rho)$ of the weak symplectic form $\omega_\rho \in \Omega^2(G/G_\rho)$ to $\mathcal{O}$ has the expression

$$\omega_{\mathcal{O}}(\mathrm{Ad}^*_{g^{-1}} \rho) \left(\mathrm{ad}^*_{\mathrm{Ad}_g \xi} \mathrm{Ad}^*_{g^{-1}} \rho, \mathrm{ad}^*_{\mathrm{Ad}_g \eta} \mathrm{Ad}^*_{g^{-1}} \rho \right) = \langle \rho, [\xi, \eta] \rangle, \qquad (6.18)$$

for $g \in G$, $\xi, \eta \in \mathfrak{g}$, and $\rho \in \mathfrak{g}_*$. Relative to this symplectic form the connected components of the coadjoint orbit $\mathcal{O}$ are symplectic leaves of the Banach Lie-Poisson space $\mathfrak{g}_*$.

Theorem 6.7 *Let the Banach Lie group G and the element $\rho \in \mathfrak{g}_*$ satisfy the hypotheses of Theorem 6.5. The following conditions are equivalent:*

(i) *$\iota : G/G_\rho \to \mathfrak{g}_*$ is an injective immersion;*
(ii) *the characteristic subspace $S_\rho := \{\mathrm{ad}^*_\xi \rho \mid \xi \in \mathfrak{g}\}$ is closed in $\mathfrak{g}_*$;*
(iii) *$S_\rho = \mathfrak{g}^\circ_\rho$, where $\mathfrak{g}^\circ_\rho$ is the annihilator of $\mathfrak{g}_\rho$ in $\mathfrak{g}_*$.*

Endow the coadjoint orbit $\mathcal{O} := \{\mathrm{Ad}^*_g \rho \mid g \in G\}$ with the manifold structure making ι a diffeomorphism. Then, under any of the hypotheses (i)–(iii), the two-form given by (6.18) is a strong symplectic form.

Example: Embedded coadjoint orbits. Consider the coadjoint orbit of the Banach Lie group $GL^\infty(\mathcal{H})$ through the element

$$\rho := \sum_{k=1}^{N} \lambda_k P_k \in L^1(\mathcal{H}),$$

where $\lambda_k \in \mathbb{C}$, $\lambda_k \neq \lambda_n$ if $k \neq n$, P_k are orthogonal projectors, and N is a natural number or infinity. If $N = \infty$, then assume that $\mathrm{rank}\, P_k < \infty$ for all $k \in \mathbb{N}$. In this case, any values of λ_i are allowed, including $\lambda_i = 0$. If N is finite, assume that $\lambda_1 = 0$ and that $\mathrm{rank}\, P_k < \infty$ for $k \geq 2$. In this case, rank P_1 could be infinite. The map $R : L^1(\mathcal{H}) \to L^1(\mathcal{H})$ defined by

$$R(\rho) := \sum_{k=1}^{N} P_k \rho P_k, \qquad \text{for} \qquad \rho \in L^1(\mathcal{H})$$

is a projector and has adjoint projector $R^* : L^\infty(\mathcal{H}) \to L^\infty(\mathcal{H})$ given by

$$R^*(X) = \sum_{k=1}^{N} P_k X P_k \qquad \text{for} \qquad X \in L^\infty(\mathcal{H}).$$

A direct computation shows that the coadjoint stabilizer Banach Lie algebra $L^\infty(\mathcal{H})_\rho$ coincides with $\mathrm{im}(R^*)$ and that the coadjoint stabilizer group is $GL^\infty(\mathcal{H})_\rho = GL^\infty(\mathcal{H}) \cap \mathrm{im}(R^*)$. As the range of a projector, $L^\infty(\mathcal{H})_\rho$ is a closed subspace of $L^\infty(\mathcal{H})$. Since the Banach space $L^1(\mathcal{H})$ is an ideal in $L^\infty(\mathcal{H})$, it is invariant under the coadjoint action of $GL^\infty(\mathcal{H})$ and so the assumptions of Theorems 6.5 and 6.6 are satisfied. This shows that the coadjoint orbit $\mathcal{O}_\rho$ is a weakly immersed weak symplectic manifold in $L^1(\mathcal{H})$.

The annihilator $(L^\infty(\mathcal{H})_\rho)^\circ \subset L^1(\mathcal{H})$ of the stabilizer Lie subalgebra equals $\ker R$ and the characteristic subspace $S_\rho\{[X,\rho] \mid X \in L^\infty(\mathcal{H})\}$ is contained in $\ker R$. If N is finite, then $S_\rho = \ker R = (L^\infty(\mathcal{H})_\rho)^\circ$, so one concludes from Theorem 6.7 that the coadjoint orbit $\mathcal{O}_\rho$ is immersed in $L^1(\mathcal{H})$ and that it is strongly symplectic.

In the special case when the rank of ρ is one, the orbit $\mathcal{O}_\rho$ can be identified with the projectivized Hilbert space $\mathbb{CP}(\mathcal{H})$ and the orbit symplectic form coincides with the Fubini-Study symplectic form $\omega([\psi]) = i\partial\bar{\partial}\langle\psi|\psi\rangle$, for $\psi \in \mathcal{H}$. In this case, it can be shown directly that the immersion of the orbit $\mathcal{O}_\rho$ is actually an embedding. The same is true if the rank of ρ is finite as shown by Bona [2].

Acknowledgments

We want to thank the Erwin Schrödinger Institute in Vienna for its hospitality during the time this paper was written. The second author was partially supported by the European Commission and the Swiss Federal Government through funding for the Research Training Network *Mechanics and Symmetry in Europe* (MASIE) as well as the Swiss National Science Foundation.

Bibliography

[1]   Bona, P.: Extended quantum mechanics. *Acta Physica Slovaca*, **50** (1), 2000, 1–198.

[2]   Bona, P.: Some considerations on topologies of infinite dimensional unitary coadjoint orbits. Comenius University, Bratislava, preprint, 2003.

[3]   Chernoff, P. R. and Marsden, J. E.: *Properties of Infinite Dimensional Hamiltonian Systems*. Lecture Notes in Mathematics, **425**, Springer Verlag, New York, 1974.

[4]   Libermann, P. and Marle, C.-M.: *Symplectic Geometry and Analytical Mechanics*, Kluwer Academic Publishers, 1987.

[5]   Murphy, G.J.: *C^*-algebras and Operator Theory*, Academic Press, San Diego, 1990.

[6]   Odzijewicz, A. and Ratiu, T.S.: Banach Lie-Poisson spaces and reduction. *Comm. Math. Phys.*, **243**, 1–54, 2003, to appear

[7]   Odzijewicz, A. and Ratiu, T.S.: The Banach Poisson geometry of the infinite Toda lattice. 2003, ESI preprint.

[8]   Odzijewicz, A. and Ratiu, T.S.: Extensions of Banach Lie-Poisson spaces, 2003, ESI preprint.

[9]   Sakai, S.: *C^*-Algebras and W^*-Algebras*. Ergebnisse der Mathematik und ihrer Grenzgebiete, **60**, 1998 reprint of the 1971 edition, Springer-Verlag, New York, 1971.

[10]   Takesaki, M.: *Theory of Operator Algebras I*, Springer-Verlag, New York, 1979.

[11]   Weinstein, A. The local structure of Poisson manifolds, *Journ. Diff. Geom* **18**, 1983, 523–557.

Chapter 7

Spectra of Operators Associated with Dynamical Systems: From Ergodicity to the Duality Principle

A.B. Antonevich[1], V.I. Bakhtin[2], A.V. Lebedev[1]

Abstract: This contribution is an overview of the ergodic, the entropy and the stochastic properties of weighted shift operators associated with their spectral characteristics. The presented material forms in essence an overview of a series of lectures given by the authors at several Białowieża conferences.

Contents

---

[1] Belarus State University / University of Bialystok
[2] Belarus State University

7.1   Introduction. Operators Associated with Dynamical Systems: Shifts and Weighted Shifts. Some Historical Notes and Problems

The article is devoted to the discussion of a number of problems related in this or that way to the operators acting on the spaces $L^p(X, \mu)$ and having the form

$$T_\gamma u(x) = u(\gamma(x)), \quad u \in L^p(X, \mu), \tag{7.1.1}$$

and

$$Bu(x) = a(x)u(\gamma(x)), \tag{7.1.2}$$

where X is a space with a given measure μ, $\gamma : X \to X$ is a given mapping and a is a given function. The triple (X, μ, γ) is called a (metrical) dynamical system, the operator T_γ is called a *shift* (or composition) operator and B is called a *weighted shift* operator (or a transfer-operator). One can come across the operators (7.1.1) and (7.1.2) in various fields of mathematics and applications and the literature treating this or that trend of investigations involving them is vast.

The properties of these operators are determined first of all by the dynamics of the mapping γ and naturally the investigation of such operators is closely related to the theory of dynamical systems. As a rule (but far from being always) in applications X is a phase space of a system and the mapping γ describes the motion law of the particles forming the system. The function u can be interpreted as a distribution function of the particles on the space X and therefore the operator T_γ describes the *evolution* (during a unit lapse of time) of the ensemble of particles considered. In examples in the theory of dynamical systems the measure μ is often invariant with respect to γ that is $\mu(\gamma^{-1}(\Omega)) = \mu(\Omega)$ for any measurable set. This property is a formulation of the conversation laws or the stationary property in the case when the dynamical system arise from a stochastic process. Under this assumption the operator T_γ is an isometry and one of the principle problems is the description of the asymptotic (in different senses) behaviour of its iterations, that is $\lim_{n \to \infty} T_\gamma^n u$.

The operators (7.1.2) spring out in the description of more complicated situations. The coefficient $a(x)$ in the expression (7.1.2) can be interpreted as a manifestation of an external influence: along with the motion of particles on the phase space at a point x we also have the process of creation with the coefficient $a(x)$ (if $a(x) > 1$) or annihilations (if $a(x) < 1$). In this case the question on the behaviour of $B^n u$ (that is the question about the evolution of the initial state u) can be divided into two questions: what is the behaviour of the sequence $\|B^n u\|$ (that is the asymptotics of the total 'quantity' of particles)? And what is the limit distribution of the sequence $\frac{B^n u}{\|B^n u\|}$ (that is the behaviour of the normalized distributions)? As a rule the behaviour of the first sequence is characterized by means of the *Lyapunov exponents*:

$$\chi(u) = \overline{\lim} \; \frac{1}{n} \; \ln \|B^n u\|. \qquad (7.1.3)$$

In particular if $\chi(u) > 0$ then the number of particles increases exponentially (that is we have an explosion) and if $\chi(u) < 0$ then the process is fading. As is known for a bounded operator we have the inequality

$$\chi(u) \le \ln r(B)$$

(where $r(B)$ is the spectral radius of B) and the equality

$$\chi(u) = \ln r(B) \qquad (7.1.4)$$

is true for 'almost all' u, where by 'almost all' we mean that the set of us that do not satisfy (7.1.4) is thin. Therefore the question about the character of the process (explosive, fading, stationary) reduces to the question about the value of the spectral radius of the operator B. The main theme of the article is the discussion of the problems, objects and results arising in the process of the answer to this question.

We would like to stress at once that the spectral analysis of the shift operators is a well developed field and a considerable attention is paid to this subject almost in every book on dynamical systems theory (and there are dozens of them). On the other hand the investigation of dynamical systems by means of weighted shift operators is not so popular although (and the material of the article shows this) the role of the spectral analysis of these operators in the dynamical systems theory and related subjects is even higher. So one of the aims of this paper is to draw more attention to this promising and important (from our point of view) branch of research.

To start with let us recall and discuss the properties of the mapping $\gamma : X \to X$ that are related to the properties of operators (7.1.1) and (7.1.2).

A measure preserving mapping γ is called *ergodic* if the condition $\gamma^{-1}(\Omega) = \Omega$ (almost everywhere) implies that $\mu(\Omega) = 0$ or $\mu(X \setminus \Omega) = 0$ that is all the invariant sets are trivial. This property has a clear operator sense: in the case when $\mu(X) < +\infty$ the ergodicity of γ is equivalent to the condition that the number $\lambda = 1$ is an eigen value of the multiplicity 1 for the operator T_γ. Moreover it follows from the statistic ergodic theorem of Von Neumann that

γ is ergodic iff for any $u \in L^2(X,\mu)$ the following equality takes place

$$\lim_{n\to\infty} \frac{1}{n}\left(u + T_\gamma u + ... T_\gamma^{n-1} u\right) = P_1 u \equiv \int_X u\, d\mu$$

where P_1 is the orthogonal projection onto the subspace of constants.

That is γ is ergodic iff $\frac{1}{n}\left(u + T_\gamma u + ... T_\gamma^{n-1} u\right)$ tends strongly to P_1.

One more property that has a clear operator sense is the so called *mixing* of γ which by definition lies in the equality

$$\lim_{n\to\infty} \int_X T_\gamma u \cdot v\, d\mu = \int_X u\, d\mu \cdot \int_X v\, d\mu$$

for any $u, v \in L^2(X,\mu)$. In other words this means that the sequence of operators T_γ^n tends weakly to the operator P_1.

Isomorphism problem. Shifts, entropy and weighted shifts. Among the problems in the dynamical systems theory where the spectral analysis of shift operators occupies a substantial place is the so-called isomorphism problem. Let us recall this problem.

Two dynamical systems (X_1, μ_1, γ_1) and (X_2, μ_2, γ_2) (where γ_i are μ_i-preserving transformations and $\mu_i(X_i) = 1$) are said to be *isomorphic* iff there exists an invertible measure preserving mapping $\phi : X_1 \to X_2$ such that $\phi \circ \gamma_1 = \gamma_2 \circ \phi$ (almost everywhere).

In fact when dealing with measure spaces it is quite natural to work with set maps (that do not 'feel' sets of measure zero) rather than with point maps as the mentioned ϕ. It is clear that the mapping ϕ induces an **-algebraic*(!) isomorphism $\tilde{\phi}$ between the functional algebras $L^\infty(X_1, \mu_1)$ and $L^\infty(X_2, \mu_2)$: $\tilde{\phi}(f) = f \circ \phi$, $f \in L^\infty(X_2, \mu_2)$ (evidently $\tilde{\phi}(f \cdot g) = \tilde{\phi}(f) \cdot \tilde{\phi}(g)$ and $\tilde{\phi}(\bar{f}) = \overline{\tilde{\phi}(f)}$ where $\bar{f}$ is the complex conjugate to f function). On the other hand on the maximal ideals level any algebraic isomorphism of semisimple commutative algebras is induced by a certain point mapping (see in this connection Theorem 7.3.1 of this article). Therefore it is convenient to involve this 'algebraic ideology' into the notion of isomorphism and in this way we arrive at the notion of *conjugacy*: dynamical systems (X_1, μ_1, γ_1) and (X_2, μ_2, γ_2) are said to be *conjugate* iff there exists a *-algebraic isomorphism $\varphi : L^\infty(X_2, \mu_2) \to L^\infty(X_1, \mu_1)$ such that $\varphi(f) \circ \gamma_1 = \varphi(f \circ \gamma_2)$, $f \in L^\infty(X_2, \mu_2)$.

Remark. It is known that in the case of Lebesgue spaces (X_i, μ_i) the conjugation of the dynamical systems implies the corresponding (point) isomorphism not only on the maximal ideals space level but also on the level of the spaces (X_i, μ_i) themselves (see, for example, [1], §2.3) so in this situation the notions of conjugacy and isomorphism do coincide.

The *isomorphism problem* is the problem of finding total dynamical invariants that are invariants that determine a dynamical system up to isomorphism (conjugation).

In a number of situations the spectral character of shift operators do resolve this problem which is stated in the next

Theorem 7.1.1 (see, for example, [1], Chapter 3) *Let γ_i, $i = 1, 2$ be ergodic transformations of the probability spaces (X_i, μ_i) and suppose that the corresponding operators T_{γ_i} acting on $L^2(X_i, \mu_i)$ have the discrete spectra then the following are equivalent:*

(i) T_{γ_1} and T_{γ_2} are spectrally isomorphic,

(ii) T_{γ_1} and T_{γ_2} have the same eigenvalues,

(iii) γ_1 and γ_2 are conjugate.

However as is well known in general there exist properties of dynamical systems that are not reflected by the properties of operators T_γ. For example, let $X = \{z \in \mathbf{C} : |z| = 1\}$ be the unit circle on the complex plane with the normalized Lebesgue measure (normalized length on the circle) and $\gamma_1(z) = z^2$ $\gamma_2(z) = z^3$. In this case the operators T_{γ_i}, $i = 1, 2$ are unitary equivalent as the operators in Hilbert space. Indeed, each of them acts identically on the one dimensional subspace consisting of constant functions and its restriction onto the orthogonal complement to this subspace is a direct sum of a countable number of operators each of which is unitary equivalent to the operator of the classical one-sided shift in the space l^2. But the mappings γ_1 and γ_2 have radically different properties (they have different entropies, the mapping γ_1 has two inverse images for each point and γ_2 has three inverse images) and they are not conjugate.

Namely in the process of solution of the isomorphism problem for the mappings of this sort A.N. Kolmogorov and Ja.G. Sinai invented the entropy of a dynamical system and D.S. Ornstein found the complete solution to the Kolmogorovs conjecture (Bernoulli shifts are conjugate iff they have equal entropies) which is stated in the next result.

Theorem 7.1.2 (Ornstein) *Let γ_i, $i = 1, 2$ be Bernoulli shifts whose* *state spaces are Lebesgue spaces then they are conjugate iff their entropies are equal.*

Of course the 'entropy answer' to the isomorphism problem is rather fascinating since in this case one determines a dynamical system by a *single* number (its entropy). However it should be mentioned at once that the entropy is far from being a complete invariant of a dynamical system in general. For example, if we take again the unit circle $X = \{z \in \mathbf{C} : |z| = 1\}$ and consider the rotations $\gamma_\varphi(z) = ze^{i\varphi}$ then for any φ the entropy of γ_φ is zero, so the entropy cannot distinguish between these mappings. On the other hand if $\frac{\varphi}{2\pi} = \frac{m}{n}$ is a rational number then $\gamma_\varphi^n(z) = z$ that is γ_φ is a periodic mapping while if $\frac{\varphi}{2\pi}$ is an irrational number then γ_φ is not periodic and therefore these mappings are essentially different.

So neither shift operators nor entropy can give a complete solution to the isomorphism problem. It turns out that the 'complete' answer here can be given in *operator algebraic* terms exploiting *weighted shift* operators.

Before passing to the description of this answer we would like first to return back to the shift operators generated by dynamical systems and stress that the difference between the conjugation and unitary equivalence is linked with the (already mentioned) fact that the unitary operator T_ϕ generated by the isomorphism mapping ϕ (that is the mapping satisfying the relation $\gamma_1 = \phi\gamma_2\phi^{-1}$) along with the unitary property possesses also the multiplicativity:

$$T_\phi(u \cdot v) = T_\phi(u) \cdot T_\phi(v) \tag{7.1.5}$$

(if the multiplication of the functions $u \cdot v$ is defined).

Remark. In the above considered example of the mappings of the circle $\gamma_1(z) = z^2 \quad \gamma_2(z) = z^3$ the operator implementing the unitary equivalence of course *does not* possess this property.

One can look at the property (7.1.5) from the following point of view. In the space $L^2(X, \mu)$ we can consider the multiplication by elements of $L^\infty(X, \mu)$ that is $L^2(X, \mu)$ is an $L^\infty(X, \mu)$-module and (7.1.5) means the compatibility of the mapping T_ϕ with this module structure of $L^2(X, \mu)$. In fact this property reduces to the (already mentioned) observation that T_ϕ is an *automorphism* of the algebra $L^\infty(X, \mu)$. In reality this is a characteristic property of the operators generated by (set) mappings of X. Namely the following statement is true.

Theorem 7.1.3 (see, for example, [1], § 2.1) *A unitary operator* $V :$ $L^2(X, \mu) \to L^2(X, \mu)$ *is induced by an automorphism of the algebra of mea-*

surable subsets of X iff it is an automorphism of the algebra $L^\infty(X, \mu)$.

So the properties of the dynamical systems should be naturally linked with the structure of $L^\infty(X, \mu)$-module on the space $L^2(X, \mu)$. But weighted shift operators are linked with this structure due to their construction. Therefore one can expect that these operators reflect wider spectrum of properties of dynamical system in comparison with the shift operators. And this is really true. In fact the presentation of this 'wider spectrum' is one of the main goals of the article.

Now we give the first example belonging to this 'spectrum' — the operator algebraic solution of the isomorphism problem in terms of weighted shifts. This solution was found by W.B. Arveson [2].

Let γ be a measure preserving *invertible* mapping of (X, μ)

In the space $L^2(X, \mu)$ we consider the operator algebra $\mathcal{B}(L^\infty, \gamma)$ — the uniform closure of the set of finite sums of weighted shift operators:

$$\sum_{k=0}^{n} a_k T_\gamma^k, \quad n \in \mathbf{N}$$

where by a_k we denote the operator of multiplication by a function $a_k \in L^\infty(X, \mu)$.

Theorem 7.1.4 [2] *Let γ_i, $i = 1, 2$ be ergodic invertible mappings of Lebesgue space (X, μ). Then γ_1 and γ_2 are conjugate iff there exists a unitary operator U such that $U\mathcal{B}(L^\infty, \gamma_1)U^* = \mathcal{B}(L^\infty, \gamma_2)$.*

So the operator algebras generated by weighted shifts 'know everything' about the dynamical system they arise from. In the subsequent part of the paper we shall show that not only the algebras but even a *single* weighted shift operator (in fact its spectrum) knows much more than a shift operator, namely it accumulates the ergodic, entropy and stochastic properties of the initial dynamical system. In Section 7.2 we uncover the interrelation between the spectrum of weighted shift operator with *invertible* shifts and ergodic measures. In Section 7.3 we shall find out that if the mapping γ is a *topological Markov chain* then the spectral radius of the corresponding weighted shift operator is linked with the topological pressure and the entropy while in the general situation of an *arbitrary* γ the spectral radius is linked with the *new* entropy-stochastic nature invariant (this is shown in Section 7.4) and in fact the spectral radius calculation brings us to the *duality principle* establishing Fenchel-Legendre duality between the spectral characteristics of weighted shift operators and stochastic behaviour of the dynamical system (described by this new invariant — T-entropy). Finally in Section 7.5 we examine the interrelation between T-entropy and the action functional for Perron-Frobenius dynamical systems.

7.2 Dynamical Nature of the Spectrum of Operators with Invertible Shifts: Spectral Radius and Ergodic Measures

Let X be a compact topological space, $\gamma : X \to X$ be a continuous invertible mapping and μ be a γ-invariant measure on X whose support coincides with X (that is any open set has a non zero measure).

The results of this section in essence can be extracted from the results presented in [3; 4].

Theorem 7.2.1 *Let $a \in C(X)$ and aT_γ be a weighted shift operator acting in one of the spaces $L^p(X, \mu)$, $1 \le p \le \infty$ or $C(X)$ then the following formula for the spectral radius $r(aT_\gamma)$ is true:*

$$\ln[r(aT_\gamma)] = \max_{\nu \in M_\gamma} \int_X \ln |a(x)| \, d\nu = \max_{\nu \in M_{\gamma,e}} \int_X \ln |a(x)| \, d\nu \qquad (7.2.1)$$

where M_γ is the set of all Borel probability γ-invariant measures on X and $M_{\gamma,e}$ is a subset of M_γ consisting of all ergodic measures.

From the 'ideological' point of view this result shows that the spectral properties of weighted shift operators depend on *all* γ-invariant measures while the properties of T_γ depend only on the measure μ.

Abstract weighted shift with invertible shift. The above mentioned conditions that X is a compact topological space and γ is a continuous mapping in fact are not restrictive which can be approved by the following argument. Let us consider a probability measure space (Y, μ) and a measure preserving invertible mapping $\alpha : Y \to Y$. Routine computation shows that for any $a \in L^\infty(Y, \mu)$ we have

$$\left[(T_\alpha a T_\alpha^{-1}) f \right] (y) = a(\alpha(y)) f(y), \quad f \in L^p(Y, \mu), \ \ 1 \le p \le \infty \qquad (7.2.2)$$

which means that the operator T_α generates *an automorphism* $\hat{T}_\alpha$ of the algebra $A = L^\infty(Y, \mu)$: $\hat{T}_\alpha(a) = T_\alpha a T_\alpha^{-1}$. The Gelfand transform establishes an isomorphism between A and $C(X)$, where X is the maximal ideal space of $L^\infty(Y, \mu)$ and under this isomorphism the automorphism $\hat{T}_\alpha$ has the form

$$\hat{T}_\alpha(a)(x) = a(\gamma(x)) \qquad (7.2.3)$$

where γ is a homeomorphism of X (in this formula and henceforth in this section we identify an element $a \in A$ with its Gelfand transform — the element of $C(X)$).

Therefore in an axiomatic way it is reasonable to consider the following object.

Definition 7.2.2 Let $A \subset L(H)$ be a subalgebra of the algebra $L(H)$ (of all linear continuous operator acting on a Banach space H) that is isometrically isomorphic to $C(X)$ for a certain compact X; and let $T : H \to H$ be an invertible isometry such that $TAT^{-1} = A$ then an operator of the form aT, $a \in A$ will be called an (abstract) *weighted shift operator* (with an invertible shift). Algebra A is called the algebra of (abstract) *weights* and T is an (abstract) *shift* operator (with an invertible shift).

For the spectral radius of an abstract weighted shift operator we have the following generalization of Theorem 7.2.1.

Theorem 7.2.3 *Let aT be an abstract weighted shift operator described in Definition 7.2.2 and γ be a homeomorphism of the maximal ideal space X of generated by the automorphism $\hat{T}$ (see (7.2.3)) then*

$$\ln[r(aT)] = \max_{\nu \in M_\gamma} \int_X \ln|a(x)|\, d\nu = \max_{\nu \in M_{\gamma,e}} \int_X \ln|a(x)|\, d\nu \qquad (7.2.4)$$

The statements of this type are called in the dynamical systems theory and related fields of analysis the *variational principles* and we shall come across many of them in the subsequent part of the article.

The variational principle (7.2.4) was stated by A.B. Antonevich and for a number of concrete situations it has been proved for example in [6; 7] where one can also find the corresponding explicit calculation of the set M_γ. In the general form (for *an arbitrary homeomorphism* γ) the principle was established by A.V. Lebedev [8] and A.K. Kitover [9]. The applications of the formula (7.2.4) to the calculation of the spectral radii of various weighted shift operators are given in [5; 3; 4].

In view of (7.2.3), the isometric property of T and the equality

$$\|a\|_{L(H)} = \max_{x \in X} |a|(x)$$

we have the next simple formula for the norms of the powers of aT:

$$\|(aT)^n\|_{L(H)} = \max_{x \in X} \left(\prod_{k=0}^{n-1} |a| \circ \gamma^k \right)(x). \qquad (7.2.5)$$

Namely the usage of this formula is one of the key moments in the proof of Theorem 7.2.3. As we shall see in the subsequent sections in the situations when the shift operator is generated by an *irreversible* mapping the expression for the norm of the powers of the weighted shift operator is essentially more complicated (it contains some mean values of sums of analogous products) and as a result the arising variational principles will take into account also the entropy and stochastic nature of γ.

Remark. It is clear from the observations preceding Theorem 7.2.3 that as the algebra A one can take for example any closed subalgebra of $L^\infty(Y,\mu)$ which is invariant with respect to T_α and T_α^{-1}. If we take $A = L^\infty(Y,\mu)$ then its maximal ideal space is extremely complicated and so there is no hope to find an explicit description of the set M_γ. Thus Theorem 7.2.3 takes into account the complexity of the algebra A: 'simpler' algebra A implies 'simpler' its maximal ideal space structure and therefore one obtains an easier description of the set M_γ.

Let us discuss now some other interrelations between the spectral properties of weighted shifts and the dynamical properties of γ.

Recall that a mapping $\gamma : X \to X$ is called *strictly ergodic*, if there exists the only one probability γ-invariant measure. Theorems 7.2.1, 7.2.3 imply that this property is in fact the property of a weighted shift.

Corollary 7.2.4 *A homeomorphism $\gamma : X \to X$ is strictly ergodic iff the spectrum of any operator aT_γ, $a \in C(X)$ with invertible weight a lies on a certain circle (here by aT_γ one can mean either the concrete weighted shift operator mentioned in Theorem 7.2.1 or an abstract weighted shift operator mentioned in Theorem 7.2.3).*

A mapping γ is called *topologically aperiodic* if for any $n \geq 1$ the set $F_n = \{x : \gamma^n(x) = x\}$ has an empty interiour.

Theorem 7.2.5 *A homeomorphism $\gamma : X \to X$ is topologically aperiodic iff the spectrum of any (abstract) weighted shift operator aT_γ, $a \in C(X)$ is invariant with respect to rotations around zero.*

It is known that if a measure μ is ergodic with respect to a mapping γ and the support of this measure is the whole of X then the spectrum of an operator aT_γ, $a \in C(X)$ acting say in $L^2(X,\mu)$ is connected (if X is not a finite set) [10]. This statement is not invertible: there exist weighted shift operators whose spectrum is connected while the corresponding measure on X is not ergodic. The connectedness of the spectrum turns to be linked with a different property of the mapping γ. The space X is called to be

γ-*connected* if there does not exist a nontrivial partition of X consisting of nonempty γ-invariant open and closed sets. In particular if μ is γ-ergodic having X as its support then X is γ-connected.

Theorem 7.2.6 *X is γ-connected iff the spectrum of any (abstract) weighted shift operator aT_γ, $a \in C(X)$ is connected.*

7.3 Dynamical Nature of the Spectral Radius of Operators with Irreversible Shifts — First Steps: Spectral Radius and Topological Pressure

It is known that in natural situations when the shift is irreversible then the spectrum of the corresponding weighted shift operator is a disk (see, for example [11]) therefore the principal moment in this case is the spectral radius calculation.

Remark. Here the 'calculation' is realized as derivation of the expression for the spectral radius in terms of other quantities — in this case the dynamical invariants. Originally the problem has been understood precisely as calculation — in order to apply formula (7.2.4) one has to know the γ-invariant measures and the problem was assumed to be solved when the description of these measures was obtained. At the next step of generalization the problem is divided into two problems: 1) the derivation of the general formula (like (7.2.4)); and 2) the calculation of invariant measures and dynamical invariants involved in it. It should be noted that in many examples the second problem does not have an *explicit* solution but once one obtains the complete answer to the first mentioned part of the problem the problem on the calculation of the spectral radius is agreed to be solved.

In the situation when the shift γ is *not invertible* the results related to the calculation of the spectral radius of a weighted shift operator can be divided into two classes. The first class contains the results referred to the situation when the weighted shift operators act in the spaces of $C(X)$ or $L^\infty(X)$ type. In this case formula (7.2.4) preserves its form (see, for example, [13]). The second class contains the results referred to the spaces $L^p(X)$, $1 \le p < \infty$. Here the situation is *qualitatively* different. As we shall see at this point the deep 'entropy' and 'stochastic' nature of the spectrum of weighted shift operator springs out. Namely the description of this phenomena is the goal of the subsequent part of the paper.

To start with we consider a model example clarifying the naturalness of the objects introduced hereafter.

Model example. Let (X, Ω, m) be a measurable space with a probability measure m and γ be a measurable, measure preserving mapping of X into itself, that is $m(\gamma^{-1}\omega) = m(\omega)$ for any $\omega \in \Omega$ (note that this condition implies in particular the equality $m(\gamma(X)) = 1$). Let us consider the space $L^p(X, m)$, $1 \le p < \infty$ and the algebra A of operators of multiplication by functions $a \in L^\infty(X, m)$. Set the *shift operator* T, (generated by the mapping γ) by the formula

$$(Tf)(x) = f(\gamma(x)), \quad f \in L^p(X, m). \tag{7.3.1}$$

We shall call the operator aT a *weighted shift* operator (with the weight a). From (7.3.1) it follows that

$$Ta = (a \circ \gamma)T, \quad a \in A. \tag{7.3.2}$$

Note that the mapping (7.3.1) defines an *endomorphism* of the algebra $L^\infty(X, m)$ that is $T(a \cdot b) = T(a) \cdot T(b)$, $a, b \in L^\infty(X, m)$. Let ξ be the partition X formed by the inverse images of γ that is

$$\xi = \{\gamma^{-1}(y)\}_{y \in X}.$$

We denote by $\xi(x)$ the element ξ containing x. Consider the *canonical factor space* $(X_\xi, \Omega_\xi, m_\xi)$ corresponding to the partition ξ and the set of *canonical conditional measures* $m^\tau(y)$, where $\tau = \xi(x)$ for some x. These objects are defined by the equality

$$\int_X f(x)\, dm(x) = \int_{X_\xi} dm_\xi(\tau) \int_\tau f(y)\, dm^\tau(y), \quad f \in L^p(X, m)$$

(the details see for example in [12], 1.5.).

We set the *conditional expectation* operator $\mathbf{E}$ in the space $L^\infty(X, m)$ by the formula

$$\mathbf{E}(a)(x) = \int_{\gamma^{-1}(x)} a(y)\, dm^{\gamma^{-1}(x)}(y), \tag{7.3.3}$$

which is defined for almost every x.

It is clear that $\mathbf{E}T(a) = a$, $a \in L^\infty(X, m)$ which means that $\mathbf{E}$ is a left inverse to T. Note that $\mathbf{E}(1) = 1$ and if $a \ge 0$ then $\mathbf{E}(a) \ge 0$.
The direct computation shows that

$$\|(aT)^n\|_{L^p(X, m)} = \left\| \mathbf{E}^n \left(\prod_{k=0}^{n-1} |a|^p \circ \gamma^k \right) \right\|_{L^\infty(X, m)}^{\frac{1}{p}}. \tag{7.3.4}$$

and therefore the problem of computation of the spectral radius of operator aT reduces to the calculation of the limit

$$r(aT) = \lim_{n\to\infty} \left\| \mathbf{E}^n\Big(\prod_{k=0}^{n-1} |a|^p \circ \gamma^k\Big) \right\|^{\frac{1}{pn}}.$$

The principal objects — abstract weighted shifts. Axiomatization. The considerations presented in the example make it natural the introduction and investigation of a number of objects that describe in an axiomatic way the weighted shift operators acting in L^p type spaces. To start with recall that if A is a semisimple commutative Banach algebra with an identity then the Gelfand transform establishes an isomorphism of this algebra with a certain subalgebra of the algebra $C(X)$ of continuous functions on the maximal ideal space X of the algebra A. We shall identify A with its image under the Gelfand transform. Recall also the following description of endomorphisms of these algebras.

Theorem 7.3.1 [13] *If $F : A \to A$ is an endomorphism of a semisimple commutative algebra A with an identity then there exists an open-closed subset $Y \subset X$ and a continuous mapping $\gamma : Y \to X$ such that*

$$(Fa)(x) = \chi_Y(x)a(\gamma(x)), \quad a \in A, \quad x \in X,$$

where χ_Y is a characteristic function of Y.
   In particular if $F(1) = 1$ then $Y = X$ and

$$(Fa)(x) = a(\gamma(x)). \tag{7.3.5}$$

Now let F be an endomorphism of the algebra $C(X)$ having the form (7.3.5) where $\gamma : X \to X$ is *onto* and $\mathbf{E}$ be a certain positive left inverse to F (that is $\mathbf{E}(a) \geq 0$, if $a \geq 0$), satisfying the condition $\|\mathbf{E}\| \leq 1$ (clearly under this condition we shall have $\|\mathbf{E}\| = 1$).
It is worth mentioning the following description of the operator $\mathbf{E}$.
For any point $x \in X$ let us consider the functional

$$\phi_x(a) = (\mathbf{E}a)(x). \tag{7.3.6}$$

Evidently ϕ_x is a positive functional and

$$\|\phi_x\| = 1, \tag{7.3.7}$$

which means that ϕ_x defines a probability measure ν_x on X. Since $\mathbf{E}(a \circ \gamma) = a$, it follows that $\nu_x(a \circ \gamma) = a(x)$ and therefore

$$\operatorname{supp} \nu_x \subset \gamma^{-1}x. \qquad (7.3.8)$$

Clearly the mapping $x \to \phi_x$ is weakly continuous on X.

It is also clear that every weakly continuous mapping $x \to \phi_x$, where ϕ_x are positive functional satisfying the conditions (7.3.7) and (7.3.8) defines a certain left inverse to F (7.3.5), that acts according to formula (7.3.6).

In particular, every operator $\mathbf{E}$ of this type satisfies the *homological* equation

$$\mathbf{E}(a \cdot (b \circ \gamma)) = b \cdot \mathbf{E}(a), \quad a, b \in A. \qquad (7.3.9)$$

Remark. In general a positive left inverse to F (if it exists) is not defined in a unique way.

Example. Let $X = \mathbf{R} \,(\mathrm{mod}\,1)$ and the mapping $\gamma : X \to X$ is given by the formula $\gamma(x) = 2x \,(\mathrm{mod}\,1)$. Take any continuous function ρ on X having the properties

$$0 \le \rho(x) \le 1, \quad x \in X,$$

$$\rho(x + \frac{1}{2}) = 1 - \rho(x), \quad 0 \le x \le \frac{1}{2}.$$

Set the operator $\mathbf{E}_\rho$ on $C(X)$ by the formula

$$(\mathbf{E}_\rho a)(x) = \sum_{t \in \gamma^{-1}(x)} a(t)\rho(t) = a(\frac{x}{2})\rho(\frac{x}{2}) + a([\frac{x}{2} + \frac{1}{2}])\rho([\frac{x}{2} + \frac{1}{2}]),$$

where $[x] = x \,(\mathrm{mod}\,1)$.

Clearly $\mathbf{E}_\rho$ is a positive left inverse to F.

It should be noted that in a concrete situation the form of the operator $\mathbf{E}$ is dictated by the functional space where the operator is considered (that is by the initial measure m).

Now the model example presented above and the foregoing notes make it natural the next

Definition 7.3.2   Let $A \subset L(B)$ be a subalgebra of the algebra $L(B)$ (of all linear continuous operator acting in a Banach space B) that is isometrically isomorphic to $C(X)$ for a certain compact X; and let F be an endomorphism of A of the form (7.3.5) and $\mathbf{E}$ be a certain fixed positive left inverse to F.

We shall say that an operator $T \in L(B)$ is an (abstract) *shift operator* (generated by F) and aT, $a \in A$ is an (abstract) *weighted shift operator (in a space of L^p, $1 \le p < \infty$ type)* if
1) the equality (7.3.2) holds,
and
2)

$$\|(aT)^n\|_{L(B)} = \left\| \mathbf{E}^n \left(\prod_{k=0}^{n-1} |a|^p \circ \gamma^k \right) \right\|_{C(X)}^{\frac{1}{p}}. \tag{7.3.10}$$

Remark 7.3.3   1) The model example presented above is a special case of a general scheme and the *complexity* of the spectral radius calculation in the general scheme is *equal* to that of a model example (cf. (7.3.10) (7.3.4)).

2) If γ is a homeomorphism (that is F is an automorphism) then $\mathbf{E}(a) = a \circ \gamma^{-1}$ and formula (7.3.10) transforms (for any p !!) into the formula

$$\|(aT)^n\|_{L(B)} = \left\| \left(\prod_{k=0}^{n-1} |a| \circ \gamma^k \right) \right\|_{C(X)}. \tag{7.3.11}$$

Precisely according to this formula there was calculated the norm $\|(aT)^n\|$ (cf. (7.2.5)) in the process of deducing the variational principle (7.2.4).

3) If $p \to \infty$ then formula (7.3.10) also transforms into (7.3.11) (for any γ !!) and this agrees with the fact (that has been already noted) that in the spaces of $C(X)$ and $L^\infty(X)$ type the variational principle (7.2.4) preserves its form.

4) Observe that for *any* given left inverse $\mathbf{E}$ to F there exists a realization of the objects mentioned in Definition 7.3.2 (for $p = 1$). Indeed, let $B = C(X)^*$, $T = \mathbf{E}^*$ and for any $a \in C(X)$ we define the operator $a : C(X)^* \to C(X)^*$ by the formula

$$(a\xi)f = \xi(a \cdot f), \quad \xi \in C(X)^*, f \in C(X),$$

where in the right-hand part $(a \cdot f)(x) = a(x) \cdot f(x)$.

Routine check shows that for a and T defined in this way all the conditions of Definition 7.3.2 (for $p = 1$) are satisfied.

Cocycles associated with endomorphisms. The principle properties. It turns out that in the situation when γ is not invertible the calculation of the spectral radius of weighted shift operators is *qualitatively* different from that in the above described situation of invertible γ. Namely the arising variational principles should contain along with the integrals with respect to invariant measures (like in Theorem 7.2.3) additional summands of entropy type nature.

The first essential progress in the calculation of the spectral radius of weighted shift operators with irreversible shifts has been achieved in the work by Ju.D. Latushkin and A.M. Stepin under a rather special assumption on the nature of the mapping γ. The rest of this section is devoted to the description and discussion of their results.

7.3.4 Assumption. Henceforth in this section we shall presuppose that the mapping γ mentioned in Definition 7.3.2 possesses the following property.

For every $x \in X$ the inverse image $\gamma^{-1}(x)$ consists of not more than a countable number of points.

7.3.5 *Under this assumption the set of functionals ϕ_x, $x \in X$ (7.3.6) defines a positive function ρ on X by the equation*

$$\phi_x(a) = \sum_{y \in \gamma^{-1}(x)} a(y)\rho(y). \tag{7.3.12}$$

The function ρ is called the *cocycle* (of the dynamical system (X, γ)).

Clearly if γ is a homeomorphism then $\rho \equiv 1$.

In the dynamical systems theory in the case when we have $|\gamma^{-1}(x)| \equiv$ const $= n$ for every $x \in X$ and $\rho = \frac{1}{n}$ the operator

$$(\mathbf{E}a)(x) = \phi_x(a) = \frac{1}{n} \sum_{y \in \gamma^{-1}(x)} a(y) \tag{7.3.13}$$

is customary called *Perron-Frobenius operator*. Following this motivation we shall also call the operator

$$(\mathbf{E}a)(x) = \sum_{y \in \gamma^{-1}(x)} a(y)\rho(y) \tag{7.3.14}$$

arising in the general situation (7.3.12) and even the abstract operator $\mathbf{E}$ mentioned in Definition 7.3.2 — Perron-Frobenius operators.

In fact the cocycle ρ has rather specific properties. Some of these properties are presented below in Lemmas 7.3.4 - 7.3.8. One can find the proofs of these results in [14].

Lemma 7.3.4   *If $\rho(x_0) = 0$ then ρ is continuous at x_0.*

A point x will be called a *local homeomorphism point* for γ iff there exists a neighbourhood $O(x)$ such that $\gamma : O(x) \to \gamma(O(x))$ is a homeomorphism.

Lemma 7.3.5   *If $\rho(x_0) \neq 0$ then ρ is continuous at x_0 if and only if x_0 is a local homeomorphism point.*

If every $x \in X$ is a local homeomorphism point then we say that γ is a *local homeomorphism*. Recall that the following general property is true.

Lemma 7.3.6   *If γ is a local homeomorphism then $|\gamma^{-1}(x)|$ is a locally constant function.*

Lemmas 7.3.5 and 7.3.6 imply the following

Corollary 7.3.7   *If $\rho(x)$ is a continuous nonvanishing function on X and X is a connected set then $|\gamma^{-1}(x)|$ is a constant function.*

Lemma 7.3.5 also shows that for the cocycle ρ the property of being continuous is valid only in rather special cases. The next Lemma 7.3.8 shows that fortunately at the points where ρ does not vanish the mapping γ behaves not 'too pathologically'.

We shall say that $x \in X$ is an $O-point$ ('open' point) if for any neighbourhood $O(x)$ its image $\gamma(O(x))$ contains some neighbourhood of $\gamma(x)$.

Lemma 7.3.8   *If $\rho(x_0) \neq 0$ then x_0 is an $O-point$.*

Spectral exponent.   Now let a, T be those described in Definition 7.3.2 and γ satisfies Assumption 7.3. In this situation (in view of (7.3.10), (7.3.6) and (7.3.12)) the spectral radius of aT is given by

$$
r(aT) = \lim_{n \to \infty} \|(aT)^n\|^{\frac{1}{n}} = \lim_{n \to \infty} \left\| \mathbf{E}^n \left(\prod_{k=0}^{n-1} |a|^p \circ \gamma^k \right) \right\|_{C(X)}^{\frac{1}{np}}
$$

$$
= \lim_{n \to \infty} \max_{x \in M} \left(\sum_{y \in \gamma^{-n}(x)} \prod_{i=0}^{n-1} (|a|^p \rho)(\gamma^i(y)) \right)^{\frac{1}{pn}}
$$

$$= \exp\left(\frac{1}{p}\ln \lim_{n\to\infty} \max_{x\in M}\left(\sum_{y\in\gamma^{-n}(x)}\prod_{i=0}^{n-1}(|a|^p\rho)(\gamma^i(y))\right)^{\frac{1}{n}}\right).$$

Set $C = |a|^p\rho$. The number $\lambda(\gamma, C)$ given by the formula

$$\lambda(\gamma, C) = \ln[r(aT)] = \frac{1}{p}\ln \lim_{n\to\infty} \sup_{x\in M}\left(\sum_{y\in\gamma^{-n}(x)}\prod_{i=0}^{n-1}C(\gamma^i(y))\right)^{\frac{1}{n}} \qquad (7.3.15)$$

(where we set $\ln 0 = -\infty$) is called the *spectral exponent*.

In the situation when γ is invertible we have that $\rho \equiv 1$ and $C \equiv |a|^p$. Therefore in this case the variational principle (7.2.4) for the spectral radius and (7.2.5) imply the variational principle for the spectral exponent $\lambda(\gamma, C)$ which looks as

$$\lambda(\gamma, C) = \frac{1}{p}\max_{\mu\in M_\gamma}\int_X \ln|C|\,d\mu = \max_{\mu\in M_\gamma}\int_X \ln|a|\,d\mu \qquad (7.3.16)$$

In the subsequent part of this section we shall deal with the variational principle for the spectral exponent (7.3.15). It turns out that in a number of situations the variational principle for the spectral exponent is closely related to the classical variational principles for the entropy and topological pressure. Therefore now we recall these notions and the corresponding principles.

Topological and metrical entropy. The variational principle linking the topological and metrical entropies was established by Dinaburg [15] and Goodman [16] and has the form

$$h(\gamma) = \sup_{\mu\in M_\gamma} h_\mu(\gamma) \qquad (7.3.17)$$

where on the left $h(\gamma)$ is the topological entropy of a topological dynamical system (X, γ) (X is a compact metric space and $\gamma : X \to X$ is a continuous mapping) and on the right $h_\mu(\gamma)$ is the metrical entropy for the metrical dynamical system (X, μ, γ) where μ belongs to the set M_γ of all probability measures that are invariant with respect to γ.

We do not present here the classical definition of the metrical entropy (the reader could find a lot of information on it in numerous sources (see, for example [12]) while we do recall the definition of the topological entropy since certain objects involved in this definition will be exploited in the subsequent part of the paper.

The definition of $h(\gamma)$ needs the introduction of the so-called (n, ε) spanning and (n, ε) separated subsets of X which are described in the following way.

For every $n \in N$ we consider the metric on X given by

$$d_n(x, y) = \max_{i=0, n-1} d(\gamma^i(x), \gamma^i(y))$$

where d is the initial metric on X. Thus with the help of d_n we are measuring the 'distance' between the segments of the trajectories of the points x and y of the length n.

If $\varepsilon > 0$ then we say that $E \subset X$ is (n, ε) *spanning* if it is an ε net for X with respect to the metric d_n that is

for any $x \in X$ there exists $x_0 \in E$ such that $d_n(x, x_0) < \varepsilon$.

A set $F \subset X$ is called (n, ε) *separated* if

for each pair of points $x, y \in F, x \neq y$ we have $d_n(x, y) > \varepsilon$.

Now the definition of the *topological entropy* looks as follows

$$h(\gamma) = \ln \lim_{\varepsilon \to 0} \overline{\lim_{n \to \infty}} \inf \left\{ |E|^{\frac{1}{n}} : E - (n, \varepsilon) spanning \right\} =$$

$$= \ln \lim_{\varepsilon \to 0} \overline{\lim_{n \to \infty}} \sup \left\{ |F|^{\frac{1}{n}} : F - (n, \varepsilon) separated \right\}. \tag{7.3.18}$$

Topological pressure.   The next classical variational principle which will be exploited here deals with the so-called topological pressure.

The topological pressure is a generalization of the notion of the topological entropy. Namely for every function $C(x)$ which is positive and continuous on X the *topological pressure* $P(\gamma, \ln C)$ is given by the formula (see [21]):

$$P(\gamma, \ln C)$$

$$= \ln \lim_{\varepsilon \to 0} \overline{\lim_{n \to \infty}} \inf \left\{ \left(\sum_{y \in E} \prod_{i=0}^{n-1} C(\gamma^i(y)) \right)^{\frac{1}{n}} : E - (n, \varepsilon) spanning \right\}$$

$$= \ln \lim_{\varepsilon \to 0} \overline{\lim_{n \to \infty}} \sup \left\{ \left(\sum_{y \in F} \prod_{i=0}^{n-1} C(\gamma^i(y)) \right)^{\frac{1}{n}} : F - (n, \varepsilon) separated \right\}.$$

$$(7.3.19)$$

Remark.

$$h(\gamma) = P(\gamma, 0) \qquad (7.3.20)$$

Ruelle [20] and Walters [21] proved the following variational principle for the topological pressure:

$$P(\gamma, a) = \sup_{\mu \in M_\gamma} \left[\int_X a \, d\mu + h_\mu(\gamma) \right]. \qquad (7.3.21)$$

Spectral exponent and topological pressure. Looking at the formulae (7.3.15) and (7.3.19) defining the spectral exponent and the topological pressure one could notice their certain similarity and indeed in a number of situations there can be established an interrelation between these objects. Namely Latushkin and Stepin [17; 18; 19] proved the next

Theorem 7.3.9 *If γ is the shift on a unilateral topological Markov chain (in particular, if γ is an expanding m-sheeted cover of a manifold X) then*

$$\lambda(\gamma, C) = P(\gamma, \ln |C|). \qquad (7.3.22)$$

Thus in this situation (in view of (7.3.21)) we have the following variational principle for the spectral exponent $\lambda(\gamma, C)$:

$$\lambda(\gamma, C) = \ln[r(aT)] = \frac{1}{p} \sup_{\mu \in M_\gamma} \left[\int_X \ln |C| \, d\mu + h_\mu(\gamma) \right] =$$

$$\sup_{\mu \in M_\gamma} \left[\int_X \ln |a| \, d\mu + \frac{1}{p} \int_X \ln \rho \, d\mu + \frac{h_\mu(\gamma)}{p} \right] \qquad (7.3.23)$$

Remark 7.3.10 1) In general (that is for any γ satisfying Assumption 7.3) the equality (7.3.22) is not true and (7.3.23) is *not* a generalization of the variational principle (7.2.4). For example if we consider an invertible mapping γ and set $C \equiv 1$ then

$$\lambda(\gamma, 1) = 0$$

while (recall (7.3.20))

$$P(\gamma, 0) = h(\gamma)$$

and in general $h(\gamma)$ could be equal to any nonnegative number.

2) In fact Latushkin and Stepin in their papers [17; 18; 19] considered only the operators with the cocycles of a *concrete* form and *did not* use the notion of the cocycle at all. Theorem 7.3.9 in its present form was obtained in [22; 23] by means of newly introduced topological invariants that also gave a number of estimates for the spectral radius. In addition it was shown in [22; 23] that the variational principle (7.2.4) and Latushkin-Stepin result are in a way the 'extreme points' of the situations one could come across when dealing with the calculation of the spectral exponent $\lambda(\gamma, C)$.

3) If $p \to \infty$ then formula (7.3.23) transforms into (7.3.11) and this agrees with the fact (that has been already noted) that in the spaces of $C(X)$ and $L^\infty(X)$ type the variational principle (7.2.4) preserves its form.

7.4 Dynamical Nature of the Spectral Radius of Operators with Irreversible Shifts — Present State: The Duality Principle

As it has been already observed the Latushkin-Stepin theorem does not solve totally the problem of the spectral radius calculation. In fact if the mapping γ does not satisfy Assumption 7.3 then even the cocycle ρ does not exist. So all the previous methods fail and we are faced with a completely new situation. Namely the investigation of this case is the theme of the present section. We shall find out that a new dynamical invariant of the 'entropy-stochastic' nature comes into play to solve the problem of the spectral radius calculation in full and the solution will be established by means of the duality principle linking this invariant and the spectral exponent.

The main object here is an abstract weighted shift operator described in Definition 7.3.2 and we do not presume any constraints on the form of γ or $\mathbf{E}$.

In essence the results of this section are taken from [24; 25].

We start with the introduction and discussion of a new entropy-stochastic type dynamical invariant.

T-entropy. Let M be the set of all Borel probability measures on X and as above we denote by M_γ the set of all γ-invariant measures from M.

Definition 7.4.1   Let $\mu \in M$. We shall call a finite Borel partition $\mathcal{D} = \{D_1, ..., D_k\}$ of X μ−*proper* if every its element has a non empty interiour and the border of every element has μ-measure zero.

Definition 7.4.2   The T-*entropy* $\tau(\mu)$ of a measure $\mu \in M$ with respect to a given operator $\mathbf{E}$ introduced in Definition 7.3.2 is a number

$$\tau(\mu) = \inf_{n,\mathcal{D}} \frac{\tau_n(\mu, \mathcal{D})}{n}, \quad \text{where } \tau_n(\mu, \mathcal{D}) = \sup_{\nu \in M} \sum_{D \in \mathcal{D}} \mu(D) \ln \frac{\mathbf{E}^{*n}\nu(D)}{\mu(D)}.$$

$$(7.4.1)$$

Here the infimum is taken over all μ-proper partitions $\mathcal{D}$. We assume that if $\mu(D) = 0$ then the summand in (7.4.1) corresponding to the set D is zero regardless the value of $\mathbf{E}^{*n}\nu(D)$.

The principal properties of T-entropy are listed in the next

Theorem 7.4.3   1) $\tau(\mu) \leq 0$ *for any measure* $\mu \in M$.
2) *If* γ *is a homeomorphism then* $\tau(\mu) = 0$ *for any measure* $\mu \in M$.
3) $\tau(\mu)$ *is a concave and upper semicontinuous function on* M.

In some particular situations we can obtain the description of $\tau(\mu)$ in a simpler way.

Theorem 7.4.4   *Let* γ *and* $\mathbf{E}$ *be such that the cocycle* ρ *introduced in 7.3 is a continuous and nonvanishing function (in particular this means that* γ *is a local homeomorphism). Then for every* $\mu \in M_\gamma$ *we have*

$$\tau(\mu) = \inf_{n,\mathcal{D}} \sum_{D \in \mathcal{D}} \mu(D) \ln \frac{\mathbf{E}^{*n}\mu(D)}{\mu(D)}. \qquad (7.4.2)$$

So here to 'calculate' $\tau(\mu)$ we do not need any measures apart from μ itself.

Remark. In the proof of this result along with the authors of the paper E.I. Vatkina also took part.

Conjecture. *Formula (7.4.2) is true for* $\mu \in M_\gamma$ *without any constraint on* γ *and* $\mathbf{E}$.

In the concrete case considered in Model example (Section 7.3) one can also obtain some additional information on $\tau(\mu)$. Namely let m be a Borel probability γ-invariant measure on X whose support is equal to X and $\mathbf{E}$ be the operator given by (7.3.3). It is clear from (7.4.1) that in this situation we have $\tau(m) = 0$. Thus according to statement 1) of Theorem 7.4.3 the function $\tau(\cdot)$ attains its maximum on M at the point m.

Theorem 7.4.5 *Let X, γ, m and $\mathbf{E}$ be those described above. If m is a unique maximum point of the function τ on M then m is ergodic.*

The duality principle. The next result is the main statement of this section — it establishes the duality between T-entropy and the spectral exponent of the weighted shift operator.

Theorem 7.4.6 (Duality principle) *For the spectral radius of the operator aT, introduced in Definition 7.3.2, the following formula holds*

$$\ln[r(aT)] = \max_{M_\gamma} \left(\int_X \ln|a| \, d\mu + \frac{\tau(\mu)}{p} \right). \qquad (7.4.3)$$

Remark 7.4.7 1) In view of the statement 2) of Theorem 7.4.3 in the situation when γ is a homeomorphism the duality principle established coincides with formula (7.2.4).

2) Let $\varphi = \ln|a|$ and $\lambda(\varphi) = \ln[r(aT)]$. Then formula (7.4.3) means that the spectral exponent $\lambda(\varphi)$ is the Fenchel-Legendre transform of the function $-\frac{\tau(\mu)}{p}$. So $\lambda(\varphi)$ and $-\frac{\tau(\mu)}{p}$ are (Fenchel-Legendre) dual objects.

In fact the duality established in this theorem is much deeper than simply the introduction of the Fenchel-Legendre convex objects into the procedure of the spectral radius calculation. Much more important (from our point of view) is the observation that formula (7.4.3) links the spectral characteristics of the weighted shift operator (the left-hand part) with the stochastic characteristics ($\tau(\mu)$ in the right-hand part) of the dynamical system. This ideology will be developed further in the next section where in particular the interrelation between $\tau(\mu)$ and the *action functional* will be examined.

3) Formula (7.4.3) reveals the partition of the process of calculation of the spectral radius into the *static component* (the first summand in the right hand part depends only on the weight a) and the *dynamical component* (the second summand depends only on the shift γ).

4) If $p \to \infty$ then the formula (7.4.3) transforms into formula (7.2.4) and this restores the variational principle (7.2.4) for the spaces of $C(X)$ and $L^\infty(X)$ type for an *arbitrary* γ!

5) Let $\mathbf{E}$ be the (Perron-Frobenius) operator mentioned in Definition 7.3.2 (see also 7.3) and $a \in C(X)$. The operator $\mathbf{E}_a : C(X) \to C(X)$ of the

form

$$\mathbf{E}_a(f) = \mathbf{E}(af), \quad f \in C(X) \tag{7.4.4}$$

is reasonable to call the *weighted Perron-Frobenius* operator.

It is evident from Definition 7.3.2 that the spectral radii of the operators aT and $\mathbf{E}_{|a|}$ coincide (for $p = 1$) therefore the duality principle means that

$$\ln[r(\mathbf{E}_{|a|})] = \max_{M_\gamma} \left(\int_X \ln|a|\, d\mu + \tau(\mu) \right). \tag{7.4.5}$$

The convex duality established in Theorem 7.4.6 and the formalism developed in [26] leads to natural introduction of the thermodynamical 'ideology' into the spectral analysis of weighted shift operators. Having in mind this motivation it is reasonable to call the measures μ at which the maximum in the right-hand part of (7.4.3) or (7.4.5) is attained the *equilibrium states*. We recall in this connection that in accordance with a common physical point of view the equilibrium states are the states at which the system 'exists in reality'. From this point of view the duality principle adds dialectics to the spectral analysis of weighted shift operators: since $\tau(\mu)$ describes the measure of the 'most typical' empirical trajectories (see, in particular, the next section) and the value $\int_X \ln|a|\, d\mu$ calculates the 'living conditions' (recall the Introduction to this article) then the duality principle tells us that that the process realizes at a point (equilibrium state) having the best combination of these components.

Theorem 7.4.8　*Let $\varphi = \ln|a| \in C(X)$ be a fixed function. The spectral exponent $\lambda(\cdot)$ (as a function on $C(X)$) possesses the Gateux derivative λ' at φ iff an equilibrium state μ in (7.4.3) is unique. In this case*

$$\lambda'(\varphi)(h) = \int_X h\, d\mu, \quad h \in C(X)$$

This theorem along with Proposition 7.4.5 imply the next

Corollary 7.4.9　*Let X, γ, m and $\mathbf{E}$ be those considered in Proposition 7.4.5. If $\lambda(\cdot)$ possesses the Gateux derivative at $\varphi = 0$ then m is ergodic.*

7.5 Perron–Frobenius Dynamical Systems

It is clear that the spectral exponent $\lambda(\varphi) = \ln[r(aT)]$ (where $\varphi = \ln|a|$) characterize the first approximation to the asymptotics of the semigroup generated by the weighted shift operator aT. On the other hand as we shall see further T-entropy describes the rough asymptotics of the distribution of the set of *empirical measures* that are defined in the following way.

Let X be a compact topological space and $\gamma : X \to X$ be a continuous mapping. Take any point $x \in X$ and consider its trajectory $\{\gamma^n(x)\}_{n \in \mathbf{N}}$. It is customary to call the measure $\delta_{x,n}$, $n = 1, 2, \ldots$ of the form

$$\delta_{x,n} = \frac{1}{n}\left(\delta_x + \delta_{\gamma(x)} + \cdots + \delta_{\gamma^{n-1}x}\right).$$

empirical. For any $f \in C(X)$ we have

$$\int_X f \, d\delta_{x,n} = \delta_{x,n}(f) = \frac{1}{n}\left(f(x) + f(\gamma(x)) + \cdots + f(\gamma^{n-1}(x))\right).$$

Let us consider any probability Borel measure μ on X and denote by $O(\mu)$ a (small) neighbourhood of μ (in *-weak topology on the set of measures). It turns out that T-entropy $\tau(\mu)$ is linked with the asymptotics of the expression $m\{x \in X \mid \delta_{x,n} \in O(\mu)\}$, $m \in M$ when $n \to \infty$. The mentioned asymptotics is simply the asymptotics of the measure m of the set of initial conditions for which the empiric measures are near μ. In fact it has been proved (and this is the basic moment in the proof of the duality principle) that the measure $m\{x \in X \mid \delta_{x,n} \in O(\mu)\}$ for large n does not exceed $e^{n(\tau(\mu)+\varepsilon)}$, where a positive number ε can be taken arbitrary small by diminishing the neighbourhood $O(\mu)$. This implies in particular that if $\tau(\mu) < 0$ then empiric measures that are close to μ can be observed only with exponentially small probability.

Of course one would like T-entropy to give not only the upper estimate for empirical measures but also the lower one. In other words one would like to have the rough asymptotics (of the probability of observing empirical measures that are close to μ) of the form $e^{n\tau(\mu)}$. In the probability theory the functional $\tau(\mu)$ possessing the mentioned property is customary to call the *action functional* (for empirical measures). It has to be stressed at once that T-entropy falls far short of coinciding with the action functional for any dynamical system. For example in the case of invertible dynamical systems we have (see Theorem 7.4.3) $\tau(\mu) \equiv 0$ although as a rule for the majority of measures μ the corresponding asymptotics is exponentially small. Nevertheless it has been found a relatively wide class of dynamical systems for which T-entropy really determines in a proper way the asymptotics we are looking for. True enough that this is correct not for all empirical measures but only for those that are near to the *equilibrium measures* (recall that the equilibrium measures are the measures at which the maximum in the duality principle (7.4.3) is attained at least for some function $\varphi = \ln |a|$).

The systems of this class are called Perron-Frobenius dynamical systems.

Before passing to their definition we would like to recall that as a rule in L^p spaces the spectrum of a weighted shift operator is invariant with respect to rotations around zero and therefore the point $r(aT)$ (spectral radius) does not differ in any special way from other points on the boundary of the spectrum, on the other hand in the spaces of smooth functions it is quite often that this point turns to be an isolated eigen value (see, for example, [3], §13) and the same is true for weighted Perron-Frobenius operators (whose spectral radius do coincide with the spectral radius of weighted shift operators (see part 5) in Remark 7.4.7)). The following definition of Perron-Frobenius dynamical systems do exploits the mentioned effect.

Let γ be a continuous mapping of a compact topological space X into itself and $\mathcal{C}$ be a γ-invariant Banach algebra consisting of continuous functions on X and such the $\mathcal{C}$ is uniformly dense in $C(X)$.

Definition 7.5.1   The dynamical system (X, γ) is called *Perron-Frobenius dynamical system* if there exist a Banach module $\mathcal{F}$ over $\mathcal{C}$ (the so called *state module*) and a linear bounded operator $W : \mathcal{F} \to \mathcal{F}$ (*evolution operator*) such that

   I. For any function $f \in \mathcal{C}$ and any state $v \in \mathcal{F}$ the following homological identity is valid: $W(f \circ \gamma \cdot v) = f W v$ (or for any function $f \in \mathcal{C}$ and any state $v \in \mathcal{F}$ the identity $W(fv) = f \circ \gamma \cdot W v$ is valid).

   II. There exist a linear functional $\nu : \mathcal{F} \to \mathbf{R}$, a state $h \in \mathcal{F}$ and a number $\lambda \in \mathbf{R}$ such that

     a) the following equalities are true: $Wh = e^\lambda h$, $\nu \circ W = e^\lambda \nu$, $\nu(h) = 1$;

     b) the spectral radius of the restriction of W onto $\ker \nu$ is less than e^λ;

     c) if a function $f \in \mathcal{C}$ is nonnegative then $\nu(fh) \geq 0$.

It is clear from a) and b) of part II of the definition that e^λ is the spectral radius of the operator W (in fact we have more — e^λ is an eigen value bounded away from the other part of the spectrum).

Remark 7.5.2   As state modules $\mathcal{F}$ there can emerge various functional spaces on X or the spaces of real-valued measures on X or the spaces of linear functionals on certain operator algebras (containing W as an element) as well as more complicated objects. As the typical examples of W one can take the shift operator T, the conditional expectation operator $\mathbf{E}$ (7.3.3) (or its abstract analogue considered in Definition 7.2.2) and in particular the classical Perron-Frobenius operator (7.3.13).

Examples. 1) Let γ be a locally expanding mapping of a compact connected Riemannian manifold X. In this case one can take as $\mathcal{F}$ the space of Holder functions (or the space of continuously differentiable functions) and as W — Perron-Frobenius operator (7.3.13) (or (7.3.14)).

2) Let γ be the shift on the space X of sequences of Bernoulli tests or the shift on the space of sequences of states of finite ergodic Markov chain. Here as $\mathcal{F}$ one can take the space of Holder functions and as W — the conditional expectation operator (7.3.3) or Perron-Frobenius operator (7.3.13). This example will be examined further in the subsequent part of this section.

3) If γ is 'bakers transform' or a topologically mixing Anosov diffeomorphism then the structure of $\mathcal{F}$ is essentially more complicated. In this case $\mathcal{F}$ consists of the so-called *foliated functions* (see [27; 28]) who look like ordinary functions along the stable foliation and as measures along the unstable one. As the evolution operator W one can take here the ordinary shift.

Let (X, γ) be a Perron-Frobenius dynamical system. By means of the linear functional ν mentioned in Definition 7.5.1 we can define a linear functional $\mu(f) = \nu(fh), \quad f \in \mathcal{C}$. Clearly the functional μ is positive (takes nonnegative values on nonnegative functions) and normalized (that is $\mu(1) = 1$). It can be expanded by the continuity from $\mathcal{C}$ onto $C(X)$. And in view of the Riesz theorem any positive normalized functional μ on $C(X)$ can be identified with a Borel probability measure μ on X so that $\mu(f) = \int_X f \, d\mu$ for any function $f \in C(X)$. The homological identity implies

$$\mu(f) = \nu(fe^{-\lambda}Wh) = \nu\big(e^{-\lambda}W(f \circ \gamma \cdot h)\big) = \mu(f \circ \gamma).$$

Which means that μ is γ-invariant.

Along with the evolution operator W for any function $\varphi \in C$ we can define the *weighted evolution* operator $W_\varphi : \mathcal{F} \to \mathcal{F}$ by the formula

$$W_\varphi v = W(e^\varphi v), \quad v \in \mathcal{F} \tag{7.5.1}$$

Remark. We would like to note that in view of the foregoing observations the weighted evolution operator can be considered as a generalization of a weighted shift operator so also as a generalization of a weighted Perron-Frobenius operator (recall part 5) in Remark 7.4.7) and in this connection we have to stress once more that the spectral radii of weighted shift operators and of the corresponding weighted Perron-Frobenius operators do coincide.

The foregoing results in essence are presented in [29; 30].

Theorem 7.5.3 *For any Perron-Frobenius dynamical system (X, γ) there exists a neighbourhood of zero $\mathcal{U} \subset C$ such that for every function $\varphi \in \mathcal{U}$ the weighted evolution operator W_φ (7.5.1) satisfies Definition 7.5.1 (that is one can take in this definition W_φ instead of W) and one can choose the corresponding linear functional ν_φ, state h_φ, number $\lambda_\varphi = \lambda(\varphi)$ and measure μ_φ to be analytically dependent on $\varphi \in C$. Moreover the function $\lambda(\varphi)$ depends convexly on $\varphi \in \mathcal{U}$ and the following identity for the measure μ_φ is valid:*

$$\mu_\varphi(f) = \nu_\varphi(f h_\varphi) = \left.\frac{d\lambda(\varphi + tf)}{dt}\right|_{t=0}, \qquad f \in C.$$

Thus the measure μ_φ coincides with the derivative $\lambda'(\varphi)$ and is an equilibrium measure corresponding to the function φ.

Let us define for any function $\varphi \in \mathcal{U}$ and measure $m \in M$ (M is the set of all Borel probability measures on X) the *action functional*

$$\tau_\varphi(m) = \inf_{\psi \in \mathcal{U}} \left\{ \lambda(\psi) - \lambda(\varphi) - \int_X (\psi - \varphi)\, dm \right\}. \tag{7.5.2}$$

Observe that if $W = \mathbf{E}$ is the conditional expectation operator (7.3.3) and $\varphi \equiv 0$ then this functional coincides with T-entropy of the Perron-Frobenius dynamical system (X, γ). In the general case the functional $\tau_\varphi(m)$ differs only by sign from the Fenchel-Legendre transform of the function $\Lambda(\psi) = \lambda(\varphi + \psi) - \lambda(\varphi)$.

Evidently $\tau_\varphi(m) \leq 0$. It turns out that for the equilibrium measures m the infimum in (7.5.2) can be easily calculated and the action functional takes especially simple form.

Theorem 7.5.4 *If for a Perron-Frobenius dynamical system a measure m coincides with an equilibrium measure $\mu_\psi = \lambda'(\psi)$ then*

$$\tau_\varphi(\mu_\psi) = \lambda(\psi) - \lambda(\varphi) - \mu_\psi(\psi - \varphi) = \lambda(\psi) - \lambda(\varphi) - \lambda'(\psi)(\psi - \varphi). \quad (7.5.3)$$

A measure $m \in M$ will be called a *limit measure* (and we denote this by $m \in M_{\lim}$) if this measure is the limit (in *-weak topology on M) of some sequence of empirical measures δ_{x_k, n_k} where $n_k \to \infty$. Evidently the set $M_{\lim}$ of limit measures is closed. In addition any limit measure is γ-invariant. This follows from the fact that for any function $f \in C(X)$

$$\lim_{n_k \to \infty} \left(\delta_{x_k, n_k}(f \circ \gamma) - \delta_{x_k, n_k}(f) \right) = \lim_{n_k \to \infty} \frac{1}{n_k} \left(f(\gamma^{n_k} x_k) - f(x_k) \right) = 0.$$

Returning back to the discussion of the interrelation between asymptotics of empirical measures and T-entropy we shall examine the following problem. Let $\mu_\varphi = \lambda'(\varphi)$ be an equilibrium measure and m be an arbitrary Borel probability measure on X. Let us consider a (small) neighbourhood $O(m)$ in *-weak topology on M. What is the asymptotics of the value $\mu_\varphi\{ x \in X \mid \delta_{x,n} \in O(m) \}$ when $n \to \infty$? The answer is given in the following three theorems.

Theorem 7.5.5 *Let (X, γ) be an arbitrary dynamical system on a compact space X. Then for any nonlimit measure $m \in M \setminus M_{\lim}$ there exists a (small) neighbourhood $O(m) \subset M$ and a (large) natural N such that $\delta_{x,n} \notin O(m)$ for all $n \geq N$ and $x \in X$.*

Theorem 7.5.6 *Let (X, γ) be a Perron-Frobenius dynamical system and μ_φ be an equilibrium measure corresponding to a function $\varphi \in \mathcal{U}$. Then for any limit measure $m \in M_{\lim}$ and any $\varepsilon > 0$ there exists a (small) neighbourhood $O(m) \subset M$ such that*

$$\limsup_{n \to \infty} \frac{1}{n} \ln \mu_\varphi\{ x \mid \delta_{x,n} \in O(m) \} < \tau_\varphi(m) + \varepsilon.$$

Theorem 7.5.7 *Let (X, γ) be a Perron-Frobenius dynamical system and μ_φ and μ_ψ be two equilibrium measures corresponding to functions $\varphi, \psi \in \mathcal{U}$ respectively. Suppose that the corresponding states h_φ, h_ψ and functionals ν_φ, ν_ψ from Definition 7.5.1 satisfy the enequalities $\nu_\psi(h_\varphi) > 0$ and $\nu_\varphi(h_\psi) > 0$. Then for any neighbourhood $O(\mu_\psi) \subset M$ we have*

$$\liminf_{n \to \infty} \frac{1}{n} \ln \mu_\varphi\{ x \mid \delta_{x,n} \in O(\mu_\psi) \} \geq \tau_\varphi(\mu_\psi).$$

Corollary 7.5.8 *Any equilibrium measure is a limit measure.*

Thus for any equilibrium measure m we have obtained the rough asymptotics of the probability (in the sense of μ_φ) to observe an empirical measure that is close to m. This probability is equal to $e^{n\tau_\varphi(m)}$ with an accuracy of subexponentially growing factor. If m is not a limit measure then this probability is zero. If m is a limit but not equilibrium measure then we have only the upper estimate for the probability considered.

Example. As an example of Perron-Frobenius dynamical system let us consider the sequence of Bernoulli independent tests. Let X be the space of all sequences $x = \{x_1, x_2, \dots\}$ where every coordinate x_i can take values from 1 to n. Define the shift γ on X by mapping a sequence $\{x_1, x_2, \dots\}$ into the sequence $\{x_2, x_3, \dots\}$. Let us fix any number $\Lambda \in (0, 1)$ and introduce the metric $\rho(x, y)$ on X by setting $\rho(x, y) = \Lambda^{i(x,y)}$ where $i(x, y) = \min\{j \mid x_j \neq y_j\}$. Let $\mathcal{C}$ be the Banach algebra consisting of bounded functions on X that satisfy the Lipschitz condition. For any function $\varphi \in \mathcal{C}$ we define the (weighted) Perron-Frobenius operator W_φ on $\mathcal{C}$ (recall part 5) in Remark 7.4.7) by the formula:

$$[W_\varphi f](x_1, x_2, \dots) = \sum_{x_0=1}^{n} e^{\varphi(x_0, x_1, x_2, \dots)} f(x_0, x_1, x_2, \dots).$$

This operator satifies the homological identity $W_\varphi(f \cdot g \circ \gamma) = g W_\varphi f$.

Let us consider the functions φ of the form $\varphi(x) = \varphi(x_1)$ that is the functions depending only on the first coordinate of the sequence x. In fact these functions are determined by the set of n values $\varphi_i = \varphi(i)$, $i = 1, \dots, n$ (i is a possible value of x_1). Observe that for the operator W_φ with the mentioned φ all the conditions of Definition 7.5.1 are satisfied (if we take $\mathcal{F} = \mathcal{C}$).

Indeed. For these φ the operator W_φ possesses the eigenfunction $h \equiv 1$ with the eigenvelue $\sum_{i=1}^{n} e^{\varphi_i}$. Let us introduce the notation

$$\lambda_\varphi = \ln\left\{ \sum_{i=1}^{n} e^{\varphi_i} \right\} \qquad p_i = \frac{d\lambda(\varphi)}{d\varphi_i} = \frac{e^{\varphi_i}}{\sum_{i=1}^{n} e^{\varphi_i}}. \tag{7.5.4}$$

Using this notation we can write the operator $e^{-\lambda(\varphi)} W_\varphi$ in the form

$$[e^{-\lambda(\varphi)} W_\varphi f](x_1, x_2, \dots) = \sum_{x_0} p_{x_0} f(x_0, x_1, x_2, \dots),$$

and its k-th power has the form

$$\left[(e^{-\lambda(\varphi)}W_\varphi)^k f\right](x_{k+1}, x_{k+2}, \dots) =$$

$$\sum_{x_1=1}^{n} \cdots \sum_{x_k=1}^{n} p_{x_1} \cdots p_{x_k} f(x_1, \dots, x_k, x_{k+1}, \dots). \quad (7.5.5)$$

It follows easily from (7.5.5) that if $f \in \mathcal{C}$ then the sequence $\left(e^{-\lambda(\varphi)}W_\varphi\right)^k f$ tends to the constant function $\int_X f \, d\mathbf{P}_\varphi$ where P_φ is the Bernoulli probability measure on X corresponding to the probabilities of the elementary outcomes: $p_1, \dots, p_n$. One can prove that the sequence of operators $\left(e^{-\lambda(\varphi)}W_\varphi\right)^k$ on the space $\mathcal{C}$ (of Lipschitz functions) tends in norm to the projection $\bar{W}_\varphi f = \int_X f \, d\mathbf{P}_\varphi$. Thus W_φ satisfies all the conditions of Definition 7.5.1 (with $\mathcal{F} = \mathcal{C}$), and the Bernoulli measure P_φ coincides with the equilibrium measure μ_φ.

Taking $\varphi_i = \ln p_i$ we make sure that any Bernoulli measure P on X (with positive probabilities of the elementary outcomes) coincides with the equilibrium measure μ_φ.

Let us consider one more Bernoulli measure Q with positive probabilities of the elementary outcomes: $q_1, \dots, q_n$. Then $Q = \mu_\psi$, where $\psi_i = \ln q_i$. Let us calculate the action functional $\tau_\varphi(\mu_\psi)$. According to (7.5.3), (7.5.4) we have

$$\tau_\varphi(\mu_\psi) = \lambda_\psi - \lambda_\varphi - \mu_\psi(\psi - \varphi) = -\mu_\psi(\psi - \varphi)$$

$$= -\sum_{i=1}^{n} q_i(\ln q_i - \ln p_i) = \sum_{i=1}^{n} q_i \ln \frac{p_i}{q_i}.$$

The expression obtained characterize the asymptotics of the probability (in the sense of P) to observe an empiric measure that is close to Q. Note that the same asymptotics follows from the local limit theorem for Bernoulli tests [31, Chapter 5, §2, Theorem 6].

Remark 7.5.9 We finish the paper with the note that the spectral analysis of weighted shift operators is not only of great value in the dynamical systems theory (which was the theme of the present article) but also in many other fields of mathematics and its applications and in particular in the theory of solvability and Fredholm solvability of functional differential equations (see [32; 33]).

Bibliography

[1] P. Walters, *An introduction to ergodic theory* // Springer-Verlag, 1982.

[2] W.B. Arveson, Operator algebras and measure preserving automorphisms // Acta. Math., **118** 1967, 95-109.

[3] A. Antonevich, *Linear functional equations. Operator approach* // Minsk, Universitetskoe, 1988 (Russian); English transl. Birkhauser Verlag, 1996.

[4] A. Antonevich, A. Lebedev, *Functional differential equations: I. C^* - theory* // Longman Scientific & Technical, 1994.

[5] A. Antonevich, A. Lebedev, On spectral properties of operators with a shift // Izv. AN SSSR. Ser. Mat., **47** 1983, No 5, 915-941 (Russian).

[6] A. Antonevich, On a class of pseudodifferential opearators with deviating argument on the torus // Diff. Uravnenija, **11** 1975, No 9, 1550-1557 (Russian).

[7] A. Antonevich, Operators with a shift generated by the action of a compact Lie group // Sibirsk. Mat. Zh., **20** 1979, No 3, 467-478 (Russian).

[8] A. Lebedev, On the invertibility of elements in C^*−algebras generated by dynamical systems // Uspekhi Mat. Nauk., **34** 1979, No 4, 199-200 (Russian).

[9] A. Kitover, On the spectrum of an automorphism with a weight and Kamowitz-Sheinberg theorem // Funktsion. Anal. i Prilozhen, **13** (1979) No 1, 70-71 (Russian).

[10] S. Parrot, Weighted translation operators: Ph. D. dis., Univ. of Michigan, Ann Arbor Mich., 1965.

[11] A. Lebedev, O. Maslak, Why the spectrum of a noninvertible weighted shift is a disk // Spectral and evolutionary problems (Proceedings of the Eight Crimean Autumn Mathematical School-Symposium Simferopol), **9** 1999, 26 - 32.

[12] N. Martin, J. England, *Mathematical theory of entropy* // Addison-Wesley 1981.

[13] S. Lo, Weighted shift operators in certain Banach spaces of functions: Diss. kand. fiz.-mat. nauk, Minsk, 1981 (Russian).

[14] A.V. Lebedev, On certain properies of cocycles associated with irreversible mappings of compacts // Operators and operator equations (collection of scientific papers), Novocherkassk university, 1995, 31-38 (Russian).

[15] E. I. Dinaburg, The relation between different entropy characteristics of dynamical systems // Izv. AN SSSR, Ser. Mat. **35** (1971), No 2, 324-366 (Russian).

[16] T. N. T. Goodman, Relating topological entropy and measure entropy // Bull. London Math. Soc., (1971), No 3, 176-180.

[17] Ju.D. Latushkin, On the integral functional operators with a nonbijective shift // Izv. AN SSSR. Ser. Mat., **45** 1981, No 6, 1241-1257 (Russian).

[18] Ju.D. Latushkin, A.M. Stepin, Weighted shift operators on a topological Markov chain // Funktsion. Anal. i Prilozhen, **22** 1988, No 4, 86-87 (Russian).

[19] Ju. D. Latushkin, A. M. Stepin, Weighted shift operators and linear extensions of dynamical systems // Uspekhi. Mat. Nauk., **46** (1991), No 2, 85-143 (Russian).

[20] D. Ruelle, Statistical mechanics on a compact set with Z^ν action satisfying expansiveness and specification // trans. Amer. Math. Soc., **185** (1973), 237-252.

[21] P. Walters, A variational principle for the pressure on continuous transformations // Amer. J. Math. **97** 1975, No 4, 937-971.

[22] A. Lebedev, O. Maslak, the variational principles for the spectral characteristics of the operators generated by dynamical systems // Proceedings of the Fifth Annual Seminar NPCS'96, Minsk, 1997, 165-170.

[23] A. Lebedev, O. Maslak, The spectral radius of a weighted shift operator, variational principles, entropy and topological pressure // Spectral and evolutionary problems (Proceedings of the Eight Crimean Autumn Mathematical School-Symposium Simferopol), **8** 1998. 26-34.

[24] A.B. Antonevich, V.I. Bakhtin, A.V. Lebedev, Variational principle for the spectral radius of weighted composition and weighted mathematical expectation operators // Doklady NAN Belarusi **44** 2000, No 6. 7–10 (Russian).

[25] A.B. Antonevich, V.I. Bakhtin, A.V. Lebedev, Variational principle for the spectral radius of weighted composition and Perron-Frobenius operators // Trudy Instituta Matematiki NAN Belarusi **5** 2000, 13-17 (Russian).

[26] A.B. Antonevich, V.I. Bakhtin, A.V. Lebedev, D.S. Sarzhinsky, Legendre analysis, thermodynamical formalizm and spectrums of Perron-Frobenius operators // Doklady RAN (to appear) (Russian).

[27] V.I. Bakhtin, Random processes generated by a hyprbolic sequence of mappings, I, II // Izvestija RAN, Ser. matem., **58** 1994, No 2 40-72, No 3 184-195 (Russian).

[28] V.I. Bakhtin, D.S. Sarzhinsky, Perron-Frobenius operator for backers transform // Vestsi NAN Belarusi. Ser. fiz.-mat. havuk. 2002, No 3, 14-20 (Russian).

[29] V.I. Bakhtin, Perron-Frobenius dynamical systems // Doklady NAN Belarusi **45** 2001, No 2. 8-11 (Russian).

[30] V.I. Bakhtin, Action functional for Perron-Frobenius cascades // Doklady NAN Belarusi **46** 2002, No 3. 28–33 (Russian).

[31] A.A. Borovkov, *Probability theory* // M..: Nauka, 1986.

[32] A. Antonevich, M. Belousov, A. Lebedev, *Functional differential equations: II. C^*- applications Part 1 Equations with continuous coefficients* // Longman, 1998.

[33] A. Antonevich, M. Belousov, A. Lebedev, *Functional differential equations: II. C^*- applications Part 2 Equations with discontinuous coefficients and boundary value problems* // Longman, 1998.

Chapter 8

An Ergodic Arnold–Liouville Theorem for Locally Symmetric Spaces

Joachim Hilgert[1]

The geodesic flow on the cotangent bundle of a locally symmetric space of negative curvature and finite volume is ergodic when restricted to the sphere bundle. When one relaxes the condition on the curvature to non-positive, i.e. when one considers quotients of non-compact Riemannian symmetric spaces of higher rank, the geodesic flow even when restricted to the sphere bundle is no longer ergodic. In fact, if we identify the tangent bundle with the cotangent bundle via the metric and in this way turn it into a symplectic manifold, we observe that the higher the rank the more Poisson commuting functions one can find, raising the degree of integrability.

The Arnold–Liouville Theorem is a classical result for completely integrable hamiltonian systems on compact symplectic manifolds which asserts that the manifold can be fibered by tori via the moment map and on these tori the flow can be linearized. This could be reformulated by saying that there is a Poisson commuting family of functions which fibers the manifold into level sets and acts transitively on each level set via its hamiltonian vector fields.

It turns out that for the cotangent bundle of a locally symmetric space of finite volume something very similar happens. Here we also have a family of Poisson commuting functions fibering the manifold into level sets and acting on these level sets via their hamiltonian vector fields. The only difference is that these actions are no longer transitive, but for generic orbits they are still ergodic. This is the result I call the ergodic Arnold–Liouville Theorem for locally symmetric spaces.

[1]Mathematics Institute, Paderborn University, D-33095 Paderborn, Germany, hilgert@math.uni-paderborn.de

I would like to thank E.B. Vinberg for useful discussions on the equivariant symplectic geometry of cotangent bundles. In particular, he showed me how simple it is to reduce Lemma 8.8 to the elementary Lemma 8.7.

8.1   Invariant Vector Fields on Homogeneous Spaces

Let $\mathcal{N}$ be any manifold and G a Lie group (possibly of dimension 0, i.e. discrete). We assume that G acts on $\mathcal{N}$ by diffeomorphisms such that the map

$$G \times \mathcal{N} \to \mathcal{N}, \quad (g,x) \mapsto g \cdot x$$

is smooth. The action is called *free* if the *stabilizer groups* $G_x = \{g \in G \mid g \cdot x = x\}$ are all trivial and *proper* if the map

$$G \times \mathcal{N} \to \mathcal{N} \times \mathcal{N}, \quad (g,x) \mapsto (x, g \cdot x)$$

is a proper map. If the action is free and proper the space $G\backslash\mathcal{N}$ of G-orbits carries a smooth manifold structure such that the canonical projection $\mathcal{N} \to G\backslash\mathcal{N}$ is a submersion (cf. [tD91], §I.5).

The action of G on $\mathcal{N}$ induces natural actions of G on the tensor bundles of $\mathcal{N}$. Viewing $g \in G$ as a diffeomorphism on $\mathcal{N}$, the action on the tangent bundle $T\mathcal{N}$ is given by the derivative $g' \colon T\mathcal{N} \to T\mathcal{N}$. Then the action on the cotangent bundle $T^*\mathcal{N}$ is defined via the canonical pairing

$$\langle g \cdot \xi, v \rangle_{\mathcal{N}} = \langle \xi, g^{-1} \cdot v \rangle_{\mathcal{N}} = \langle \xi, (g^{-1})'(x)(v) \rangle_{\mathcal{N}} \quad \forall \xi \in T_x^*\mathcal{N}, v \in T_x\mathcal{N}.$$

The action on the higher tensor bundles $T_s^r\mathcal{N}$ is then defined via tensor products. Note that in this way all the canonical projections are G-equivariant and G acts on all bundles by bundle maps covering the original action, so we obtain induced actions on the spaces of sections. For example G acts on the space $\mathcal{X}(\mathcal{N})$ of smooth vector fields via

$$(g \cdot X)(x) = g \cdot (X(g^{-1} \cdot x)). \tag{8.1}$$

It is now clear what is meant by a G-invariant vector field, metric, symplectic form, and so on.

Let now G be a Lie group and H a closed subgroup of G. We denote the Lie algebras of G and H by $\mathfrak{g}$ and $\mathfrak{h}$. We have an identification $T(G/H) \cong G \times_H (\mathfrak{g}/\mathfrak{h})$, where $G \times_H (\mathfrak{g}/\mathfrak{h})$ is the fiber product of G and $\mathfrak{g}/\mathfrak{h}$ over H. More precisely, two pairs $(g, X + \mathfrak{h})$ and $(g', X' + \mathfrak{h})$ in $G \times (\mathfrak{g}/\mathfrak{h})$ are called equivalent, if they belong to the same orbit under the right action of H on $G \times (\mathfrak{g}/\mathfrak{h})$ given by

$$(g, X + \mathfrak{h}) \cdot h = (gh, h^{-1} \cdot X + \mathfrak{h}),$$

where $h \cdot X$ denotes the adjoint action. Then $G \times_H (\mathfrak{g}/\mathfrak{h})$ is the set of equivalence classes. We denote the equivalence class of $(g, X + \mathfrak{h})$ by $[g, X + \mathfrak{h}]$. Then

$$g_1 \cdot [g_2, X + \mathfrak{h}] := [g_1 g_2, X + \mathfrak{h}]$$

defines a left action of G on $G \times_H (\mathfrak{g}/\mathfrak{h})$. Set

$$\lambda_g \colon G/H \to G/H, \quad g_1 H \mapsto g g_1 H.$$

Then the derivative $\lambda_g'(x) \colon T_x(G/H) \to T_{g \cdot x}(G/H)$ is an isomorphism and we identify $G \times_H (\mathfrak{g}/\mathfrak{h})$ with $T(G/H)$ via the equivariant correspondence

$$[g, X + \mathfrak{h}] \longleftrightarrow \lambda_g'(o)(X + \mathfrak{h}), \tag{8.2}$$

where $o = H \in G/H$ is the canonical base point of G/H.

The dimensions in the stratification of $T(G/H)$ by G-orbits are

$$\begin{aligned}
\dim(G \cdot [g, X + \mathfrak{h}]) &= \dim(G/H) + \dim(gHg^{-1} \cdot [g, X + \mathfrak{h}]) \\
&= \dim(G/H) + \dim(H \cdot [\mathbf{1}, X + \mathfrak{h}]) \\
&= \dim(G/H) + \dim(H \cdot X + \mathfrak{h})
\end{aligned}$$

Moreover, $[g, X + \mathfrak{h}]$ and $[\tilde{g}, \tilde{X} + \mathfrak{h}]$ belong to the same G-orbit in $G \times_H (\mathfrak{g}/\mathfrak{h})$ if and only if $X + \mathfrak{h}$ and $\tilde{X} + \mathfrak{h}$ belong to the same H-orbit in $\mathfrak{g}/\mathfrak{h}$.

Let $\pi \colon G \to G/H$ be the canonical projection whose derivative $\pi' \colon TG \to T(G/H)$ is surjective. Under the above identifications the map π' is given by

$$\pi'(g, X) = [g, X + \mathfrak{h}].$$

The vector fields on G/H correspond to smooth functions $\overline{X} \colon G \to \mathfrak{g}/\mathfrak{h}$ satisfying

$$\overline{X}(gh) = h^{-1} \cdot \overline{X}(g) \quad \forall g \in G, h \in H.$$

The vector field is then given by $gH \mapsto [g, \overline{X}(g)]$. A vector field $gH \mapsto [g, \overline{X}(g)]$ is G-invariant if and only if $\overline{X}$ is a constant map:

$$
\begin{aligned}
[x, \overline{X}(x)] &= \left(g \cdot [\bullet, \overline{X}(\bullet)]\right)(x) \\
&\overset{(8.1)}{=} g \cdot [g^{-1}x, \overline{X}(g^{-1}x)] \\
&= [x, \overline{X}(g^{-1}x)].
\end{aligned}
$$

In that case the value of $\overline{X}$ has to be a H-fixed point in $\mathfrak{g}/\mathfrak{h}$. Therefore the map

$$
\mathcal{X}(G/H)^G \to (\mathfrak{g}/\mathfrak{h})^H, \quad \overline{X} \mapsto \overline{X}(o),
$$

where $\mathcal{X}(G/H)^G$ denotes the G-invariant vector fields and $(\mathfrak{g}/\mathfrak{h})^H$ the H-fixed points, is a linear isomorphism. If $\overline{X}(o) = X + \mathfrak{h}$ with $X \in \mathfrak{g}$, then the flow of $\overline{X}$ is given by

$$
\Phi^t(g \cdot o) = g(\exp tX) \cdot o. \tag{8.3}
$$

In particular, the invariant vector fields are complete.

Let $\mathfrak{n} := \mathfrak{n}_{\mathfrak{g}}(\mathfrak{h})$ be the normalizer of $\mathfrak{h}$ in $\mathfrak{g}$. Then $(\mathfrak{g}/\mathfrak{h})^H \subseteq \mathfrak{n}/\mathfrak{h}$. In fact, $H \cdot (X + \mathfrak{h}) = X + \mathfrak{h}$ is equivalent to $\mathrm{Ad}(H)X \subseteq X + \mathfrak{h}$ and this implies $\mathrm{ad}(\mathfrak{h})X \subseteq \mathfrak{h}$. For $X \in \mathfrak{n}$ and $Y \in \mathfrak{h}$ we have

$$
\mathrm{Ad}(\exp Y)X = e^{\mathrm{ad}Y} X \in X + \mathfrak{h},
$$

so that $X + \mathfrak{h}$ is fixed under the identity component of H. Thus, if H is connected, we have $(\mathfrak{g}/\mathfrak{h})^H = \mathfrak{n}_{\mathfrak{g}}(\mathfrak{h})$.

8.2   Tangent Bundles of Symmetric Spaces

Let $(\mathcal{M}, \omega)$ be a symplectic manifold with tangent bundle $T\mathcal{M}$ and cotangent bundle $T^*\mathcal{M}$. Then for each smooth function $f \in C^\infty(\mathcal{M})$ one defines its *symplectic gradient* or *hamiltonian vector field* H_f via

$$
-\omega(H_f, X) = \langle df, X \rangle_{\mathcal{M}},
$$

where $X \in \mathcal{X}(\mathcal{M})$ is a vector field on $\mathcal{M}$, and $\langle \cdot, \cdot \rangle_{\mathcal{M}}$ again denotes the canonical pairing between $T^*\mathcal{M}$ and $T\mathcal{M}$ (for the basic facts of symplectic geometry see e.g. [AM87]). The hamiltonian vector fields satisfy

$$
H_{f_1 f_2} = f_2 H_{f_1} + f_1 H_{f_2}. \tag{8.4}
$$

For $f_1, f_2 \in C^\infty(\mathcal{M})$ one has the *Poisson bracket* $\{f_1, f_2\} \in C^\infty(\mathcal{M})$ defined by

$$\{f_1, f_2\} = \omega(H_{f_1}, H_{f_2}).$$

It defines a Lie algebra structure on $C^\infty(\mathcal{M})$ and the equality $\{f_1, f_2\} = df_2(H_{f_1})$ shows that for two Poisson commuting functions the one function (and hence its level sets) is invariant under the flow of the other's hamiltonian vector field.

Suppose that ω is a G-invariant symplectic form on $\mathcal{M}$ and $f \in C^\infty(\mathcal{M})$ is also G-invariant. Then the hamiltonian vector field $H_f \in \mathcal{X}(\mathcal{M})$ is again G-invariant. As a consequence we see that the flow $\Phi^t(x)$ of H_f is G-equivariant, i.e.

$$\Phi^t(g \cdot x) = g \cdot \Phi^t(x) \quad \text{for } (x,t) \in \mathcal{M} \times \mathbb{R},$$

where (x,t) is restricted to the domain of definition for the flow.

Let $\mathcal{N}$ be any manifold. Recall the *canonical symplectic form* $\omega = d\theta$ on $T^*\mathcal{N}$. If $p_{T^*\mathcal{N}} \colon T^*\mathcal{N} \to \mathcal{N}$ is the canonical projection and $p'_{T^*\mathcal{N}} \colon T(T^*\mathcal{N}) \to T\mathcal{N}$ its derivative, then θ is given by the formula

$$\langle \theta(\xi), v \rangle = \langle \xi, p'_{T^*\mathcal{N}}(v) \rangle \quad \forall v \in T_\xi(T^*\mathcal{N}), \forall \xi \in T^*\mathcal{N}.$$

For a diffeomorphism $h \colon \mathcal{M} \to \mathcal{N}$ the induced map $h^* \colon T^*\mathcal{N} \to T^*\mathcal{M}$ is a canonical transformation w.r.t. to the standard symplectic structures on the two cotangent bundles. Thus the induced map

$$C^\infty(T^*\mathcal{M}) \to C^\infty(T^*\mathcal{N}), \quad f \mapsto h_* f := f \circ h^*$$

is a Poisson isomorphism. In particular we have

$$h_* H_f = H_{h_* f}$$

so that the push-forward of vector fields $h_* \colon \mathcal{X}(\mathcal{M}) \to \mathcal{X}(\mathcal{N})$ maps hamiltonian vector fields to hamiltonian vector fields.

If now G acts smoothly on $\mathcal{N}$, these considerations show that the hamiltonian vector fields of G-invariant functions on $T^*\mathcal{N}$ are G-invariant hamiltonian vector fields on $T^*\mathcal{N}$. Moreover, the action of G on $T^*\mathcal{N}$ is *hamiltonian* with *moment map* $J \colon T^*\mathcal{N} \to \mathfrak{g}^*$ given by

$$\langle J(\eta), X \rangle = \langle \eta, \tilde{X} \circ p_{T^*\mathcal{N}}(\eta) \rangle_{\mathcal{N}},$$

where $X \in \mathfrak{g}$ and $\tilde{X} \in \mathcal{X}(\mathcal{N})$ is the vector field on $\mathcal{N}$ induced by the derived action of G on $\mathcal{N}$ via

$$\tilde{X}(x) = \frac{d}{dt}\big|_{t=0}(\exp tX) \cdot x.$$

Now suppose that $(\mathcal{N}, B)$ is a Riemannian manifold. The Riemannian metric B allows us to identify $T\mathcal{N}$ and the cotangent bundle $T^*\mathcal{N}$ of $\mathcal{N}$. Then the formula for θ reads

$$\langle \theta(\xi), v \rangle = \langle \xi, p'_{T^*\mathcal{N}}(v) \rangle = B\big(\xi, p'_{T\mathcal{N}}(v)\big)$$

for $\xi \in T^*\mathcal{N} \cong T\mathcal{N}$ and $v \in T_\xi(T^*\mathcal{N}) \cong T_\xi(T\mathcal{N})$. The set

$$\{\xi \in T(T^*\mathcal{N}) \mid p'_{T^*\mathcal{N}}(\xi) = 0\}$$

is called the *vertical subbundle* of $T(T^*\mathcal{N})$. Analogously one has a vertical subbundle of $T(T\mathcal{N})$. It is clear that the identification $T\mathcal{N} \cong T^*\mathcal{N}$ preserves the vertical bundles. Since $T_v(T\mathcal{N})$ is canonically identified with $T_{p_{T\mathcal{N}}(v)}\mathcal{N} \times T_{p_{T\mathcal{N}}(v)}\mathcal{N}$, the metric B induces a Riemannian metric on $T(T\mathcal{N})$. Therefore it makes sense to speak about the *horizontal subbundle*, i.e. the bundle of elements orthogonal to the elements of the vertical subbundle.

For the remainder this section we suppose that $\mathcal{N} = G/K$ is a Riemannian symmetric space and $\mathfrak{g} = \mathfrak{k} + \mathfrak{p}$ the corresponding Cartan decomposition of $\mathfrak{g}$. Then one identifies $T_o(G/K) \cong \mathfrak{g}/\mathfrak{k}$ with $\mathfrak{p}$. Thus we obtain an identification $T(G/K) \cong G \times_K \mathfrak{p}$ via

$$[g, X] \longleftrightarrow \lambda'_g(o)X$$

and see that $[g, X]$ and $[\tilde{g}, \tilde{X}]$ belong to the same G-orbit in $G \times_K \mathfrak{p}$ if and only if X and $\tilde{X}$ belong to the same K-orbit in $\mathfrak{p}$. Moreover, the derivative $\pi' \colon TG \to T(G/K)$ of the canonical projection is given by

$$\pi'(g, X) = [g, \mathrm{pr}_\mathfrak{p}(X)],$$

where $\mathrm{pr}_\mathfrak{p} \colon \mathfrak{g} = \mathfrak{k} + \mathfrak{p} \to \mathfrak{p}$ is the projection with kernel $\mathfrak{k}$.

Using $T^*\mathcal{N} \cong G \times_K \mathfrak{p}^*$ and identifying $\mathfrak{p}$ and $\mathfrak{p}^*$ via the Killing form B gives

$$T^*\mathcal{N} \cong G \times_K \mathfrak{p} \cong T\mathcal{N}.$$

Under these identifications the moment map is given by

$$J([g, X]) = \mathrm{Ad}(g)X.$$

For the following remarks see also [Th81] and [GS84], §45. Set $P :=$ $\exp \mathfrak{p}$ and note that the Cartan decomposition $G = PK$ yields a retraction

$$j \colon \mathcal{N} = G/K \to P, \quad pK \mapsto p$$

for the canonical quotient map $\pi \colon G \to G/K$. We identify $\mathcal{N} = G/K$ with $P \subseteq G$. Then the derivative $\exp' \colon T\mathfrak{p} \cong \mathfrak{p} \times \mathfrak{p} \to TP \subseteq TG \cong G \times \mathfrak{g}$ yields a coordinate system for $T\mathcal{N}$. More precisely, for $(X, Y) \in T\mathfrak{p} = \mathfrak{p} \times \mathfrak{p}$ we have

$$
\begin{aligned}
\exp'(X, Y) &= (\exp X, \exp'(X)(Y)) \\
&= (\exp X, \lambda'_{\exp X}(\mathbf{1}) \tfrac{\mathrm{id} - e^{-\mathrm{ad}X}}{\mathrm{ad}X} Y) \\
&\overset{(8.2)}{=} (\exp X, \sum_{k=1}^{\infty} \tfrac{(-\mathrm{ad}X)^{k-1}}{k!} Y)
\end{aligned}
$$

(see [He78] for the standard facts concerning the geometry of Lie groups). We want to determine the derivative

$$f' \colon T(T\mathfrak{p}) = (\mathfrak{p} \times \mathfrak{p}) \times (\mathfrak{p} \times \mathfrak{p}) \to T(TG) \cong (G \times \mathfrak{g}) \times (\mathfrak{g} \times \mathfrak{g})$$

of the map $f = \exp'$. To this end we set

$$\Psi \colon \mathfrak{p} \to \mathrm{End}(\mathfrak{g}), \quad X \mapsto \sum_{k=1}^{\infty} \frac{(-\mathrm{ad}X)^{k-1}}{k!}, \tag{8.5}$$

so that

$$f \colon \mathfrak{p} \times \mathfrak{p} \to G \times \mathfrak{g}, \quad (X, Y) \mapsto (\exp X, \Psi(X)Y).$$

Therefore we have

$$f'(X, Y)(A, B) \cong (\exp X, \Psi(X)Y, \Psi(X)A, (\Psi'(X)A)Y + \Psi(X)B).$$

Note that $\Psi(0) = \mathrm{id}$ and $\Psi'(0) = -\tfrac{1}{2}\mathrm{ad}$ so that for $X = 0$ the formula for f' simplifies to

$$f'(0, Y)(A, B) = (\mathbf{1}, Y, A, B - \tfrac{1}{2}[A, Y]).$$

Thus $T\mathcal{N}$ can be viewed as a subset of $TG \cong G \times \mathfrak{g}$. Under the identification of the tangent bundle $T(TG)$ with $(G \times \mathfrak{g}) \times (\mathfrak{g} \times \mathfrak{g})$ and the corresponding identification of $T(T\mathcal{N})$ as a subset of $(G \times \mathfrak{g}) \times (\mathfrak{g} \times \mathfrak{g})$ for $Y = (0, Y) \in T_o\mathcal{N} \cong \mathfrak{p}$ we have

$$T_Y(T\mathcal{N}) = \{(\mathbf{1}, Y, A, -\tfrac{1}{2}[A, Y] + B) \mid A, B \in \mathfrak{p}\}. \tag{8.6}$$

The other tangent spaces are then determined via G-invariance.

We describe the vertical and the horizontal subbundle of $T(T\mathcal{N})$ under the above identifications.

Lemma 8.1　*Viewing $T_Y(T\mathcal{N})$ as a subset of $T(TG) = (G \times \mathfrak{g}) \times (\mathfrak{g} \times \mathfrak{g})$ we find:*

(i) *The space of vertical vectors is $\{(\mathbf{1}, Y, 0, B) \mid B \in \mathfrak{p}\}$.*

(ii) *The space of horizontal vectors is $\{(\mathbf{1}, Y, A, -\frac{1}{2}[A, Y]) \mid A \in \mathfrak{p}\}$.*

Proof.

(i) Using $\mathcal{N} \cong P \cong \mathfrak{p}$ and $T\mathcal{N} = TP \cong \mathfrak{p} \times \mathfrak{p}$ we see that the map $p_{T\mathcal{N}}$ corresponds to the projection onto the first factor. The same is true if we view $T\mathcal{N}$ as a subset of $G \times \mathfrak{g}$. The derivative of $\mathrm{pr}_1 : G \times \mathfrak{g} \to G$ is given by

$$\mathrm{pr}_1' : (G \times \mathfrak{g}) \times (\mathfrak{g} \times \mathfrak{g}) \to G \times \mathfrak{g}, \quad ((g, Y), (A, B)) \mapsto (g, A).$$

(ii) Note that the metric on $\mathfrak{p}$ is just the Killing form B. A vector $v \in T(T\mathcal{N})$ horizontal if it is orthogonal to the vertical vectors $w \in T(T\mathcal{N})$ with $p_{T(T\mathcal{N})}(v) = p_{T(T\mathcal{N})}(w)$, where $p_{T(T\mathcal{N})} : T(T\mathcal{N}) \to T\mathcal{N}$ is the canonical projection. Thus in our example the horizontal vectors are the elements of the form $((X, Y), (A, 0))$ so that the space of horizontal vectors in $T_Y(T\mathcal{N})$ viewed as subset of $T(TG) = (G \times \mathfrak{g}) \times (\mathfrak{g} \times \mathfrak{g})$ is

$$\{(\mathbf{1}, Y, A, -\tfrac{1}{2}[A, Y]) \mid A \in \mathfrak{p}\}$$

as asserted.

$\square$

We also describe the G-actions on $\mathcal{N}$ and $T\mathcal{N}$ in our various identifications. To this end we consider the projection

$$p_P : G = PK \to P, \quad pk \mapsto p$$

and note that $p^2 = (pk)\theta(pk)^{-1}$, where $\theta : G \to G$ is the Cartan involution corresponding to the Cartan decomposition $G = PK$. The G-action on $\mathcal{N} = G/K$ by left translation becomes

$$g \cdot p = p_P(gp) \quad \forall g \in G, p \in P$$

under the above identification of $\mathcal{N}$ with P. Thus

$$(g \exp X)\theta(g \exp X)^{-1} = g(\exp X)\theta(\exp X)^{-1}\theta(g)^{-1} = g(\exp 2X)\theta(g)^{-1}$$

for $g \in G$ and $X \in \mathfrak{p}$ shows that the induced action on $\mathfrak{p}$ is given by

$$g \cdot X = \tfrac{1}{2} \log(g(\exp 2X)\theta(g)^{-1}). \tag{8.7}$$

Lemma 8.2   *The action of G on $TN \cong \mathfrak{p} \times \mathfrak{p}$ is given explicitly by*

$$k \cdot (X,Y) = (\mathrm{Ad}(k)X, \mathrm{Ad}(k)Y)$$
$$p \cdot (X,Y) = \left(\tfrac{1}{2} \log(p(\exp 2X)p), \left(\mathrm{Ad}(p) \circ \Psi(\log(p(\exp 2X)p))\right)^{-1} \Psi(2X)Y \right)$$

for $k \in K$, $p \in P$ and $X, Y \in \mathfrak{p}$.

Proof.   For $k \in K$ we have $\theta(k)^{-1} = k^{-1}$, so that the action (8.7) reduces to

$$k \cdot X = \mathrm{Ad}(k)X \quad \forall k \in K, X \in \mathfrak{p}.$$

Thus the induced action of K on $TN \cong \mathfrak{p} \times \mathfrak{p}$ is given by

$$k \cdot (X,Y) = (\mathrm{Ad}(k)X, \mathrm{Ad}(k)Y).$$

For $p \in P$ the calculation of $p \cdot X$ with $X \in \mathfrak{p}$ is more involved. We write $f(X) := \tfrac{1}{2} \log(p(\exp 2X)p)$ and

$$f = \mu_{\frac{1}{2}} \circ \log \circ \rho_p \circ \lambda_p \circ \exp \circ \mu_2,$$

where $\mu_c(X) = cX$ and $\rho_p(g) = gp$. Then for $X \in \mathfrak{p}$ and $q \in P$ we obtain the derivatives $\mu_c'(X)Y = cY$ and

$$\exp'(X)Y = \lambda'_{\exp X}(\mathbf{1}) \circ \Psi(X)Y, \quad \log'(q)Z = (\lambda'_q(\mathbf{1}) \circ \Psi(\log q))^{-1}Z.$$

Note that $\lambda^{-1}_{p(\exp 2X)p} \circ \rho_p \circ \lambda_{p\exp 2X} = \lambda_{p^{-1}} \circ \rho_p$, so that

$$\lambda'_{p(\exp 2X)p}(\mathbf{1})^{-1} \circ \rho_p'(p\exp 2X) \circ \lambda'_{p\exp 2X}(\mathbf{1}) = \mathrm{Ad}(p)^{-1}.$$

Now the chain rule gives

$$\begin{aligned}
f'(X)Y &= \mu'_{\frac{1}{2}}(\log \circ \rho_p \circ \lambda_p \circ \exp \circ \mu_2(X)) \circ (\log \circ \rho_p \circ \lambda_p \circ \exp \circ \mu_2)'(X)Y \\
&= \Psi(\log(p(\exp 2X)p))^{-1} \circ \mathrm{Ad}(p)^{-1} \circ \Psi(2X)Y \\
&= \left(\mathrm{Ad}(p) \circ \Psi(p(\exp 2X)p)\right)^{-1}\Psi(2X)Y
\end{aligned}$$

and this yields the desired formula for $p \cdot (X,Y)$. $\square$

Lemma 8.3   *Consider the identification map*

$$T_{(0,Y)}(\mathfrak{p} \times \mathfrak{p}) = \mathfrak{p} \times \mathfrak{p} \to \mathfrak{g} \times \mathfrak{g} \cong T_{(1,Y)}(G \times \mathfrak{g})$$
$$(A,B) \mapsto (A, B - \tfrac{1}{2}[A,Y]).$$

Using this identification we find:

(i) *The vector fields $\tilde{Z} \in \mathcal{X}(T\mathcal{N})$ induced by the action of $\exp \mathbb{R}Z$ are given by*

$$\tilde{Z}(\mathbf{1}, Y) = (\mathbf{1}, Y, 0, [Z, Y]) \quad \forall Z \in \mathfrak{k}, Y \in \mathfrak{p}$$
$$\tilde{Z}(\mathbf{1}, Y) = (\mathbf{1}, Y, Z, -\tfrac{1}{2}[Z, Y]) \quad \forall Z \in \mathfrak{p}, Y \in \mathfrak{p}.$$

In particular, $\tilde{Z}$ is a section of the vertical bundle for $Z \in \mathfrak{k}$ and a section of the horizontal bundle for $Z \in \mathfrak{p}$.

(ii) *The tangent space $T_{(\mathbf{1}, Y)}(G \cdot (\mathbf{1}, Y))$ is given by*

$$\{(\mathbf{1}, Y, A, [B, Y] - \tfrac{1}{2}[A, Y]) \mid A \in \mathfrak{p}, B \in \mathfrak{k}\} \cong \mathfrak{p} \times \mathrm{ad}(Y)\mathfrak{k}.$$

It contains all the horizontal vectors in $T_{(\mathbf{1}, Y)}(T\mathcal{N})$.

Proof. Part (ii) is an immediate consequence of (i), so we only have to prove the first part. Writing $k(t) = \exp tZ$ with $Z \in \mathfrak{k}$ and taking the derivative w.r.t. t in 0, we see that the vector field $\tilde{Z} \in \mathcal{X}(T\mathcal{N})$ induced by the action of $\exp \mathbb{R}Z$ takes the value

$$\tilde{Z}(0, Y) = (0, [Z, Y]) \quad \forall Z \in \mathfrak{k}, Y \in \mathfrak{p}.$$

We rewrite the formula for $\tilde{Z}$ as $\tilde{Z}(\mathbf{1}, Y) = (\mathbf{1}, Y, 0, [Z, Y])$. In particular, G-invariance implies that $\tilde{Z}$ is a section of the vertical bundle.

For the special case of $X = 0$ the formula for $p \cdot (X, Y)$ from Lemma 8.2 simplifies to

$$p \cdot (0, Y) = \left(\log p, \left(\mathrm{Ad}(p) \circ \Psi(\log p^2) \right)^{-1} Y \right).$$

Now we set $p = \exp Z$ and observe

$$\mathrm{Ad}(p) \circ \Psi(\log p^2) = e^{\mathrm{ad}Z} \circ \Psi(2Z) = \frac{e^{\mathrm{ad}Z} - e^{-\mathrm{ad}Z}}{2\mathrm{ad}Z}.$$

Therefore

$$\exp tZ \cdot (0, Y) = (tZ, (\mathrm{id} + O(t^2))Y) \quad \forall t \in \mathbb{R},$$

so in this case we obtain

$$\tilde{Z}(0, Y) = (Z, 0) \quad \forall Z \in \mathfrak{p}.$$

Again we use the identification map to find $\tilde{Z}(\mathbf{1}, Y) = (\mathbf{1}, Y, Z, -\tfrac{1}{2}[Z, Y])$ for $Z \in \mathfrak{p}$. In particular, by G-invariance, $\tilde{Z}$ takes its values in the horizontal bundle. $\qquad\square$

For the proofs of the following facts see [Th81]. The natural symplectic form on $T(T\mathcal{N})$ is given by

$$\omega_{(1,Y)}\big((A_1, B_1 - \tfrac{1}{2}[A_1, Y]), (A_2, B_2 - \tfrac{1}{2}[A_2, Y])\big) = B(A_1, B_2) - B(A_2, B_1).$$
(8.8)

Let $f \in C^\infty(T\mathcal{N})$ be G-invariant and $h \in C^\infty(\mathfrak{p})$ the restriction of f to $T_o\mathcal{N} = \mathfrak{p}$ (which determines f uniquely). Then the hamiltonian vector field H_f is determined by

$$H_f(X) = \big(\operatorname{grad} h(X), -\tfrac{1}{2}[\operatorname{grad} h(X), X]\big),$$
(8.9)

where the gradient is taken with respect to the Killing form on $\mathfrak{p}$.

In particular, H_f is horizontal, so by Lemma 8.3 the flow of H_f preserves G-orbits. According to (8.3) the flow $\Phi_f \colon \mathbb{R} \times T(G/K) \to T(G/K)$ of H_f on $T(G/K)$ is given by

$$\Phi_f^t([g, X]) = [g \exp(t \operatorname{grad} h(X)), X]$$
(8.10)

and preserves G-orbits. Therefore the flow can be studied on these orbits separately. Note further that (8.8) and (8.9) show that any two G-invariant functions $f_1, f_2 \in C^\infty(T\mathcal{N})$ commute under the Poisson bracket, i.e.

$$\{f_1, f_2\} = 0 \quad \forall f_1, f_2 \in C^\infty(T\mathcal{N})^G.$$
(8.11)

8.3 Commuting Vector Fields and Flows

Let $(\mathcal{M}, \omega)$ be a symplectic manifold and recall that for two Poisson commuting functions the one function (and hence its level sets) is invariant under the flow of the other's hamiltonian vector field.

Let $\mathcal{A} \subseteq C^\infty(\mathcal{M})$ be a finitely generated associative subalgebra consisting of Poisson commuting functions. Let $f_1, \ldots, f_k$ be a set of generators and define $F \colon \mathcal{M} \to \mathbb{R}^k$ by $F = (f_1, \ldots, f_k)$. Then an arbitrary element of $\mathcal{A}$ can be written as $f = \sum_{\alpha \in \mathbb{N}_0^n} c_\alpha F^\alpha$, where we use the usual multi-index notation. For any $v \in \mathbb{R}^k$ we consider the closed subset $\mathcal{M}_{(v)} := F^{-1}(v)$ of $\mathcal{M}$. Since the f_j Poisson-commute we see that all the $\mathcal{M}_{(v)}$ are stable under all the flows of the H_{f_j}'s.

If now $\mathcal{M}_{(v)}$ is a *submanifold* of $\mathcal{M}$, then the hamiltonian vector fields H_{f_j} are all tangent to $\mathcal{M}_{(v)}$, i.e. restrict to vector fields on $\mathcal{M}_{(v)}$. Note that all elements of $\mathcal{A}$ are constant on $\mathcal{M}_{(v)}$. But then the identity (8.4) implies that the restriction of any H_f with $f \in \mathcal{A}$ is a linear combination of the restrictions of the H_{f_j}.

In general, the $\mathcal{M}_{(v)}$ will not be manifolds. It is, however, always possible to find submanifolds of $\mathcal{M}$ contained in $\mathcal{M}_{(v)}$ such that all the vector fields H_{f_k} restrict to vector fields of these submanifolds. To this end we recall the notion of a $\mathcal{D}$-orbit for a family $\mathcal{D}$ of vector fields from [Su73]. We set

$$\mathcal{D} := \{H_f \mid f \in \mathcal{A}\} \tag{8.12}$$

and note that according to (8.4) the linear spans of $\{H_f(x) \mid f \in \mathcal{A}\}$ and $\{H_{f_1}(x), \ldots, H_{f_k}(x)\}$ in $T_x\mathcal{M}$ agree for all $x \in \mathcal{M}$. We denote this linear span by $\Delta(x)$. Then $x \mapsto \Delta(x)$ is a smooth distribution in the sense of [Su73]. It is involutive since $\mathcal{D}$ is a commutative family of vector fields. Moreover, this shows that Δ satisfies the condition (e) from [Su73], Theorem 4.2. Therefore the $\mathcal{D}$-orbits $\mathcal{D} \cdot x$ are maximal integral manifolds of Δ. They satisfy

$$\dim(\mathcal{D} \cdot x) = \dim \Delta(x). \tag{8.13}$$

¿From the construction of the $\mathcal{D}$-orbits it is clear that they are invariant under the flows of the $H_f \in \mathcal{D}$. Since these flows preserve the $\mathcal{M}_{(v)}$ we see also that the $\mathcal{D}$-orbits are contained in single $\mathcal{M}_{(v)}$'s. This, finally, shows that the restriction of H_f with $f \in \mathcal{A}$ to any $\mathcal{D}$-orbit is a linear combination (with coefficients depending on v) of the restrictions of the $H_{f_1}, \ldots, H_{f_k}$.

Suppose that y is a regular value of F, i.e. the derivative $F'(x) \colon T_x\mathcal{M} \to \mathbb{R}^k$ is surjective for all $x \in F^{-1}(y)$. Then $\mathcal{M}_{(F(x))}$ is a closed submanifold of $\mathcal{M}$ with $\dim \mathcal{M}_{(F(x))} = \dim \mathcal{M} - k$. On the other hand, k is equal to

$$\dim \operatorname{span}\{f_1'(x), \ldots, f_k'(x)\} = \dim \operatorname{span}\{H_{f_1}(x), \ldots, H_{f_k}(x)\} = \dim(\mathcal{D} \cdot x)$$

so that

$$\dim(\mathcal{D} \cdot x) + \dim \mathcal{M}_{(F(x))} = \dim \mathcal{M}. \tag{8.14}$$

The commuting vector fields $H_{f_1}, \ldots, H_{f_k}$ define a (local) action of $\mathbb{R}^k$ on $\mathcal{M}$ which leaves the $\mathcal{D}$-orbits invariant.

Let $\mathfrak{a}$ be a *Cartan subspace*, i.e. a maximal abelian subspace of $\mathfrak{p}$. Further let $N_K(\mathfrak{a})$ be the *normalizer* of $\mathfrak{a}$ in K and $M := Z_K(\mathfrak{a})$ the *centralizer* of $\mathfrak{a}$ in K. Then $Z_K(\mathfrak{a})$ is normal in $N_K(\mathfrak{a})$. The quotient group $W(\mathfrak{a}, K) := N_K(\mathfrak{a})/Z_K(\mathfrak{a})$ is the *Weyl group*.

Lemma 8.4 *Let G act on the left of $G/M \times \mathfrak{a}$ via $h \cdot (gM, X) = (hgM, X)$.*

Then the map

$$\phi\colon G/M \times \mathfrak{a} \to T(G/K) \cong G \times_K \mathfrak{p}, \quad (gM, X) \mapsto \lambda'_g(o)X = [g, X]$$

is G-equivariant, surjective, and the following two statements are equivalent

(1) $\phi(gM, X) = \phi(\tilde{g}M, \tilde{X})$.
(2) There exists $k \in K$ with $gk = \tilde{g}$ and $\mathrm{Ad}(k)X = \tilde{X}$.

In particular, $\tilde{X}$ is contained in the Weyl group orbit of X and we have a bijection of orbit spaces

$$G\backslash T(G/K) \longleftrightarrow W(\mathfrak{a}, K)\backslash \mathfrak{a}.$$

Proof. Let $Y \in T_x(G/K)$ and $x = g \cdot o$. Then $\tilde{Y} := (\lambda'_g(o))^{-1}Y \in \mathfrak{p}$ and there exists a $k \in K$ with

$$X := k^{-1} \cdot \tilde{Y} = \lambda'_k(o)^{-1}\tilde{Y} \in \mathfrak{a}.$$

For this $X \in \mathfrak{a}$ we have $\lambda'_{gk}(o)X = \lambda'_g(o)\lambda'_k(o)X = \lambda'_g(o)\tilde{Y} = Y$. On the other hand, if $\lambda'_g(o)X = \lambda'_{\tilde{g}}(o)\tilde{X}$, then $g \cdot o = \tilde{g} \cdot o$. Thus there is a $k \in K$ with $gk = \tilde{g}$. Then we calculate

$$\lambda'_{\tilde{g}}(o)\tilde{X} = \lambda'_{gk}(o)\tilde{X} = \lambda'_g(o)\lambda'_k(o)\tilde{X} = \lambda'_g(o)\mathrm{Ad}(k)(\tilde{X})$$

to find $X = \mathrm{Ad}(k)(\tilde{X}) \in \mathrm{Ad}(K)\tilde{X} \cap \mathfrak{a} = W(\mathfrak{a}, K)\tilde{X}$, where the last equality follows from [He78], Prop. VII.2.2. $\qquad\square$

Theorem 8.1 *Consider the algebra $\mathcal{A}$ of G-invariant functions in $C^\infty(T^*\mathcal{N})$ which restrict to polynomials on $\mathfrak{p}$. Then $\mathcal{A}$ is finitely generated, commutative under the Poisson bracket, and the joint level sets of these functions are precisely the G-orbits in $T^*(G/K)$.*

Proof. The Poisson commutativity follows from (8.11). To prove the other claims we note that the restriction $\mathfrak{p}^* \to \mathfrak{a}^*$ induces an isomorphism between the K-invariant polynomials on $\mathfrak{p}$ and the W-invariant polynomials on $\mathfrak{a}$. Moreover, this space is isomorphic to the space of real polynomials in $\dim \mathfrak{a}$ variables (see [Hu90], §3.5) and each of these polynomials is the restriction of a smooth G-invariant function on $C^\infty(T^*\mathcal{N})$ (see [He84], Cor. II.5.12, for the extension to a K-invariant polynomial h on $\mathfrak{p}$ and then set $f([g, X]) = h(X)$.

Now the G-orbit structure of $T(G/K) \cong T^*(G/K)$ described in Lemma 8.4 shows that it suffices to prove that W-orbits in $\mathfrak{a}$ can be separated by W-invariant polynomials. To do that, given two (finite) orbits, one chooses a polynomial taking positive values on one and negative values on the other orbit. Averaging over W then yields the desired W-invariant separating polynomial. $\qquad\square$

Even though we will not deal with quantizations in this paper the following remark seems appropriate here.

Remark 8.1 *Let $\mathbb{D}(G/K)$ be the (commutative) algebra of invariant differential operators on G/K. Then using the results of Chevalley and Harish-Chandra we find isomorphisms*

$$\mathbb{D}(G/K) \cong \mathcal{E}'_K(G/K) \cong U(\mathfrak{a})^W \cong I(\mathfrak{a}^*_{\mathbb{C}}) \cong \mathbb{C}[X_1,\ldots,X_k],$$

where $k = \dim_{\mathbb{R}} \mathfrak{a}$ is the rank of G/K, $\mathcal{E}'_K(G/K)$ denotes the space of distributions supported in the base point, $U(\mathfrak{a})$ is the universal enveloping algebra of $\mathfrak{a}$, and $I(\mathfrak{a}^*_{\mathbb{C}})$ is the algebra of W-invariant polynomials. In fact, the generators of $I(\mathfrak{a}^*_{\mathbb{C}})$ can be chosen real, i.e. as real polynomials on $\mathfrak{a}$, which then come from differential operators with real valued coefficients (see [Hu90], §3.5). In particular, the associative subalgebra of $C^\infty(G/K)$ generated by the principal symbols of elements of $\mathbb{D}(G/K)$ coincides with the complexification of the algebra $\mathcal{A}$ introduced in Theorem 8.1. $\qquad\square$

Lemma 8.5 *Let $h\colon \mathfrak{p} \to \mathbb{R}$ be a K-invariant smooth function. Then $\operatorname{grad} h(X) \in \mathfrak{a}$ for all $X \in \mathfrak{a}$. In particular, we have $\operatorname{grad} h(X) = \operatorname{grad} h|_{\mathfrak{a}}(X)$.*

Proof. Consider the K-orbit $K \cdot X$ of X in $\mathfrak{p}$ and its tangent space $T_X(K \cdot X) = \operatorname{ad} X(\mathfrak{k})$ at X. The function h is constant on K-orbits, therefore we have

$$0 = h'(X)\,(\operatorname{ad} X(Z)) = B(\operatorname{grad} h(X), [X,Z]) = B([\operatorname{grad} h(X), X], Z)$$

for all $Z \in \mathfrak{k}$, so that $[\operatorname{grad} h(X), X] \subseteq \mathfrak{p} \cap \mathfrak{k} = \{0\}$. In other words, $\operatorname{grad} h(X) \in \mathfrak{z}_{\mathfrak{p}}(X)$. If X is regular, this means that $\operatorname{grad} h(X) \in \mathfrak{a}$. But $\operatorname{grad} h$ is continuous and the regular elements are dense in $\mathfrak{a}$. Thus $\operatorname{grad} h(X) \in \mathfrak{a}$ for all $X \in \mathfrak{a}$. $\qquad\square$

Lemma 8.5 together with (8.10) shows that we have a map

$$\mathcal{F}\colon \mathcal{A} \times \mathfrak{a} \to \mathfrak{a}, \quad (f,X) \mapsto \operatorname{grad} f|_{\mathfrak{a}}(X)$$

such that the flow $\Phi_f\colon \mathbb{R} \times T(G/K) \to T(G/K)$ of H_f on $T(G/K)$ is given by

$$\Phi_f^t([g,X]) = [g\exp(t\,\mathcal{F}(f,X)), X]. \tag{8.15}$$

We let $A = \exp \mathfrak{a}$ act on the right of $G/M \times \mathfrak{a}$ via

$$(gM, X) \cdot a := \rho(a)(gM, X) := (gaM, X).$$

Then we obtain the following equivariance property of $\phi \colon G/M \times \mathfrak{a} \to T(G/K)$ under Φ_f:

$$\phi \circ \rho \left(e^{t \operatorname{grad} f(X)} \right) = \Phi_f^t \circ \phi. \tag{8.16}$$

Example 8.1 Recall the geodesic flow Φ^t on the tangent bundle of a Riemannian manifold $\mathcal{N}$: with each tangent vector $v \in T_x\mathcal{N}$ one associates the tangent vector $\gamma_v'(t)$ to the (uniquely determined) geodesic γ_v with $\gamma_v'(0) = v$. In the case of $\mathcal{N} = G/K$ the geodesic γ in $T(G/K)$ with $\gamma(0) = gK$ and $\gamma'(0) = \lambda_g'(o)X$ is given by

$$\gamma(t) = g \exp(tX)K$$

Thus the geodesic flow on $G \times_K \mathfrak{p}$ can be written

$$\Phi^t([g, X]) = [g \exp tX, X].$$

Each geodesic, viewed as curve in $T(G/K)$, i.e. as orbit of the geodesic flow, is completely contained in a G-orbit. This follows from (8.10), but can be verified directly: It suffices to consider the case $x = o$. Since the derivative $\gamma_X'(t)$ of $\gamma_X(t)$ is given by $\lambda_{\exp tX}'(o)X$ (see [He78, S.226]), we have

$$\lambda_{g_t}'(o)(\gamma_X'(0)) = \lambda_{g_t}'(o)X = \gamma_X'(t)$$

for $g_t = \exp tX$. This means that $\gamma_X'(t)$ is contained in the G-orbit of $\gamma_X'(0)$.

If G/K has rank 1, i.e. if $\dim \mathfrak{a} = 1$, the K-invariant polynomials on $\mathfrak{p}$ are generated by $X \mapsto B(X, X)$, so that the G-orbits in $T(G/K)$ are simply the sphere bundles (and the zero section). Thus the relevant $T^*\mathcal{N}_{(v)}$ in question is (up to scaling) the sphere bundle in $T\mathcal{N}$. The sphere bundle $S(G/K)$ of G/K can in this case be identified with G/M and the geodesic flow is given by the right multiplication of A. More precisely, in this case we have the Iwasawa decomposition $G = KAK = NAK$ with $\dim A = 1$, so that $\mathfrak{g} = \mathfrak{k} + \mathfrak{p}$ implies

$$T(G/K) \cong G \times_K \mathfrak{p} \cong NA \times \mathfrak{p}.$$

G acts transitively on the sphere bundle in $T(G/K)$ since K acts transitively on the sphere (w.r.t. the Killing form) in $\mathfrak{p}$:

$$\{X \in \mathfrak{p} \mid \|X\| = 1\} = K \cdot X_o \cong K/Z_K(A)$$

and for $X_o \cong (\mathbf{1}, X_o) \in T(G/K)$

$$nak \cdot X_o = [nak, X_o] = [na, k \cdot X_o] \cong (na, k \cdot X_o) \in S(G/K).$$

Thus the sphere bundles are indeed given by G/M. $\square$

To study the multi-parameter flow on G-orbits separately, we simply fix $X \in \mathfrak{a}$. To determine the orbits of the flow we have to determine the sets $\{\operatorname{grad} h(X) \mid h \in \mathbb{R}[\mathfrak{a}]^W\}$, where $\mathbb{R}[\mathfrak{a}]^W$ denotes the W-invariant real polynomials on $\mathfrak{a}$. To this end we introduce some more structure theory. For $0 \neq \alpha \in \mathfrak{a}^*$ set

$$\mathfrak{g}_\alpha := \{X \in \mathfrak{g} \mid (\forall D \in \mathfrak{a})\ [D, X] = \alpha(D)X\}$$

and

$$\mathfrak{k}_\alpha := \mathfrak{k} \cap (\mathfrak{g}_\alpha + \mathfrak{g}_{-\alpha}), \quad \mathfrak{p}_\alpha := \mathfrak{p} \cap (\mathfrak{g}_\alpha + \mathfrak{g}_{-\alpha}).$$

If $\mathfrak{g}_\alpha \neq \{0\}$, then α is a *restricted root* of $(\mathfrak{a}, \mathfrak{g})$. We denote the set of restricted roots of $(\mathfrak{a}, \mathfrak{g})$ by $\Delta := \Delta(\mathfrak{a}, \mathfrak{g})$. An element X of $\mathfrak{a}$ is regular if $\alpha(X) \neq 0$ for all $\alpha \in \Delta(\mathfrak{a}, \mathfrak{g})$. For each regular element X_o one can define a set of *positive roots* via

$$\Delta^+ := \Delta^+(\mathfrak{a}, \mathfrak{g}) := \{\alpha \in \Delta(\mathfrak{a}, \mathfrak{g}) \mid \alpha(X_o) > 0\}.$$

Then Δ is the disjoint union of Δ^+ and $-\Delta^+$, the set of *negative roots*. We fix some choice of positive roots.

Note that $T_e(G/M)$, where $e := M \in G/M$ is the canonical base point, can be identified with $\mathfrak{g}/\mathfrak{m}$. This in turn can be identified with $\sum_{\alpha \in \Delta^+} \mathfrak{k}_\alpha$.

Lemma 8.6

(i) *The map $\phi\colon G/M \times \mathfrak{a} \to T(G/K)$ from Lemma 8.4 is smooth and surjective. Nevertheless, ϕ is not a covering. The derivative $\phi'(gM, X)$ is bijective if and only if X is regular. More precisely,*

$$\ker\phi'(e, X) = \sum_{X \in \ker\alpha} \mathfrak{k}_\alpha \quad and \quad \operatorname{im}\phi'(e, X) = \mathfrak{p} \times (\mathfrak{a} + \sum_{X \notin \ker\alpha} \mathfrak{p}_\alpha).$$

(ii) *The additive group $\mathfrak{a}$ acts (from the right) on $G/M \times \mathfrak{a}$ via $(gM, Y) \cdot X = (g\exp(X)M, Y)$.*

(iii) *If the action is restricted to $G/M \times \{X\}$ and $\exp \mathbb{R}X$, then one obtains the geodesic flow in the image of ϕ.*

Proof. In view of Example 8.1 only part (i) needs a proof. To calculate the derivative

$$\phi'(e, X) \colon \mathfrak{p} \oplus \sum_{\alpha \in \Delta^+} \mathfrak{k}_\alpha \times \mathfrak{a} \to \mathfrak{p} \times \mathfrak{p} \cong T_{\phi(e,X)}(G/K)$$

of ϕ in (e, X) we use Lemma 8.3 which says that under the identification $T_{\phi(e,X)}(G/K) \cong \mathfrak{p} \times \mathfrak{p}$ we have

$$\phi'(e, X)(Y_p + Y_k, W) = (Y_p, [Y_k, X] + W), \qquad (8.17)$$

where $Y_p \in \mathfrak{p}$, $Y_k \in \sum_{\alpha \in \Delta^+} \mathfrak{k}_\alpha$, and $W \in \mathfrak{a}$. Writing $Y_k = \sum_{\alpha \in \Delta^+}(Y_\alpha + \theta Y_\alpha)$ with $Y_\alpha \in \mathfrak{g}_\alpha$ and the Cartan involution $\theta \colon \mathfrak{g} \to \mathfrak{g}$ with respect to the decomposition $\mathfrak{g} = \mathfrak{k} + \mathfrak{p}$ we find

$$[Y_k, X] = \sum_{\alpha \in \Delta^+} \alpha(X)(Y_\alpha - \theta Y_\alpha) \in \mathfrak{p}. \qquad (8.18)$$

Thus the kernel of $\phi'(e, X)$ consists of all those $Y_k = \sum_{\alpha \in \Delta^+}(Y_\alpha + \theta Y_\alpha)$ with $Y_\alpha \in \ker \alpha$ and this proves the first claim. Since $\mathfrak{p}_\alpha = \{X - \theta X \mid X \in \mathfrak{g}_\alpha\}$ the second claim also follows from the equations (8.18) and (8.17). $\qquad \square$

Note that the centralizer $\mathfrak{k}_X$ of X in $\mathfrak{k}$ is given by

$$\mathfrak{m} + \sum_{X \in \ker \alpha} \mathfrak{k}_\alpha := \mathfrak{z}_\mathfrak{k}(X) := \{Y \in \mathfrak{k} \mid [X, Y] = 0\}$$

Analogously, given $X \in \mathfrak{a}$ consider the centralizer $\mathfrak{g}_X := \mathfrak{z}_\mathfrak{g}(X) := \{Y \in \mathfrak{g} \mid [Y, X] = 0\}$. If $\Delta_X = \{\alpha \in \Delta \mid \alpha(X) = 0\}$, then $\mathfrak{g}_X = \mathfrak{m} + \mathfrak{a} + \sum_{\alpha \in \Delta_X} \mathfrak{g}_\alpha$. Consider the center $\mathfrak{z}(\mathfrak{g}_X)$ of the reductive θ-invariant algebra $\mathfrak{g}_X$ and $\mathfrak{a}(X) := [\mathfrak{g}_X, \mathfrak{g}_X] \cap \mathfrak{a}$. Then $\mathfrak{a}(X)$ is a Cartan subspace for $[\mathfrak{g}_X, \mathfrak{g}_X]$ and $\mathfrak{a} = \mathfrak{a}(X) \oplus E_X$, where

$$E_X := \mathfrak{p} \cap \mathfrak{z}(\mathfrak{g}_X). \qquad (8.19)$$

Lemma 8.7 *Let $p, q_1, \ldots, q_k$ be different points in $\mathbb{R}^n$ and $v \in \mathbb{R}^n$. Then there exists a polynomial $h \in \mathbb{R}[x_1, \ldots, x_n]$ such that $h(p) = 0 = h(q_j)$ for $j = 1, \ldots, k$ and*

$$\operatorname{grad} h(p) = v$$
$$\operatorname{grad} h(q_j) = 0 \quad \forall j = 1, \ldots, k.$$

Proof. We use induction over k. Let $e_1, \ldots, e_n$ be the standard basis for $\mathbb{R}^n$. If $k = 1$ and $v = \sum_{j=1}^{n} \alpha_j e_j$ and $p = (p_1, \ldots, p_n)$ we set $f := \sum_{j=1}^{n} \alpha_j (x_j - p_j)$ and find:

$$f(p) = 0 \quad \text{and} \quad \operatorname{grad} f(p) = v.$$

Clearly we can find a polynomial $g_1 \in \mathbb{R}[x_1, \ldots, x_n]$ such that

$$g_1(q_1) = 0 \quad \text{and} \quad g_1(p) = 1.$$

Then $f_1 := fg_1$ satisfies $\operatorname{grad} f_1 = g_1 \operatorname{grad} f + f \operatorname{grad} g_1$ and therefore

$$\operatorname{grad} f_1(p) = \operatorname{grad} f(p) = v$$
$$f_1(p) = 0$$
$$f_1(q_1) = 0.$$

Finally we set $h_1 := f_1 g_1$ and obtain

$$\operatorname{grad} h_1(p) = \operatorname{grad} f_1(p) = v$$
$$\operatorname{grad} h_1(q_1) = 0$$
$$h_1(p) = 0$$
$$h_1(q_1) = 0.$$

using $\operatorname{grad} h_1 = g_1 \operatorname{grad} f_1 + f_1 \operatorname{grad} g_1$.

To do the induction we now assume that we have $h_k \in \mathbb{R}[x_1, \ldots, x_n]$ such that $h_k(p) = 0 = h_k(q_j)$ for $j = 1, \ldots, k$ and

$$\operatorname{grad} h_k(p) = v$$
$$\operatorname{grad} h_k(q_j) = 0 \quad \forall j = 1, \ldots, k$$

as well as a polynomial $g_{k+1} \in \mathbb{R}[x_1, \ldots, x_n]$ such that $g_{k+1}(p) = 1$ and $g_{k+1}(q_j) = 0$ for $j = 1, \ldots, k+1$. Then we set $f_{k+1} := g_{k+1} h_k$ and find

$$\operatorname{grad} f_{k+1}(p) = \operatorname{grad} h_k(p) = v$$
$$f_{k+1}(p) = 0$$
$$f_{k+1}(q_j) = 0 \quad \forall j = 1, \ldots, k+1.$$

Then $h_{k+1} := f_{k+1} g_{k+1}$ yields

$$\operatorname{grad} h_{k+1}(p) = \operatorname{grad} f_{k+1}(p) = v$$
$$\operatorname{grad} h_{k+1}(q_j) = 0 \quad \forall j = 1, \ldots, k+1$$
$$h_{k+1}(p) = 0$$
$$h_1(q_j) = 0 \quad \forall j = 1, \ldots, k+1$$

and hence the claim. $\qquad\square$

Lemma 8.8 *Let E be a real Euclidian vector space and W a finite subgroup of $\mathrm{O}(E)$. Given $v \in E$ and $\xi \in T_v E \cong E$ such that $w(\xi) = \xi$ for all w in the stabilizer W_v of v in W, one can find a W-invariant polynomial f on E with $\operatorname{grad} f(v) = \xi$.*

Proof. Lemma 8.7 shows that one can find a polynomial h on E such that $\operatorname{grad} h(\gamma v) = \gamma \xi$ for all $\gamma \in W$. Then $f = \frac{1}{|W|} \sum_{\gamma \in W} h \circ \gamma$ has the desired properties. In fact,

$$\operatorname{grad} f(v) = \frac{1}{|W|} \sum_{\gamma \in W} \operatorname{grad}(h \circ \gamma^{-1})(v) = \frac{1}{|W|} \sum_{\gamma \in W} \gamma \operatorname{grad} h(\gamma^{-1} v)$$
$$= \frac{1}{|W|} \sum_{\gamma \in W} \gamma \gamma^{-1} \xi = \xi$$

$\qquad\square$

Let $f_1, \ldots, f_k$ be a set of algebraically independent generators of $\mathcal{A}$ and $h_j := f_j|_{\mathfrak{a}}$ for $j = 1, \ldots, k$. Then $k = \dim \mathfrak{a}$ and the equality $\operatorname{grad}(uv) = u \operatorname{grad} v + v \operatorname{grad} u$ for functions on a Euclidean vector space shows that

$$\{\operatorname{grad} h(X) \mid h \in \mathbb{R}[\mathfrak{a}]^W\} = \sum_{j=1}^{k} \mathbb{R} \operatorname{grad} h_j(X) \tag{8.20}$$

since the f_j are constant on G-orbits. Set $H = (h_1, \ldots, h_j) \colon \mathfrak{a} \to \mathbb{R}^k$ and choose a basis for $\mathfrak{a}$ to view H as a map $H \colon \mathbb{R}^k \to \mathbb{R}^k$. Then [Hu90], Prop. 3.13, says that up to a non-zero scalar the Jacobian determinant of H is given by $\prod_{\alpha \in \Delta^+} \alpha$. Thus for regular X the linear map $H'(X) \colon \mathbb{R}^k \to \mathbb{R}^k$ is bijective and hence

$$\{\operatorname{grad} h(X) \mid h \in \mathbb{R}[\mathfrak{a}]^W\} = \mathfrak{a}. \tag{8.21}$$

For singular X the Jacobian vanishes, so the span $\sum_{j=1}^{k} \mathbb{R} \operatorname{grad} h_j(X)$ is a proper subset of $\mathfrak{a}$. Using Lemma 8.8 we can calculate $\sum_{j=1}^{k} \mathbb{R} \operatorname{grad} h_j(X)$.

Lemma 8.9	*Let $f_1, \ldots, f_k$ be a set of algebraically independent generators of $\mathcal{A}$ and $X \in \mathfrak{a}$. Then*

$$E_X = \sum_{j=1}^{k} \mathbb{R} \operatorname{grad} f_j(X).$$

Proof.	Since E_X is the intersection of all maximal abelian subspaces of $\mathfrak{p}$ containing X (see (8.19) and [Eb96], Lemma 2.20.9), Lemma 8.5 applied to any of these spaces shows that $\operatorname{grad} h(X) \in E_X$ for all K-invariant smooth functions h on $\mathfrak{p}$. Thus we have $\sum_{j=1}^{k} \mathbb{R} \operatorname{grad} h_j(X) \subseteq E_X$. For the converse we note that the stabilizer W_X of X in the Weyl group W acts trivially on E_X. Therefore (see Lemma 8.8) any element ξ of E_X occurs as the gradient of a W-invariant polynomial $h \in \mathbb{R}(\mathfrak{a})^W$ in X. But then equation (8.20) proves the claim.				$\square$

Summing up the results of this section we obtain

Theorem 8.2	*The smooth distribution Δ associated with the family of vector fields $\mathcal{D} = \{H_f \mid f \in \mathcal{A}\}$ introduced in (8.12) is given by*

$$\Delta([\mathbf{1}, X]) = E_X \times \{0\} \in \mathfrak{p} \times \mathfrak{p} \cong T_{[\mathbf{1},X]}(G \times_K \mathfrak{p}) \cong T_{[\mathbf{1},X]}(T(G/K))$$

for $X \in \mathfrak{a}$.

## 8.4	The Ergodic Arnold–Liouville Theorem

Fix $X \in \mathfrak{a}$. Then, according to Lemma 8.4 $[g, X] = [\mathbf{1}, X]$ is equivalent to $g \in K$ with $\operatorname{Ad}(g)X = X$. Therefore the stabilizer of $[\mathbf{1}, X] \in G \times_K \mathfrak{p} \cong T(G/K)$ in G is

$$\{g \in G \mid g \cdot [\mathbf{1}, X] = [\mathbf{1}, X]\} = \{k \in K \mid \operatorname{Ad}(k)X = X\} =: K_X. \qquad (8.22)$$

Lemma 8.10	*Suppose that Γ is a lattice in G, i.e. a discrete subgroup such that $\Gamma \backslash G$ has finite measure. Fix $X \in \mathfrak{a}$. Then the abelian group $\exp(E_X)$ acts ergodically on $\Gamma \backslash G / K_X$ if X has a non-zero component in each simple summand of $\mathfrak{g}$. If there exists a simple summand of $\mathfrak{g}$ in which X has no component, then in general $\exp E_X$ does not act ergodically on $\Gamma \backslash G / K_X$.*

Proof.	Suppose that $f \in L^2(\Gamma \backslash G / K_X)$ is invariant under $\exp E_X$. Since K_X is compact, we can lift f to an L^2-function $\tilde{f}$ on $\Gamma \backslash G$ which is invariant under $\exp E_X$ (and K_X). If X has a non-zero component in each simple summand of $\mathfrak{g}$, then the group $\exp E_X$ which contains $\exp \mathbb{R}X$ is totally non-compact in the sense of [BM00], Def.III.2.4, so that we can apply Moore's ergodicity theorem (see [BM00], Thm.III.2.5) which implies that $\tilde{f}$ is constant. Therefore f is constant and this shows that $\exp E_X$ acts ergodically on $\Gamma\backslash G/K_X$.

For the converse we note that E_X is contained in the sum of those simple summands of $\mathfrak{g}$ for which X has non-zero components. So if X has no component in some simple summand of $\mathfrak{g}$, the corresponding factor in G is pointwise fixed under $\exp E_X$ which shows that for example in the case of Γ being a product of lattices the action cannot be ergodic. $\qquad\square$

Note that the role of the lattice in the previous lemma is only to guarantee that the space on which our action lives has finite measure, so that we are in the framework of standard ergodic theory. Moore's theorem works for unitary representations and could have been applied also in the case $\Gamma = \{\mathbf{1}\}$.

Moreover, we note that the same argument shows that already the one-parameter group $\exp(\mathbb{R}X)$ acts ergodically if X has non-zero components in each simple summand of $\mathfrak{g}$.

Theorem 8.3 (Ergodic Arnold–Liouville) *Let G/K be a Riemannian symmetric space of non-compact type and $\mathcal{A}$ be the algebra of G-invariant smooth functions on $T^*(G/K) \cong T(G/K)$ which restrict to polynomials on $\mathfrak{p}$.*

(i) *As an additive group $\mathcal{A}$ acts on $T^*(G/K)$ via $f \cdot [g, X] = \Phi_f^1([g, X])$, where Φ_f^t is the flow of the hamiltonian vector field associated with f.*

(ii) *We have $f \cdot [g, X] := [g \exp (\operatorname{grad} f|_\mathfrak{a}(X)), X]$.*

(iii) *The $\mathcal{A}$-action commutes with the natural G-action on $T(G/K)$.*

(iv) *The joint level sets of $\mathcal{A}$ are precisely the G-orbits in $T(G/K)$.*

(v) *Suppose that Γ is a lattice in G. Then the $\mathcal{A}$-action factors through to an action on $\Gamma\backslash T(G/K) \cong T(\Gamma\backslash G/K)$ and the $\mathcal{A}$-action on $\Gamma\backslash G \cdot [\mathbf{1}, X]$ is ergodic if X has a non-zero component in each simple summand of $\mathfrak{g}$. If there exists a simple summand of $\mathfrak{g}$ in which X has no component, then in general $\exp E_X$ does not act ergodically on $\Gamma\backslash G \cdot [\mathbf{1}, X]$.*

Proof. Part (i) is clear. Parts (ii) and (iii) are immediate consequences of (8.15) and (8.16). Part (iv) was shown in Theorem 8.1. To prove part (v) we note first that the factorization of the $\mathcal{A}$-action is an immediate consequence of the G-equivariance of all the constructions involved. By (8.15) and Lemma 8.9 it now suffices to consider the $\exp E_X$-action on

$\Gamma \backslash G \cdot [\mathbf{1}, X]$. But then (8.16) together with (8.22) reduces the problem to characterizing the ergodicity of the right multiplication by $\exp E_X$ on $\Gamma \backslash G / K_X$ which in turn has been established in Lemma 8.10. $\qquad \square$

If G is simple, then the $\mathcal{A}$-action is ergodic on all G-orbits except the zero-section. On the zero-section the $\mathcal{A}$-action is in fact trivial since $E_0 = \{0\}$.

Bibliography

[AM87] Abraham, R., and J.E. Marsden, *Foundations of Mechanics*. Second Edition, Addison Wesley, Redwood City, 1987

[BM00] Bekka, M.B., and M. Mayer, *Ergodic Theory and Topological Dynamics of Group Actions on Homogeneous spaces*. Cambridge Univ. Press, 2000

[tD91] tom Dieck, T., *Topologie*. De Gruyter, Berlin, 1991

[Eb96] Eberlein, P. *Geometry of Nonpositively Curved Manifolds*. The University of Chicago Press, 1996

[GS84] Guillemin, V., and S. Sternberg, *Symplectic techniques in physics*. Cambridge Univ. Press, 1984

[He78] Helgason, S.: *Differential Geometry, Lie Groups, and Symmetric Spaces*. Acad. Press, Orlando, 1978

[He84] Helgason, S.: *Groups and Geometric Analysis*. Acad. Press, Orlando, 1984

[Hu90] Humphreys, J.E., *Reflection Groups and Coxeter Groups*. Cambridge Univ. Press, 1990

[Su73] Sussmann, H., "Orbits of Families of Vector Fields and Integrability of Distributions." Trans. Amer. Math. Soc. **180** (1973), 171–188

[Th81] Thimm, A., "Integrable geodesic flows on homogeneous spaces." Ergod. Th. and Dynam. Sys. **1** (1981), 495–517

Chapter 9

The Renormalization Fixed Point as a Mathematical Object

R. P. Langlands[1]

9.1 Introduction

The success of renormalization group methods in statistical mechanics and, in particular, in the study of critical phenomena is well known to be a consequence of the presence of only a small number of expanding directions, often just one or two, at the pertinent fixed point of the associated infinite-dimensional dynamical system, the other directions being contracting. What cannot be sufficiently emphasized is that the numerical success and the great robustness of the methods appear to result from the extreme rapidity with which the eigenvalues in the contracting directions descend to 0. Given the importance of this property, it is troubling that no methods have been found to establish it rigorously in important concrete cases such as percolation or the Ising model.

It is not immediately clear what is called for, some flexibility certainly. There is of course a dynamical system to define, but its relation to the given model is not prescribed. One might wish, as in the early paper [W] of K. G. Wilson to replace a model on a discrete lattice, for example the Ising model, by a family of Ginzburg-Landau models, more generally a difficult model by an easier model that is more easily imbedded in a family or, for some other reason, more easily treated. Then the appropriate strategy might be to establish the necessary dynamics in this family, enlarged if necessary, and only afterwards transfer the results to the original model, discrete or not. This second step will also be analytical and will presumably rely in

---

[1]Institute for Advanced Studies, Einstein Drive, Princeton, NJ 08540, U.S.A,
rpl@ias.edu

turn on a different form of the very characteristics used to establish the properties of the dynamical system.

The fixed point lies in the infinite-dimensional space of the dynamical system and is to be described by coordinates. Since the space is not necessarily linear, these coordinates may not be of the same nature or have the same meaning at the fixed point as they have at the models, which, I recall, are themselves to be regarded as points in the space, but perhaps in a very different part, where the coordinates have quite a different interpretation. The fixed point of the Ising model is, for example, related to a very special conformal field theory, the minimal model with central charge $c = 1/2$. The data defining this field theory must be present in the coordinates of the fixed point, either implicitly or explicitly, but are scarcely to be seen, except by inference, in the model itself.

Another possibility is to search not for an infinite-dimensional system that contains the dynamics of the renormalization but rather for a sequence of finite-dimensional approximations to it. This is largely just a matter of realizing the analytic problems concretely. The second step would then be to transfer the results for this sequence to the original model.

I have thought about these questions over the years, very often in collaboration with Yvan Saint-Aubin and with a number of students at the Université de Montréal, performing some instructive experiments but without making any real mathematical inroads. I would like to take the opportunity of this conference[2] to review the results, and to reflect – in a highly speculative way – on some of the analytic problems that I would like to see solved and on further possible numerical investigations. It is best to turn immediately to the models, for once they are defined, it will be possible to explain in a precise, concrete way what could only be intimated in these introductory remarks.

9.2 Percolation

There are many models for percolation. In some sense all the usual planar models, for example those discussed in [P2][3], are associated to the same fixed point. Some care has to be taken when interpreting this statement. The group $GL(2, \mathbb{R})$ of linear transformations of the plane operates on the

---

[2]This paper is based on notes for lectures at the Białowieża conference that I was unable, at the last minute, to attend.

[3]The papers [P1, P2, P3] are also available on the website
www.sunsite.ubc.ca/DigitalMathArchive/Langlands/

models. So it should operate on the pertinent fixed points of the renormalization dynamics. It turns out, numerically at least, that to each model is associated a conformal structure on the plane and that the appropriate fixed point is determined by this conformal structure. Since the set of conformal structures is a homogeneous space under $GL(2,\mathbb{R})$ that can be identified with the upper half-plane, so is the set of fixed points. To remove this indeterminacy, we consider only models symmetric with respect to both coordinate axes and with respect to interchange of the two axes

To explain the strategy, we fix a model, to be specific, percolation by sites on the square lattice, but any model would do. Recall that, in this model, each site (m,n), $m,n \in \mathbb{Z}$, is open with a probability p, $0 \leq p \leq 1$. One interesting value attached to the model is the probability $\pi_L(p)$ that there is a crossing (by leaping from one open site to another open site adjacent in the sense of the lattice) of a large square of side L. There is a unique critical value p_c, $0 < p_c < 1$, for which

$$0 < \liminf_{L \to \infty} \pi_L(p_c) \leq \limsup_{L \to \infty} \pi_L(p_c) < 1.$$

For $p < p_c$ both limits are 0; for $p > p_c$ they are both 1. The simplest open problems are whether the limit superior and the limit inferior are equal to each other at $p = p_c$ and whether they are both equal to .5. In comparison to other questions about critical points, they are extremely easy to state, although not necessarily easier to prove. The value .5 is supposed to be universal. It is believed to be valid for all other planar models provided they have the three basic symmetries: reflection symmetry in the two axes and symmetry under interchange of the two axes.

There is a more general form of the question ([P2]). Take, once and for all, $p = p_c$. Suppose C is a simple closed curve in the plane and α, β are two intervals on it. If L is any large positive number, we can define as in [P2] the probability $\pi_C^L(\alpha, \beta)$ of a crossing inside the dilation LC of C from $L\alpha$ to $L\beta$. Then it is believed – the belief is supported by the numerical evidence – that the limit, $\pi_C(\alpha, \beta)$ of $\pi_C^L(\alpha, \beta)$ as L approaches infinity exists and is universal. More generally, it should be possible to define a similar limiting probability $\pi_C(\alpha_1, \beta_1, \alpha_2, \beta_2, \ldots; \gamma_1, \delta_1, \gamma_2, \delta_2, \ldots)$ that there are crossings from α_1 to β_1, α_2 to β_2 and so on, but none from γ_1 to δ_1 and so on. Numerical studies give us every reason to believe that these limits, referred to as crossing probabilities, exist and that they are universal, thus independent of the model. This was implicit in [P3], where they, or rather collections of numbers,

$$\rho(E) = \rho_C(\alpha_1, \beta_1, \alpha_2, \beta_2, \ldots; \gamma_1, \delta_1 \ldots),$$

in which E stands for the *event* or *crossing* defined by C and the collection of intervals, are used as coordinates in the space in which the dynamics is defined. In particular, the collection

$$\{\pi(E)\} = \{\pi_C(\alpha_1, \beta_1, \alpha_2, \beta_2, \ldots; \gamma_1, \delta_1, \gamma_2, \delta_2, \ldots)\},$$

in which C runs over all admissible curves and the intervals in C are arbitrary, are supposed to be the coordinates of the pertinent fixed point.

Before coming to [P3], I recall that thanks to Schramm, Smirnov, and several other authors (see [SS] and the papers referred to there) a good deal more is now known than was known when [P2] and [P3] were written. Since the central problem, universality, remains unsolved, it is still possible, none the less, that [P3] has something to offer. It will no doubt be clear to the reader that to overcome the technical difficulties that arise in pursuing the strategy of that paper a much better command of the available techniques than I possess at present will be required. Although life for many of us is not so short as it once was, art too grows longer and ever more rapidly; this has to serve as my apology for presenting my reflections in a half-baked form.

In response to the studies on crossing probabilities reported in [P1], M. Aizenman suggested an hypothesis of conformal invariance for the crossing probabilities. It is still not known that the crossing probabilities are defined for any but a few very special models. It is, for example, not known that they are defined for the square lattice. It is therefore certainly not known that they are universal. Smirnov has, however, proved that the crossing probabilities are defined for percolation on the triangular lattice and that Aizenman's hypothesis of conformal invariance is valid in this case. Thus what remains to be proved is existence in general and universality. Although the papers [P1, P3] were numerical, they were also, for me at least, an attempt to create some confidence in a particular analytic strategy for establishing universality as a consequence of the existence of fixed points for an appropriate renormalizing dynamical system. Although these systems were introduced in [P3], the strategy was not explained. I would like to explain it here, even though I have not yet made any serious attempt to deal with the analytic problems that arise; they are formidable. All I can do is remind myself of them. I begin with a brief review of the definitions of [P3], referring the reader to that paper for more precision.

If conformal invariance is assumed, many of the coordinates $\pi_C(\alpha_1, \ldots; \gamma_1, \ldots)$ are redundant. In particular, it is enough to take C to be a unit square. For each positive integer l we divide each of its sides into l intervals of length $1/l$. This yields a set $\mathfrak{A}_l$ of $4l$ intervals on the boundary of the square. On the assumption of continuity of the crossing probabilities, it would be enough to know the coordinates $\pi_C(\alpha_1, \beta_1, \alpha_2, \beta_2, \ldots; \gamma_1, \delta_1 \ldots)$ for intervals α_i, β_i, γ_j and δ_j that are unions of some of the intervals in $\mathfrak{A}_l$, provided of course that l is taken larger and larger. In other words, as an approximation to the full set of crossing probabilities, we can consider only the set defined by the events associated to the unit square and intervals α_i, β_i, γ_j and δ_j each of which is a union of intervals in $\mathfrak{A}_l$.

These events can be defined by a family of basic events. We can attach to each configuration for percolation a function y on pairs (α, β) in $\mathfrak{A}_l$ that takes values in $\{0, 1\}$. The value is 1 if the configuration contains a crossing from α to β and is otherwise 0. The underlying space of the dynamical system is, in principle, the space Π_l of measures on the set $\mathfrak{Z}_l$ of such functions. The insistence in [P3] and [P4] on the FKG-inequality was somewhat of a luxury. It would, however, have been much better to take, as we ultimately did in [P3], only measures that respect the three basic symmetries. So I now add this to the definition of Π_l. It is clear that whenever we can attach crossing probabilities to a given model M of percolation, we can attach to it an element $\eta_l = \eta_l(M)$ of the space Π_l. Moreover, if $l|m$ then we can deduce $\eta_l(M)$ from $\eta_m(M)$ because the intervals in $\mathfrak{A}_l$ are unions of the intervals in $\mathfrak{A}_m$. Finally, universality amounts, at least for crossing probabilities, to the assertion that the family $\{\eta_l \,|\, l \in \mathbb{N}\}$ is independent of M.

To establish universality it would suffice to show that there is another family ν_l, defined independently of any particular model, such that, for any given model M, the point $\eta_l(M)$ is well approximated by ν_l for large l, a relation more precisely expressed by (7.1) of [P3]. It is not, by the way, supposed that ν_l can be directly deduced from ν_m if $l|m$. This will not be so.

The idea explained, or rather the wish expressed, in [P3] and in [P4] is that ν_l could be introduced as the fixed point of a transformation defined independently of any model. The transformation, which is intended to be a finite-dimensional approximation to the dynamics of renormalization, was defined in [P3] and a fixed point was exhibited numerically for $l = 2$, but no attempt was made to begin the analysis.

There is no need to consider all $l > 0$. It clearly suffices, for the purpose of establishing universality, to consider any sequence $\{l_k\}$ of integers that

approaches infinity multiplicatively, for example, the sequence 2^k, $k \geq 0$. It
was also necessary to replace $\mathfrak{Z}_l$ by a subset $\mathfrak{Y}_l$, or rather to demand that
all the probability measures in Π_l assign the measure 0 to points outside
$\mathfrak{Y}_l$. This is easy to arrange even for the measures associated to percolation
models, but begs a question that, sooner or later, will come back to haunt
anyone who attempts to apply the strategy. In essence, the observation
is that the finite model cannot be an approximation to percolation if connections between neighboring intervals are admitted indiscrimately. So we
considered only the functions in a set $\mathfrak{Y}_l$ that is defined by excluding most
such connections. This too entails possible difficulties and it has still to be
shown that they do not arise.

   To explain this question, consider the dynamical transformation $\Theta_l =
\Theta_l^{(2)} : \Pi_l \rightarrow \Pi_l$ introduced in [P3, P4]. A basic object is the square with its
boundary divided into $4l$ intervals of equal size. Suppose we fit four such
squares together to form a single large square. The intervals in the sides of
the smaller squares that lie on the boundary of the large square will divide
it into $8l$ intervals of equal size. We fuse adjacent intervals in pairs to arrive
at a division of the boundary of the large square into $4l$ equal intervals.

   If we have for each of the small squares $\sigma_{i,j}$ a point $y_{i,j}$ in $\mathfrak{Y}_l$, then
we can try, using just the crossings of $y_{i,j}$, to cross from one of the $4l$
intervals on the boundary of the large square to another, the understanding
being that we connect a crossing of $\sigma_{i,j}$ to one of $\sigma_{i',j'}$ if these two small
squares have a common side and if the two points $y_{i,j}$ and $y_{i',j'}$ both reach
a common interval in the common side. In this way, we attach to the
collection $\{y_{i,j}\}$ an element of $\mathfrak{Z}_l$. Modifying it by removing the connections
between adjacent sides, we arrive finally at a point in $\mathfrak{Y}_l$. This map of the
4-fold product of $\mathfrak{Y}_l$ with itself to $\mathfrak{Y}_l$ yields immediately the associated map
Θ_l on measures and it is this map that defines the dynamics and for which
we need to establish the existence of a fixed point ν_l.

   We need to establish not only the existence of a sequence of fixed points

$$\nu_{l_1}, \nu_{l_2}, \ldots$$

as $l_1, l_2, \ldots$ runs through a sequence of positive integers tending multiplicatively to infinity, but also that, for any given model M and for each l in the
sequence, ν_l approaches $\eta_l = \eta_l(M)$, or rather that, in the notation of [P3],

$$\lim_{k \to \infty} \Gamma_m^{l_k}(\nu_{l_k}) = \eta_m.$$

The map $\Gamma_m^{l_k}$ is the map on measures attached to the coarsening map $\mathfrak{A}_{l_k} \rightarrow$

$\mathfrak{A}_m$, thus to the coarsening map $\mathfrak{Z}_{l_k} \to \mathfrak{Z}_m$. A small, but essential, technical point aside, this is defined by the condition that the image of a function z joins two intervals in $\mathfrak{A}_m$ precisely when these two intervals contain intervals in $\mathfrak{A}_{l_k}$ joined by z.

We have therefore to show that, among other things, η_l is an approximate fixed point of the transformation Θ_l. Since we pass from $\mathfrak{Z}_l$ to $\mathfrak{Y}_l$ by suppressing connections between adjacent intervals, we will have to establish quantitative forms of the following type of assertion. Take the unit square, divide it in two parts by a central vertical line, and divide this vertical line into l equal intervals. Then, for l large, the probability that there is a horizontal crossing of the square is approximately the probability that there is a horizontal crossing without any subpath that moves from one of the l equal subintervals to an adjacent one.[4]

The existence of a sequence of fixed points will perhaps be most easily established for the indices $l_k = 2^k$. Because of the numerical results of [P3], it can, in some sense, be taken for granted that ν_2 exists and even, although this is not of much use, that it is close to η_2. Since we are now taking only measures fixed by the three basic symmetries, many of the eigenvalues appearing in Table III of [P3] are no longer pertinent. The first pertinent ones are 1.6345851, 0.4072630, 0.2445117, 0.1721207, 0.1123677. Thus there is only one that is greater than 1. It is the only eigenvalue that is relevant in the technical sense. The others are less than 1 and, apparently, rapidly decreasing to 0. The proposed strategy is to begin with the fixed point at level $k = 1$ and to establish the existence of the fixed point for larger k inductively using Newton's method. This will require, of course, showing that the eigenvalue structure, one relevant eigenvalue and a rapidly decreasing sequence of irrelevant eigenvalues, is preserved. Although I briefly outline this strategy in the following paragraphs, I stress immediately, once and for all, that I have not, perhaps to my shame, begun to think about the estimates that will be needed. They will not be easy to establish.

Given any point y_{2^k} in $\mathfrak{Y}_{2^k}$, we can construct, by the heaping $\Phi^{(2)}_{2^k}$ of [P3], a point in $\mathfrak{Y}_{2^{k+1}}$, basically by the same process as before, except that we map the 4-fold product of $\mathfrak{Y}_{2^k}$ not to $\mathfrak{Y}_{2^k}$ but to $\mathfrak{Y}_{2^{k+1}}$, as is possible because the side of each square $\sigma_{i,j}$, $i, j = 1, 2$, that lies in the boundary of the large square is already divided into 2^k sides. There is also, as in

---

[4]There are mathematicians, for example O. Schramm or S. Smirnov, who have thought much more deeply about such questions than I and who have, I believe, partial answers to them.

[P3], a coarsening $\Gamma^{2^{k+1}}_{2^k}$ that maps $\mathfrak{Y}_{2^{k+1}}$ to $\mathfrak{Y}_{2^k}$. The associated maps on measures yield a repeating sequence

$$\to \Pi_{2^k} \to \Pi_{2^{k+1}} \to \Pi_{2^k} \to \Pi_{2^{k+1}} \to \ .$$

Let $\Psi_k : \Pi_{2^{k+1}} \to \Pi_{2^{k+1}}$ be the composition $\Phi^{(2)}_{2^k} \circ \Gamma^{2^{k+1}}_{2^k}$ of the two distinct maps in this sequence and, to simplify notation, let $\Delta_k = \Theta_{2^{k+1}}$. It is $\Gamma^{2^{k+1}}_{2^k} \circ \Phi^{(2)}_{2^k}$. Suppose a fixed point ν_{2^k} of Θ_{2^k} has been found. Let ψ_k be its image in $\Pi_{2^{k+1}}$. It is clear that ψ_k is a fixed point of Ψ_k.

The map Ψ_k is a projection $\Gamma^{2^{k+1}}_{2^k}$ on Π_{2^k} followed by the map $\Phi^{(2)}_{2^k}$. The tangent map $D\Gamma^{2^{k+1}}_{2^k}$ at ψ_k and, indeed, at any point is also a projection. So the behavior of $D\Psi_k$ is, apart from a preliminary compression, that of $D\Phi^{(2)}_{2^k}$. Thus, in order to show that $D\Psi_k$ has, apart from a large number of additional very small eigenvalues, eigenvalues close to those of $D\Theta_{2^k}$, thereby beginning the induction, we have to show that $D\Gamma^{2^{k+1}}_{2^k}$ does not deform the image of Π_{2^k} in $\Pi_{2^{k+1}}$, at least not in a neighborhood of the fixed point ν_{2^k}.

We can express this in terms of matrices. Suppose we choose coordinates in Π_{2^k} and $\Pi_{2^{k+1}}$ so that

$$D\Gamma^{2^{k+1}}_{2^k} = \begin{pmatrix} I & 0 \end{pmatrix}.$$

Let

$$\begin{pmatrix} A \\ B \end{pmatrix}$$

be the matrix of $D\Phi^{(2)}_{2^k}$ at ν_{2^k}. Then

$$D\Psi_k = \begin{pmatrix} A & 0 \\ B & 0 \end{pmatrix}, \qquad D\Theta_{2^k} = A.$$

So we seem to need to show that B is of the form CA, for then

$$\begin{pmatrix} I & 0 \\ C & I \end{pmatrix} \begin{pmatrix} A & 0 \\ 0 & 0 \end{pmatrix} \begin{pmatrix} I & 0 \\ -C & 0 \end{pmatrix} = \begin{pmatrix} A & 0 \\ B & 0 \end{pmatrix}.$$

It will be important to control the size of C.

So long as we have not fixed the metrics on Π_{2^k} and $\Pi_{2^{k+1}}$, this is not a meaningful demand. What it might mean ultimately is that if α and β are intervals of length $1/2^k$ on the boundary and α_1, α_2, respectively β_1, β_2, the two subintervals of length $1/2^{k+1}$ into which they can be divided and if π' is a measure near the fixed point, π its image under $\Phi^{(2)}_{2^k}$, and π'' the image of π under $\Gamma^{2^{k+1}}_{2^k}$, then the probability, with reference to π, that α_i is connected to β_j is approximately independent of i and j and approximately determined, in a universal way, thus independently of α and β, by the probability that α is connected to β.

Our goal is not to find a fixed point of Ψ_k but of Δ_k. The method of Newton, in which one establishes that a map

$$\pi \to \pi - D^{-1}(\Delta_k \pi - \pi),$$

D being a constant approximation to the tangent map $D\Delta_k$, is contracting on some domain, is the obvious technique to apply ([L]). To make it work here, we would like to show that Δ_k is close to Ψ_k.

The maps Δ_k and Ψ_k on measures are both attached to maps

$$\mathfrak{Y}_{2^{k+1}} \times \mathfrak{Y}_{2^{k+1}} \times \mathfrak{Y}_{2^{k+1}} \times \mathfrak{Y}_{2^{k+1}} \to \mathfrak{Y}_{2^{k+1}}.$$

For the first, we heap and then coarsen; for the second, we coarsen and then heap. Were it not that some connections are suppressed upon coarsening, the second would always yield an element of $\mathfrak{Y}_{2^{k+1}}$ connecting more pairs of intervals. There would not, in general, be many more connections if we could be certain of a condition that I now attempt to explain.

Suppose we have two abutting unit squares T_1, T_2 and in their common side one of the intervals α of length $1/2^k$ into which it is divided. Let α_1 and α_2 be the two halves of α, both of length $1/2^{k+1}$. Suppose y_1 and y_2 lie in $\mathfrak{Y}_{2^{k+1}}$, the first with respect to one of the two squares and the second with respect to the other and suppose y_1 connects some interval β_1 to α_1 and y_2 connects some interval β_2 to α_2. Then we want there to be some other interval γ of length $1/2^{k+1}$ on the common side such that, for $i = 1$ and $i = 2$, y_i connects β_i to γ. We cannot expect this always to be so, but we would like it be so for most β_i and most β_j with a probability in y_1 and y_2 that is almost 1 with respect to the measure ψ_k, and thus with respect to any measure close to ψ_k.

Once again, no metric has been defined, but any metric on the measures will have to regard two measures as close not only if they assign approximately the same measure to each set but also if they assign equal measures to approximately the same set. Thus two atomic measures $c_1 \delta_{y_1}$ and $c_2 \delta_{y_2}$ will have to be regarded as close if c_1 is close to c_2 and y_1 is close to y_2. So this condition appears to be what is necessary to show that when acting upon a neighborhood of ψ_k, Δ_k is close to Ψ_k.

The measure ψ_k is the image of ν_{2^k} and ν_{2^k} is supposed to be an approximation to η_{2^k}. Thus ψ_k can be expected to be an approximation to $\eta_{2^{k+1}}$. The required property with respect to this measure is, at least roughly and intuitively, a consequence of a well known property of critical percolation on, say, the triangular lattice, which it will be necessary to establish anew for ν_{2^k} or ψ_k in the context of finite-dimensional approximations to percolation. Suppose each y_i is defined by an occupied percolation path p_i. Consider the square S_1 of side $1/2^{k+1}$ whose center is the common endpoint of α_1 and α_2. Let S_2 be a second square with the same center and a side $a/2^{k+1}$, where a is chosen as large as possible with respect to the condition that neither β_1 nor β_2 meets S_2. Then, as a result of Lemma 7.2 of [K], with a very high probability (of the order of $1 - a^{-\delta}, \delta > 0$) the square S_2 contains a path surrounding S_1. This path together with the union of the connected path in p_1 from β_1 to α_1 and the connected path in p_2 from β_2 to α_2 would, for the triangular lattice, be an occupied path joining β_1 to β_2.

Whether this strategy or something completely different will ultimately be used to establish universality in percolation, I do not know. Nor do I know whether I shall ever return to the problem in a serious way. It is nevertheless a pleasure to remind myself now and again of its depth.

9.3 The Ising Model

The paper [I] is rather long and some of the central numerical conclusions are easy to overlook. There is one in particular that I want to recall here. Since we are passing to a different topic, all notation is again free.

Various forms of the Ising model were considered in [I]. They are all defined by a graph Γ on a surface S, closed or open, with or without boundary. If Π is the set of vertices of Γ, then each model assigns a probability to each configuration $\sigma : \Pi \to \pm 1$, thus to each configuration of spins. Each configuration defines a unique partition of the vertices into the maximal subsets of constant sign that are connected within Γ. For the models of [I], it was possible to attach to each such partition a collection of simple oriented curves, the contour lines, $L_1, L_2, \ldots$ There will be, in general, several such collections attached to each σ because the orientations are arbitrary. Moreover, for most models there are configurations for which even the unoriented curves are ambiguously defined. Thus to each σ is attached the finite set Λ_σ of such collections and to each element $\lambda = \{L_1, L_2, \ldots\}$

of Λ_σ a probability. Set $\Lambda = \cup_\sigma \Lambda_\sigma$. It is a set furnished with a probability.

There was, in addition, for each model a notion of mesh ϵ and the possibility existed of taking the mesh to 0. Suppose we have on S an oriented curve C that is, at least at first, closed and smooth (even implicitly analytic) although not necessarily connected. Thus it is the union of a finite number of simple closed curves. Although some care has to be taken with perhaps degenerate intersections, it is pretty clear how to attach to each collection $\lambda = \{L_1, L_2, \dots\}$ a distribution δ_λ on C. The distribution will, in fact, be a measure, the sum of atomic measures of mass ± 1 at each of the intersections of each L_i with C, the sign being determined by the relative orientation of L_i and C at the given intersection. The map $\lambda \to \delta_\lambda$ allows us to transfer the probability measure on Λ to a probability measure on the space of distributions on C. This measure we denote μ_ϵ to emphasize the dependence on ϵ. What some of the numerical experiments of [I] demonstrate, or at least were meant to demonstrate, is that

$$\lim_{\epsilon \to 0} \mu_\epsilon = \mu = \mu_C = \mu_C^S$$

exists as a measure on the space of distributions on C and that it is conformally invariant and universal. As in percolation, the conformal invariance is with respect to a conformal structure determined by the model. Universality is, of course, also valid only within the family of models defining a given structure. It is important to observe that the measure will depend on S. It will be important to determine the extent of this dependence.

The experiments establishing the conformal invariance were not nearly so extensive as those undertaken in [P2] for percolation. Moreover, conformal invariance refers now to a structure with more components: the surface S, which may have a boundary, and the curve C, which may or may not lie in that boundary. So the conformal invariance is sometimes of a different nature than the conformal invariance for percolation: the pertinent maps refer to a surface and a curve, not simply to a curve and its interior. For the present purposes, the most important case is that of a compact S without boundary, which was experimentally the most difficult case of [I] and also the one to which the least space was given. Indeed, in that paper it is little more than an afterthought.

What I want to do here is to take the existence of μ_C for granted and suggest further properties that it might possess and that might be tested or even established, although proving that it has these properties is likely to be much harder than the problems for percolation discussed in the previous section. The measure μ_C depends strongly on the way C lies in S. We shall here be concerned primarily with compact S without boundary, thus, for example, with the plane compactified to the Riemann sphere or with an infinitely long cylinder also compactified to the sphere.[5] Numerical experiments for these examples were discussed toward the end of §3.2 of [I]; they are for me the most suggestive of the paper. The compact surface S is implicitly endowed with a conformal structure that is determined by the model – or universality class of models – with which we begin.

When considering percolation, we supposed implicitly that we were dealing with translation-invariant models in the plane. So the resulting conformal structures were parametrized by the upper half-plane and it was natural to separate the models into universality classes according to the attached conformal structure. When allowing, as now for the Ising model, models on various surface, each model defining on the surface one of the very many possible conformal structures on it, the division into universality classes is not so natural. It is perhaps somewhat better to treat all models as belonging to one universality class and to consider whatever data, parameters, or set of coordinates that define the fixed point as referring to all possible S and all possible conformal structures on them. As an example, whatever information we have to give to define the fixed point must allow us to generate all the measures μ_C. For translation-invariant percolation the matter is at first glance simpler as the symmetric fixed point generates under the action of $GL(2,\mathbb{R})$ all the others. More general models are, however, attached to less simple conformal structures ([P2]). So, at least in principle, to determine fully the universal fixed-point even for percolation requires that we be able to calculate crossing probabilities not only in the plane but also on surfaces with other conformal structures.

For percolation and for the Ising model, there is abundant evidence that there are also conformal field theories attached to the universality class. What I want to discuss in the remainder of this paper is their possible relation to the measures μ_C^S, confining myself to the Ising model for which the evidence of [I] is available. There would be several steps to the construction. First of all, a Hilbert space $\mathcal{H}$ has to be attached to the parametrized boundary of the unit disk. It will be defined as an L^2-space with respect to one of the measures μ_C^S. The space $\mathcal{H}$ introduced, there is a second, more recondite, collection of objects to be defined. If the compact surface S with

---

[5] As a consequence, the measures μ_ϵ and the limiting measure μ_C are concentrated on distributions that annihilate the constant functions. There are other possibilities that lead, in the language of conformal field theory, to different sectors. They will be ignored.

conformal structure together with two families of parametrized, smooth, oriented, simple closed curves $\{C_1, \ldots, C_m\}$ and $\{C_1', \ldots, C_n'\}$ on it, each of these $m + n$ curves disjoint from all the others and if $\Sigma \subset S$ has as oriented boundary the first set of curves as oriented together with the other set of curves oppositely oriented, then there is an operator

$$K_\Sigma : \otimes_{i=1}^m \mathcal{H} \to \otimes_{j=1}^n \mathcal{H}$$

attached to Σ. These operators are, in fact to depend only on Σ and not on the closed surface S.

They are also to be multiplicative, in the sense that if Σ' is a second surface with boundary $\{C_1', \ldots, C_n'\}$, with the opposite orientation, and $\{C_1'', \ldots, C_p''\}$, then

$$(1) \qquad K_{\Sigma'} \circ K_\Sigma = \alpha K_{\Sigma''},$$

if Σ'' is obtained by pasting Σ and Σ' along $\cup_{j=1}^n C_j'$. The relation (1) is a projective relation, valid for some constant α.

There is to be in addition an action of the circle group $e^{i\theta} \to \pi(e^{i\theta})$ on $\mathcal{H}$. Taking as Σ the annulus with inner radius 1 and outer radius e^r and setting $\pi(e^r) = K_\Sigma$, we see from (1) that $\pi(e^r)\pi(e^t) = \pi(e^{r+t})$. These two actions together will yield a representation of the semigroup $\{z \in \mathbb{C} \,|\, |z| > 1\}$. More generally, as in the papers of G. Segal ([S]), the operators K_Σ will yield an action of the semigroup of annuli with parametrized boundaries and then an action of the direct sum of two copies of the Virasoro algebra, one holomorphic, one antiholomorphic. There is a final step, the factorization of this representation into a direct sum of tensor products of an irreducible representation of the holomorphic algebra and one of the antiholomorphic algebra, and then the factorization of K_T in general into the contribution of a holomorphic conformal field theory and the contribution of an antiholomorphic field theory. This last step is very elaborate even for such a simple model as the free boson (see [CG] and the papers there referred to); there is no point at this stage, at least not for me, in speculating on it for the Ising model. I do, however, find it essential to be quite clear about those properties of the measures μ_C^S that permit the introduction of K_Σ. So I begin by reviewing them in the context of the free boson. They are simple enough theoretically and undoubtedly commonplaces for specialists, but I have no suitable reference. What is perhaps perfectly obvious to others was not always so to me.

Suppose $C = \cup_{i=1}^n C_i$ is the union of disjoint simple curves and is contained in the compact surface S. We take S to be without boundary at first, although a similar construction can be made when it has a boundary, even when some of the curves C_i lie in its boundary. If φ is a smooth function on C then we extend it to a function φ_S that is harmonic on each component of the complement of C in S with boundary values φ. The Dirichlet form

$$D(\varphi) = D(\varphi_S) = \frac{1}{2\pi} \int_S \{(\frac{\partial \varphi_S}{\partial x})^2 + (\frac{\partial \varphi_S}{\partial y})^2\} dx dy$$

is a quadratic form that depends only on the conformal structure. If g is any positive constant, we can introduce the gaussian measure on the distributions that annihilate the constant functions that is defined by

$$\exp(-gD(\varphi)).$$

This is the measure μ_C^S attached to C in S for the free boson. A property of μ_C^S that seems to be very important is that its equivalence class, in the sense of mutual absolute continuity, is independent of S, provided that S is compact without boundary.

In essence, this means that if S_1 and S_2 are two such surfaces, then the difference $D(\varphi_{S_1}) - D(\varphi_{S_2})$, defined at first only for sufficiently smooth φ, can be extended to all distributions, or at least, given our restrictions, to all distributions annihilating the constant functions, or even, the weakest possible assertion, just to all distributions outside a set of measure 0 with respect to the gaussian measure. Before sketching a proof of this, consider some examples.[6]

Consider first the unit circle C in the Riemann sphere. Let

$$(2) \qquad \varphi(z) = \sum_{k>0} a_k z^k + \sum_{k>0} a_{-k} \bar{z}^k, \quad |z| = 1.$$

The function φ extends as φ_{S_l} to $S_l = \{|z| \leq 1\}$ in the form given. It extends to $S_r = \{|z| \geq 1\}$ as

$$\varphi_{S_r} = \sum_{k>0} a_k \bar{z}^{-k} + \sum_{k>0} a_{-k} z^{-k}.$$

The form

$$(3) \qquad D(\varphi_S) = D(\varphi_{S_l}) + D(\varphi_{S_r}).$$

---

[6] In the calculations, I have suppressed the constant term of the functions φ. This makes the formulas more transparent and suffices for our purposes. The formula can easily be extended to include the constant terms ([E]).

The two terms are easily calculated as, for example, in [E]. The first and second terms on the right are both equal to

$$(4) \qquad \sum_{k=1}^{\infty} 2k a_k a_{-k}.$$

So (3) is

$$(5) \qquad \sum_{k=1}^{\infty} 4k a_k a_{-k}.$$

We can also imbed the unit circle in a torus S_A by taking the annulus S bounded by $C_1 = \{z|\,|z| = 1\}$ and by $C_2 = \{z|\,|z| = A\}$, $A > 1$, and by identifying $e^{i\theta}$ with $Ae^{i\theta}$. Then the function φ will be defined on C_1 by (2) and on C_2 by

$$(6) \qquad \varphi(z) = \sum_{k>0} \frac{a_k}{A^k} z^k + \sum_{k>0} \frac{a_{-k}}{A^k} \bar{z}^k, \quad |z| = A.$$

We apply (6.4) of [E] with $q = A^{-1}$ to obtain

$$(7) \qquad D(\varphi_S) = \sum_{k=1}^{\infty} 4k a_k a_{-k} - \sum_{k=1}^{\infty} 8 a_k a_{-k} \frac{q^k}{1 - q^{2k}}.$$

The two results are not the same, but the difference is a series that converges for the Fourier coefficients of any distribution on the circle. As a result the measure $\mu_C = \mu_C^S$ obtained from the Riemann sphere S is absolutely continuous with respect to any of the measures $\mu_C^{S_A}$ and conversely.

We make a similar calculation for the annulus T bounded by the two circles $C_i = \{z|\,|z| = A_i\}$, $A_1 < A_2$. We first of all imbed it in the Riemann sphere. The function φ on $C = C_1 \cup C_2$ is given by specifying it as φ_i on C_i. I suppose, for simplicity, that the constant terms of each φ_i are 0. Let $\{a_k/A_1^{|k|}\}$ be the Fourier coefficients of φ_1 and $\{b_k/A_2^{|k|}\}$ those of φ_2. The Riemann sphere is the union of the three regions, $S_l = \{z|\,|z| < A_1\}$, T, $S_r = \{z|\,|z| > A_2\}$. It is clear that

$$D(\varphi_S) = D(\varphi_{S_l}) + D(\varphi_T) + D(\varphi_{S_r}).$$

The first and third terms on the right are calculated from (4), the second from (6.4) of [E]. If $q = A_1/A_2$, the result, upon simplifying, is that $D(\varphi_S)$

equals

$$\sum_{k=1}^{\infty} 2k\{a_k a_{-k}\frac{2}{1-q^{2k}} - a_k b_{-k}\frac{2q^k}{1-q^{2k}} - b_k a_{-k}\frac{2q^k}{1-q^{2k}} + b_k b_{-k}\frac{2}{1-q^{2k}}\}.$$

Apart from a term that converges for the Fourier coefficients of any distribution on C this is the result that would be obtained if we took the disjoint union $\tilde{S}$ of two Riemann spheres and imbedded C_1 in the first and C_2 in the second to obtain an imbedding of C in $\tilde{S}$. So μ_C^S and $\mu_C^{\tilde{S}}$ are mutually absolutely continuous.

We can calculate μ_C with respect to yet another surface by taking $A = A_3 > A_2$ and identifying $A_1 e^{i\theta}$ with $A_3 e^{i\theta}$ to form a torus S_A. We again calculate $D(\varphi_{S_A})$ with the help of (6.4) of [E]. If $q_1 = A_1/A_2$ and $q_2 = A_2/A_3$, the result is the sum of

$$\sum_{k=1}^{\infty} 4k\{a_k a_{-k} + b_k b_{-k}\}$$

and

$$-\sum_{i=1}^{2}\sum_{k=1}^{\infty} 4k\{a_k a_{-k}\frac{q_i^{2k}}{1-q_i^{2k}} + a_k b_{-k}\frac{2q_i^k}{1-q_i^{2k}} + b_k a_{-k}\frac{2q_i^k}{1-q_i^{2k}} + b_k b_{-k}\frac{q_i^{2k}}{1-q_i^{2k}}\}.$$

Once again, this second term converges for the Fourier coefficients of any distribution.

It is, more generally, easy to see that if $C = \cup_{j=1}^{n} C_j$ is the disjoint union of n simple closed smooth curves imbedded in a compact Riemann surface S without boundary, then the absolute continuity class of the measure μ_C^S is independent of S. I present a rough argument. Suppose that S_1 and S_2 are two such surfaces to which we give metrics compatible with the Riemannian structure. I show only that $D(\varphi_{S_1}) - D(\varphi_{S_2})$ is defined for any distribution φ. We parametrize C, or more precisely we parametrize the curves C_i. Suppose that $T = \cup_{j=1}^{n} T_j$ is a disjoint union of open annuli such that the imbedding of C in S_i, $i = 1, 2$ extends to a holomorphic imbedding of T into the same S_i. Suppose ϵ is a smooth function on T that is 1 in a neighborhood of C and 0 near the boundary of T. If φ is given on C, let φ_{S_i} be the harmonic function on $S_i - C$ with boundary values φ. The function $\epsilon\varphi_{S_i}$ can be regarded as a function on T or on S_1 and can also be transported to S_2. If Δ_2 is the Laplacian on S_2, then

$$(8) \qquad \Delta_2(\varphi_{S_2} - \epsilon\varphi_{S_1}) = -\Delta_2(\epsilon\varphi_{S_1}).$$

Take, at first, S_1 to be a disjoint union of Riemann spheres. If φ is given by (2) on the curve C_i, then it is clear that, even if φ is a distribution, the function φ_{S_1} is defined outside of T and that we can bound any derivative of φ_{S_1} in the region in $\cup T_j$ where $\epsilon \neq 1$, the bound depending of course on the distribution φ. If φ is given on C_j by (2) with coefficients a_k^j, then, for an appropriate integer $m > 0$, it can be taken to be a universal constant A_m times

$$\sum_{j=1}^{n} \sum_{k=1}^{\infty} (|a_k^j| + |a_{-k}^j|)k^{-m}.$$

Moreover, the right side of (8) is 0 where $\epsilon = 1$. As a consequence, we can bound the derivatives of $\Delta_2(\epsilon \varphi_{S_1})$, which is of course 0 where $\epsilon = 0$. By the standard theory of elliptic differential equations, we can majorize the function $\varphi_{S_2} - \epsilon \varphi_{S_1}$, which vanishes on T, and its derivatives, and thus $D_2(\varphi_{S_2} - \epsilon \varphi_{S_1})$, in terms of the original distribution φ. We can also, of course, majorize the derivatives of $\varphi_{S_1} - \epsilon \varphi_{S_1}$. To majorize $D_1(\varphi_{S_1}) - D_2(\varphi_{S_2})$, we observe that it is equal to the sum of three terms: the difference

$$(9) \qquad D_1(\varphi_{S_1} - \epsilon \varphi_{S_1}) - D_2(\varphi_{S_2} - \epsilon \varphi_{S_1})$$

and the difference of the two cross-terms coming from polarization,

$$(10) \qquad \frac{1}{2\pi} \int \frac{\partial}{\partial x}(\varphi_{S_j} - \epsilon \varphi_{S_1})\frac{\partial}{\partial x}(\epsilon \varphi_{S_1}) + \frac{\partial}{\partial y}(\varphi_{S_j} - \epsilon \varphi_{S_1})\frac{\partial}{\partial y}(\epsilon \varphi_{S_1}).$$

There is now no difficulty in bounding (9) and to bound (10) all we need to do is to integrate sufficiently often by parts, inserting an appropriate partition of unity. Thus $\mu_C^{S_1}$ and $\mu_C^{S_2}$ are mutually absolutely continuous when S_1 is chosen in the way indicated. By transitivity, the assertion then remains true for any pair S_1 and S_2.

There is thus a positive, measurable function $\xi_{S_1}^{S_2} = \xi_{S_1}^{S_2}(C)$ on distributions integrable with respect to $\mu_C^{S_1}$ such that

$$\mu_C^{S_2} = \xi_{S_1}^{S_2} \mu_C^{S_1}.$$

As a consequence, there is a canonical isomorphism $f \to g = \sqrt{\xi_{S_2}^{S_1}}\, f$ from $L^2(\mu_C^{S_1})$ to $L^2(\mu_C^{S_2})$ that allows us to identify the two spaces and to define $\mathcal{H}_C$.

Taking, in particular, C to be the unit circle imbedded in the Riemann sphere, we obtain a canonical measure and a canonical L^2-space that we denote $\mathcal{H}$. Then, provided we have parametrized each C_i, we can identify $\mathcal{H}_C$ with $\otimes_{i=1}^{n}\mathcal{H}$.

Thus if Σ is any oriented Riemann surface with oriented boundary C the union of $C_l = \cup_{i=1}^{m} C_i$ and $C_r = \cup_{j=1}^{n} C_j'$, the first with the given orientation of each C_i, the second with the orientation of each C_j' reversed, then we may identify $\mathcal{H}_{C_l}$ with the tensor product of $\otimes_{i=1}^{m}\mathcal{H}$ and $\mathcal{H}_{C_r}$ with $\otimes_{j=1}^{n}\mathcal{H}$. We complete Σ to a closed surface S by parametrizing each C_i and each C_j' and then capping C_l with S_l and C_r with S_r. One possibility is to cap each C_i and C_j' separately, but this is by no means necessary, or even desirable. Let S be the resulting surface. The measure μ_C^S is absolutely continuous with respect to $\mu_{C_l}^S \times \mu_{C_r}^S$. Let

$$(11) \qquad \mu_C^S = Z(\phi_l, \phi_r)\mu_{C_l}^S \times \mu_{C_r}^S.$$

Observe that Z is a function of a pair of functions ϕ_l and ϕ_r on distributions that is integrable with respect to $\mu_{C_l}^S \times \mu_{C_r}^S$. So I shall be arguing formally with some fairly fancy notions. There is no difficulty in making them precise for the gaussian measures that arise for free bosons.

The operator K_Σ is to be defined as a map from $L^2(\mu_{C_l}^S)$ to $L^2(\mu_{C_r}^S)$ by means of a kernel. This kernel will necessarily depend on the two measures $\mu_{C_l}^S$ and $\mu_{C_r}^S$, but it will have to be shown that the map from $\mathcal{H}_{C_l}$ to $\mathcal{H}_{C_r}$ does not. The function $Z(\phi_l, \phi_r)$ will be a factor of $K_\Sigma(\phi_l, \phi_r)$ but not the only factor. We shall take

$$(12) \qquad K_\Sigma(\phi_l, \phi_r) = \eta_{rl}(\phi_l)Z(\phi_l, \phi_r)\eta_{lr}(\phi_r).$$

The two functions η_{rl} and η_{lr} are still to be defined.

The function η_{lr} is defined just as η_{rl} except that the roles of C_l and C_r are interchanged. So it suffices to define η_{rl}. The curve C_l is the boundary of two Riemann surfaces, S_l and $S_r' = \Sigma \cup S_r$. So $\mu_C^{S_l}$ and $\mu_C^{S_r'}$ are both defined, although, as follows for example from the calculation of (3), they will not be equivalent, in the sense of absolute continuity, to μ_C^S. The curve C_l is, of course, oriented, and we have attached S_r', also oriented, on the right. The surface $\bar{S}_r'$, obtained by reversing the orientation of S_r' is attached on the left.[7] Although C_l lies in the boundary of S_l and $\bar{S}_r'$, we may again define $\mu_{C_l}^{S_l}$ and $\mu_{C_l}^{\bar{S}_r'}$. Because S_l and S_r' are attached to C_l on the same side, the previous arguments can be extended to compare φ_{S_l} and $\varphi_{\bar{S}_r'}$,

---

[7] It may be better to think concretely of the unit circle and to take $S_r = \{z \mid |z| \geq 1\}$. Then $\bar{S}_r$ may be identified with $\{z \mid |z| \leq 1\}$, the point z on S_r becoming in $\bar{S}_r$ the point $\bar{z}^{-1}$.

defined by $\varphi_{\bar{S}'_r}(\bar{s}) = \bar{\varphi}_{S'_r}(s)$. Here $\bar{s}$ is simply $s \in S_r$ regarded as a point in $\bar{S}'_r$. So we can conclude that $\mu_{C_l}^{S_l}$ and $\mu_{C_l}^{S'_r} = \mu_{C_l}^{\bar{S}'_r}$ are mutually absolutely continuous. Thus we can introduce the Radon-Nikodym derivative $\xi_{S_l}^{S'_r}$ of the second with respect to the first. We define the positive function η_{rl} by the relation

$$\eta_{rl} = \sqrt{\xi_{S_l}^{S'_r}}.$$

The kernel K_Σ is not necessarily bounded. So the most that we can assert at first is that the associated operator is densely defined. This is enough for our present purposes. Indeed I want only to explain why it is well defined projectively, thus up to a constant, independently of the choice of S_l and S_r and why the multiplicative relation $K_{\Sigma'} \circ K_\Sigma = \alpha K_{\Sigma''}$ is valid. It is well to be explicit about the function of the parametrizations, for they are somewhat extrinsic to the constructions. First of all, they allow us to identify $\mathcal{H}_{C_l}$ and $\mathcal{H}_{C_r}$ with tensor products of the space $\mathcal{H}$ with itself. Secondly, the gluing of Σ to Σ' requires an identification of C_r and C'_l and this can be effected by the parametrizations.

The operator K_Σ is defined on $L^2(\mu_C^S)$ as

$$\phi_l \to g(\phi_r) = \int f(\phi_l) K_\Sigma(\phi_l, \phi_r) d\mu_{C_l}^S.$$

Suppose that $\tilde{S}$, defined by $\tilde{S}_l$ and $\tilde{S}_r$, is a second choice for S. Then

$$\tilde{f}(\phi_l) = \sqrt{\xi_{\tilde{S}}^{S}(C_l)} f(\phi_l)$$

and

$$\tilde{g}(\phi_r) = \sqrt{\xi_{\tilde{S}}^{S}(C_r)} g(\phi_r).$$

Thus, in terms of $\tilde{f}$ and $\tilde{g}$, the operator K_Σ would be given by a different kernel,

$$
\begin{aligned}
\tilde{g}(\phi_r) &= \int \tilde{f}(\phi_l) \sqrt{\xi_S^{\tilde{S}}(C_l)} K_\Sigma(\phi_l, \phi_r) \sqrt{\xi_{\tilde{S}}^{S}(C_r)} d\mu_{C_l}^S \\
&= \int \tilde{f}(\phi_l) \sqrt{\xi_{\tilde{S}}^{S}(C_l)} K_\Sigma(\phi_l, \phi_r) \sqrt{\xi_{\tilde{S}}^{S}(C_r)} d\mu_{C_l}^{\tilde{S}}.
\end{aligned}
$$
(13)

We need to verify that the kernel appearing here is equal to that given by the definition (12) applied directly to $\tilde{S}$.

As was already observed, the previous arguments that showed that $\mu_C^{S_1}$ and $\mu_C^{S_2}$ were mutually absolutely continuous can be extended to the case that S_1 and S_2 have boundaries and some of the simple closed curves forming C lie in the boundary of both S_1 and S_2. Thus the four functions $\xi_{S_l}^{\tilde{S}_l}(C_l)$, $\xi_{S_r'}^{\tilde{S}_r'}(C_l)$, $\xi_{S_r'}^{\tilde{S}_r}(C_r)$, and $\xi_{S_l'}^{\tilde{S}_l'}(C_r)$ are all defined and

$$(14) \qquad \xi_S^{\tilde{S}}(C_l) \doteq \xi_{S_l}^{\tilde{S}_l}(C_l)\xi_{S_r'}^{\tilde{S}_r'}(C_l), \quad \xi_S^{\tilde{S}}(C_r) \doteq \xi_{S_l'}^{\tilde{S}_l'}(C_r)\xi_{S_r}^{\tilde{S}_r}(C_r)$$

because, for example,

$$(15) \qquad D(\varphi_S) = D(\varphi_{S_l}) + D(\varphi_{S_r'}).$$

The dot above the equalities in (14) indicate that they are only valid projectively. From (15) we can only deduce a relation between gaussian measures up to a constant that will be determined as the quotient of two determinants.

Let $\tilde{Z}(\phi_l, \phi_r)$ be the analogue of $Z(\phi_l, \phi_r)$. Then, as a consequence of (14) and the definition (11),

$$(16) \qquad \frac{\tilde{Z}(\phi_l, \phi_r)}{Z(\phi_l, \phi_r)} \doteq \xi_S^{\tilde{S}}(C)\xi_{\tilde{S}_l}^{S_l}(C_l)\xi_{\tilde{S}_r'}^{S_r'}(C_l)\xi_{\tilde{S}_l'}^{S_l'}(C_r)\xi_{\tilde{S}_r}^{S_r}(C_r).$$

The arguments ϕ_l and ϕ_r have been omitted on the right. ¿From the analogue of (15) for the decompositions $S = S_l \cup \Sigma \cup S_r$ and $\tilde{S} = \tilde{S}_l \cup \Sigma \cup \tilde{S}_r$, we conclude that

$$(17) \qquad \xi_S^{\tilde{S}}(C) \doteq \xi_{S_l}^{\tilde{S}_l}(C_l)\xi_{S_r}^{\tilde{S}_r}(C_r).$$

Thus

$$(18) \qquad \tilde{Z}(\phi_l, \phi_r) \doteq Z(\phi_l, \phi_r)\xi_{\tilde{S}_r'}^{S_r'}(C_l)\xi_{\tilde{S}_l'}^{S_l'}(C_r).$$

The functions η_{rl} is the square root of $\xi_{S_l}^{S_r'}(C_l)$ and η_{lr} is the square root of $\xi_{S_r}^{S_l'}(C_r)$. Let $\tilde{\eta}_{rl}$ and $\tilde{\eta}_{lr}$ be the analogues of η_{rl} and η_{lr}. They are the square roots of $\xi_{\tilde{S}_l}^{\tilde{S}_r'}(C_l)$ and $\xi_{\tilde{S}_r}^{\tilde{S}_l'}(C_r)$.

For C_l there are four measures in play, those defined by S_l, $\tilde{S}_l$, S_r' and $\tilde{S}_r'$. All are absolutely continuous with respect to each other and there will be obvious relations of transitivity. Making use of (18), we write the kernel of (13) up to a constant factor as

$$\sqrt{\xi_{\tilde{S}_l}^{S_l}(C_l)\xi_{\tilde{S}_r}^{S'_r}(C_l)}\sqrt{\xi_{S_l}^{S'_r}(C_l)}\frac{1}{\xi_{\tilde{S}_r}^{S'_r}(C_l)}\tilde{Z}(\phi_l,\phi_r)$$

$$\times\sqrt{\xi_{\tilde{S}'_l}^{S'_l}(C_r)\xi_{\tilde{S}_r}^{S_r}(C_r)}\sqrt{\xi_{S_r}^{S'_l}(C_r)}\frac{1}{\xi_{\tilde{S}'_l}^{S'_l}(C_r)}\ .$$

We have to verify that the expression to the left of $\tilde{Z}(\phi_l,\phi_r)$ is equal to $\tilde{\eta}_{rl}$ and that the expression to the right is $\tilde{\eta}_{lr}$. Consider the first and apply the relations of transitivity. The desired relation is immediate.

The most important lesson to be learned from this formal argument is that the relations (14) and the analogous relations for the decompositions $S_l \cup \Sigma \cup S_r$ and $\tilde{S}_l \cup \Sigma \cup \tilde{S}_r$ are critical to the verification that the operator K_Σ is well defined. So we shall need something similar for the Ising model. We first verify the multiplicative property for the free boson. With the freedom we now have in the choice of cappings, this is easy.

After pasting Σ and Σ' together, we cap Σ on the left with S_l and Σ' on the right with S_r to obtain S, a surface that contains both Σ and Σ' and that can be used to define all three operators K_Σ, $K_{\Sigma'}$, and $K_{\Sigma''}$, $\Sigma'' = \Sigma \cup \Sigma'$.

The kernels defining the three operators are given by (12) and its variants for Σ' and Σ'', which we express by adding the appropriate number of primes. When S is chosen as above to be the same for all three of Σ, Σ' and Σ'', then $\eta_{lr}\eta'_{rl} = 1$, $\eta_{rl} = \eta''_{rl}$ and $\eta'_{lr} = \eta''_{lr}$. As a consequence, the multiplicativity becomes

$$(19)\qquad \int Z(\phi_l,\phi_m)Z'(\phi_m,\phi_r)d\mu_{C_r}^S(\phi_m) = \alpha Z''(\phi_l,\phi_r)$$

where $C_r = C'_l$ is the curve along which Σ and Σ' are glued.

Consider (19) as a function of ϕ_l for a fixed ϕ_r and multiply this function against the measure $\mu_{C_l}^S$. The result on the right side is a conditional probability of ϕ_l with respect to the measure $\mu_{C_l \cup C_r}^S$ and the given ϕ_r. The result on the left is the integral over ϕ_m of the conditional probability of ϕ_l with respect to $\mu_{C_l \cup C_r}^S$ and ϕ_m of the conditional probability of ϕ_m with respect to $\mu_{C'_l \cup C'_r}^S$. The Markov property of the gaussian measures, in essence a result of the relations (14) and (15), then yields (19).

9.4   Possible Construction of a Conformal Field Theory

After this cavalier discussion of the free boson, we return to the construction of [I] to see whether it is reasonable to hope that it offers some analogue of (15). It is best to work with the Ising model on a triangular lattice or, if we want to consider models other than translation-invariant planar models, on triangulated surfaces S. The advantage, ultimately of no importance, is that there is no ambiguity about the level curves attached to a given configuration σ. We choose barycenters for each triangle and each edge of the triangulation and join the barycenter of an edge to the barycenters of the two triangles in which it lies. This yields for each edge a broken segment crossing it. The level curve or total, unoriented contour curve attached to a given configuration is a possibly disconnected curve formed as the union of some of these broken segments, those crossing an edge whose two vertices are assigned opposite spins in σ. The elements λ of Λ_σ are obtained by assigning an orientation to each connected component of the level curve. For the triangular lattice or for triangulated surfaces the probability of $\lambda \in \Lambda_\sigma$ is more easily defined than for other models. It is the probability of the configuration σ divided by 2^{l_σ} if l_σ is the number of connected components in the level curve of σ. Thus 2^{l_σ} is also the number of elements in Λ_σ.

I only want to consider the analogue of (14) or (15) at the level of a finite triangulation. There is no question in this paper of proving anything beyond this level or even of pursuing the experiments of [I] further. So let S be closed and compact and let C be a smooth curve, thus the union of a finite number of simple closed parametrized curves, that divides S into two disjoint pieces, S_l and S_r. It is best to suppose – this is clearly a technical issue to be resolved when the time comes to pass to the limit in the mesh – that the level curves cut C transversely. Then each element λ in Λ_σ defines two sets of points on C, the set X of points at which C crosses it with positive orientation and the set Y of points at which C crosses it with negative orientation. Varying σ and choosing for each σ all possible λ, we obtain a collection of pairs (X, Y), possibly repeated and each with a probability, that of λ. The sum of the probabilities is 1. We denote by $\mu_C^{\Pi}(X, Y)$ the sum over all occurrences of (X, Y) of the probability of the individual occurrence. Then

$$\sum_{(X,Y)} \mu_C^{\Pi}(X, Y) = 1.$$

As in [I], μ_C^Π may also be considered a measure on distributions. A function f on C is sent by the distribution associated to (X,Y) to the number $\sum f(X) - \sum f(Y)$.

Each vertex σ of the graph Γ is surrounded by a star St_σ that is spanned by the barycenters of the simplices (points, edges and triangles) containing the vertex. Consider the set Π_C of vertices such that St_σ meets C. The set of vertices not in Π_C is divided into two parts, those lying in S_l and those lying in S_r. We denote the two parts by Π_l and Π_r.

The meaning of (14), which we now want to interpret at the level of the finite triangulation, is that the measure μ_C^Π is of the form

$$(20) \qquad \xi^{S_l}(C)\xi^{S_r}(C)\nu_C,$$

where ν_C is a measure that is independent of the choice of S_l and S_r and where, for example, $\xi^{S_l}(C)$ is a function of (X,Y) that depends only on S_l but not on S_r. In fact, it will depend on the collection Θ_l of edges connecting points of Π_l to other points of Π_l or to points in Π_C and the graph they define. Given (20), we could introduce the Radon-Nikodym derivatives

$$\xi^{S_l}_{S'_l}(C) = \frac{\xi^{S_l}(C)}{\xi^{S'_l}(C)}$$

and the essential factors η_{rl} and η_{lr}. There would be, by the way, nothing canonical about such a factorization since ν_C is clearly not uniquely determined. Different ν lead, however, to the same Radon-Nikodym derivatives. The measure ν may have no limit as the mesh decreases to 0, so that in the limit, when the mesh becomes zero, it is purely fictitious and, as for the free boson, only the Radon-Nikodym derivatives survive. For the Ising model there appears to be an additional complication, so that we cannot simply take the definition (12) at the finite level and then pass to the limit. Before explaining the difficulty, it is useful, as a supplement to the discussion of the free boson, to take a few lines to explain to what the definition (12) reduces at the finite level when (20) is available, not only for a curve C that divides S into a left part S_l and a right part S_r but also for a curve $C = C_l \cup C_r$ that separates S into three parts S_l, Σ and S_r as in the definition of the operator K_Σ.

For such a curve, the analogue of (20) takes the form

$$\mu_C^\Pi = \xi^{S_l}(C_l)\xi^\Sigma \xi^{S_r}(C_r)\nu_{C_l} \times \nu_{C_r}.$$

The formula (20) applied to C_l and C_r yields

$$\mu_{C_l}^{\Pi} = \xi^{S_l}(C_l)\xi^{S_r'}(C_l)\nu_{C_l}, \qquad \mu_{C_r}^{\Pi} = \xi^{S_l'}(C_r)\xi^{S_r}(C_r)\nu_{C_r}.$$

The map

$$f \to F = f\sqrt{\xi^{S_l}(C_l)}\sqrt{\xi^{S_r'}(C_l)}$$

identifies $\mathcal{H}_{C_l} = L^2(\mu_{C_l}^{\Pi})$ with $L^2(\nu_{C_l})$. The map

$$g \to G = g\sqrt{\xi^{S_l'}(C_r)}\sqrt{\xi^{S_r}(C_r)}$$

identifies $\mathcal{H}_{C_r} = L^2(\mu_{C_r}^{\Pi})$ with $L^2(\nu_{C_r})$.

The kernel K_Σ defined by (12) is

$$(21) \qquad \sqrt{\frac{\xi^{S_r'}(C_l)}{\xi^{S_l}(C_l)}}\frac{\xi^{S_l}(C_l)\xi^{\Sigma}\xi^{S_r}(C_r)}{\xi^{S_l}(C_l)\xi^{S_r'}(C_l)\xi^{S_l'}(C_r)\xi^{S_r}(C_r)}\sqrt{\frac{\xi^{S_l'}(C_r)}{\xi^{S_r}(C_r)}}.$$

We have suppressed the two variables ϕ_l and ϕ_r. Integrated against f with respect to the measure $\mu_{C_l}^{\Pi}$, this kernel yields the image g of f under the operator K_Σ. The argument of f is ϕ_l, that of g is ϕ_r. Expressed in terms of F and G and an integration with respect to ν_{C_l}, the kernel of K_Σ is the product of (21) with

$$\frac{1}{\sqrt{\xi^{S_l}(C_l)\xi^{S_r'}(C_l)}}\xi^{S_l}(C_l)\xi^{S_r'}(C_l)\sqrt{\xi^{S_l'}(C_r)\xi^{S_r}(C_r)}\ .$$

When all possible cancellations in the product are carried out, nothing is left but ξ^{Σ}. The multiplicative property then reduces to a relation similar to (19),

$$\int \xi^{\Sigma}(\phi_l, \phi)\xi^{\Sigma'}(\phi, \phi_r)d\nu_{C_r}(\phi) = \alpha\xi^{\Sigma''}.$$

We return to the Ising model and the apparent failure of (20). Given (X, Y), the σ that generate it are defined by a triple: σ_l on the vertices of Π_l, σ_r on the vertices of Π_r, and σ_C on the vertices of Π_C. The configuration σ is determined up to a global sign on each component of S by the level curve λ. The Boltzmann weight is not affected by these global signs. That factor β_C of the Boltzmann weight for σ contributed by edges joining vertices in Π_C is determined by (X, Y) alone. The Boltzmann weight is a product of this factor and two other factors, β_l and β_r, the first the product of the contributions from the edges in Θ_l crossing that part λ_l of the level curve that lies in S_l, the second the product of the contributions from the edges in Θ_r that cross λ_r. The number l_σ of connected closed curves in λ is given by $l_\sigma = l_l + l_r + l_C$, where l_l is the number of closed curves lying entirely in S_l, l_r the number lying entirely in S_r, and l_C the number that meet C. So, apart from the normalizing factor given by the partition function,

$$(22) \qquad \mu_C^\Pi(X,Y) = 2^{|S|}\beta_C \sum_{\lambda_l}\sum_{\lambda_r} \frac{\beta_l\beta_r}{2^{l_l+l_r+l_C}},$$

the sum being over all possible λ_l and λ_r compatible with the given collection (X,Y) of positively and negatively oriented crossings of C. The exponent $|S|$ is the number of connected components of S.

It is the term 2^{l_C} in the denominator of (22) that prevents the factorization of (20). So we might attempt to modify the construction. There are two pairings of X with Y associated to compatible λ and (X,Y),. If, as is implicit in the notation, we have been careful with our orientations on C, then at each $x_i \in X$, $i = 1,\ldots,N$ one component of λ crosses from S_r into S_l. Let y_i be the first point at which it crosses back into S_r. The first pairing is $W_l = \{(x_1,y_1),\ldots,(x_N,y_N)\}$. The second, $W_r = \{(y_1',x_1'),\ldots,(y_N',x_N')\}$, is defined in the same way, except that the roles of S_l and S_r are reversed. These two collections together define a distribution on $C \times C$,

$$f \to \sum_i f(x_i,y_i) - \sum_i f(y_i',x_i'),$$

f being a function on $C \times C$. Taking $f(x,y) = g(x) - g(y)$, we recover twice the original distribution,

$$f \to \sum\{g(x_i) - g(y_i) - g(y_i') + g(x_i')\} = 2\sum_i\{g(x_i) - g(y_i)\}$$

because $\{x_i\} = \{x_i'\}$ and $\{y_i\} = \{y_i'\}$. This construction does not demand the global existence of S_l and S_r; it simply requires that C be oriented.

The measure on the collection Λ of all possible level curves defines one, $\mu_{C\times C}^\Pi$, on the family of collections

$$(23) \quad W = (W_l,W_r) = (\{(x_1,y_1),\ldots,(x_N,y_N)\},\{(y_1',x_1'),\ldots,(y_N',x_N')\})$$

associated to the λ in it. To calculate $\mu_{C\times C}^\Pi(W)$, we first construct λ_l in S_l compatible with the first component W_l of W, and λ_r in S_r compatible with the second, W_r. If we join them at the points where they meet on C, they form together a λ compatible with W and all such λ are so obtained. Once the collection λ of oriented level curves is fixed, the configuration σ

is determined up to the choices of global sign. Since l_C is determined by W alone, we write $l_C = l_W$. The relation (22) now becomes

$$(24) \qquad \mu^{\Pi}_{C \times C}(W) = 2^{|S|-l_W} \beta_C \Big\{ \sum_{\lambda_l} \frac{\beta_l}{2^{l_l}} \Big\} \Big\{ \sum_{\lambda_r} \frac{\beta_r}{2^{l_r}} \Big\},$$

$|\lambda_l|$ and $|\lambda_r|$ being the number of closed curves in λ_l and λ_r respectively, and l_W being the number of closed curves in λ that meet C. It is a number that is determined by W alone. The formula (20) for the measure on the collection of W follows from (24).

Unfortunately measures on distributions on $C \times C$, and this is what might result from (24) on passage to the limit over decreasing mesh, have a number of disadvantages that make them unsuitable for the construction of a conformally invariant theory. For example, there is no possibility that when C has more than one connected component the absolute continuity class of $\mu^S_{C \times C}$ is independent of S either at the finite level or in the limit. So we have to find our way back to pairs (X, Y).

The object W is a pair (W_l, W_r) and there are maps $W_l \to (X, Y)$ and $W_r \to (X, Y)$. Consider the matrix with entries $a(W_l, W_r) = 2^{-l_W}$. If the entries could be written as

$$(25) \qquad a(W_l, W_r) = \alpha b(W_l) b(W_r),$$

α a constant, then the right side of (24) would become

$$2^{|S|} \alpha \beta_C \Big\{ \sum_{\lambda_l} \frac{b(W_l)\beta_l}{2^{l_l}} \Big\} \Big\{ \sum_{\lambda_r} \frac{b(W_r)\beta_r}{2^{l_r}} \Big\}.$$

Passage to the level of (X, Y) requires summing over all pairs (W_l, W_r) for which the image of both W_l and W_r is (X, Y). The mass of (X, Y), again apart from the normalization given by the partition function, would be

$$(26) \quad \mu^{\Pi}_C(X, Y) = 2^{|S|} \alpha \beta_C \Big\{ \sum_{\lambda_l, W_l \to (X,Y)} \frac{b(W_l)\beta_l}{2^{l_l}} \Big\} \Big\{ \sum_{\lambda_r, W_r \to (X,Y)} \frac{b(W_r)\beta_r}{2^{l_r}} \Big\}.$$

This is the factorization required.

There is no possibility that the simple representation (25) of $a(W_l, W_r)$ exists, but as we are passing to a limit so much is not required. The relation (25) has presumably only to hold approximately except for a collection of (X, Y) whose measure tends to zero as the mesh does. This statement will be easier to understand once we examine the matrix $A(X, Y) = (a(W_l, W_r))$ more carefully. It depends on (X, Y), X and Y being two sets of points with the same number n of elements. We label them both as $\{1, \ldots, n\}$. It is likely that the probability that n is less than any given bound is going to zero with the mesh. The family W_l is nothing but a permutation r of $\{1, \ldots, n\}$ and W_r a permutation s^{-1}. The number l_W is the number $\gamma(rs^{-1})$ of cycles in rs^{-1}. Thus $A(X, Y)$ is the matrix of

$$(27) \qquad R = R_n = \sum_r 2^{-\gamma(r)} r$$

in the regular representation of the symmetric group on n symbols with its standard basis.

We can decompose the regular representation into irreducible constituents τ. Since (27) is clearly a central element in the group algebra, it is represented as a scalar matrix in each of these constituents. If absolute precision is not demanded, then the approximate form of (25) is that, as n increases, the eigenvalues of $\tau(R)$, all of which are equal, divided by the eigenvalue for the trivial representation approach zero, provided that τ is not trivial. The vector $(b(W))$ would then have all its components equal. As a somewhat unexpected conclusion to a lecture on percolation and the Ising model, I therefore briefly describe the necessary representation theory of the symmetric group, which I take from [J].

9.5 Calculations for the Symmetric Group

We begin with the calculation of $\tau(R)$ for the trivial representation. It is given by a generating function, the coefficient of x^n being the eigenvalue for the symmetric group on n symbols divided by $n!$. For a given n we give the cycle lengths as i_j cycles of length j, for $j = 1, 2, \ldots$, with $\sum_j i_j j = n$. The number of cycles corresponding to these lengths is

$$n! \prod_{j=1}^{\infty} \left(\frac{(j-1)!}{j!}\right)^{i_j} \frac{1}{i_j!} = n! \prod_{j=1}^{\infty} \left(\frac{1}{j}\right)^{i_j} \frac{1}{i_j!}.$$

Observe that when $i_j = 0$ the corresponding factor is 1, so that it is permissible to write an infinite product. It is to be multiplied by

$$\prod_j \frac{1}{2^{i_j}}.$$

Multiplying by x^n, dividing by $n!$, and summing over all possible families of nonnegative integers i_j, we obtain for the generating function the result

$$\prod_{j=0}^{\infty} \exp(x^j/2j) = \exp(-\ln(1-x)/2) = \frac{1}{(1-x)^{1/2}}.$$

Thus the eigenvalue of R_n for the trivial representation is

$$(28) \qquad \frac{1}{2}\frac{3}{2} \ldots (n - \frac{1}{2}).$$

We can expect that this is the largest eigenvalue of R_n.

A similar calculation for the one-dimensional representation $\tau(r) = \text{sgn}(r)$, but with division by $(-1)^n n!$, yields the series of $(1-x)^{1/2}$. The quotient of the eigenvalues of R_n in the two representations is thus

$$(-1)^{n-1} \frac{\frac{1}{2}\frac{1}{2}\frac{3}{2} \cdots \frac{2n-3}{2}}{\frac{1}{2}\frac{3}{2} \cdots \frac{2n-1}{2}} = \frac{(-1)^{n-1}}{2n-1},$$

confirming our hopes.

The general irreducible representation τ is given as τ^μ, where μ is a partition of the set $\{1, \ldots, n\}$. The properties of τ^μ are described in [J]. First of all, there is an ordering on partitions, the partition $\lambda = \{l_1 \geq l_2 \ldots\}$ dominating the partition $\mu = \{k_1 \geq k_2 \geq \ldots\}$ if and only if $\sum_{i=1}^{j} l_i \geq \sum_{i=1}^{j} k_i$ for all j. The representations τ^μ have the property that τ^μ is contained exactly once in the representation ι^μ induced from the trivial representation of the subgroup fixing the partition μ and every other representation contained in this induced representation is equivalent to a τ^λ, $\lambda \geq \mu$.

The trace of $\iota^\mu(R_n)$ is readily calculated. Let $\mu = \{k_1, \ldots, k_s\}$. Then the trace of $\iota^\mu(R)$ is obtained by taking the sum over all decompositions of $\{1, \ldots, n\}$ into s subsets of respectively $k_1, \ldots, k_s$ elements of the product of the traces of R_{k_l} with respect to the trivial representation of the symmetric group on k_l elements, thus

$$\text{tr}(\iota^\mu(R_n)) = \frac{n!}{k_1! \ldots k_s!} \prod_{j=1}^{s} \frac{1}{2}\frac{3}{2} \ldots (k_j - \frac{1}{2}).$$

To compute the trace of τ^μ, we use the *determinantal form* appearing on p. 74 of [J]. Writing formally $\iota^\mu = [k_1] \ldots [k_s]$ when $\mu = \{k_1, \ldots, k_s\}$, we can treat any formal determinant

$$(29) \qquad \big| [m_{i,j}] \big|$$

that is of size $s \times s$ and in which $\sum_i m_{i,r(i)} = n$ is independent of the permutation r as a linear combination of induced representations,

$$\sum_r \operatorname{sgn}(r)[m_{1,r(1)}][m_{2,r(2)}] \ldots [m_{s,r(s)}].$$

If in (29) $m_{i,j} = 0$, then $[m_{i,j}]$ is a multiplicative identity, whereas if $m_{i,j} < 0$ then $[m_{i,j}]$ is the multiplicative zero.

The determinantal form described and proved in [J] expresses τ^μ in this way. If μ is given by $k_1 \geq k_2 \geq \cdots \geq k_s$, then

$$(30) \qquad \tau^\mu = \big| [k_i - i + j] \big|.$$

Since the dimension of ι^μ, $\mu = \{k_1, \ldots, k_s\}$, $n = \sum k_i$, is $n!/k_1! \ldots k_s!$, this yields a simple formula for the dimension of τ^μ. It is

$$(31) \qquad n! \left| \frac{1}{(k_i - i + j)} \right|.$$

It also yields a simple formula for the trace of $\tau^\mu(R_n)$ as

$$n! \left| \frac{\frac{1}{2}\frac{3}{2} \ldots (k_i - i + j - \frac{1}{2})}{(k_i - i + j)!} \right|.$$

The eigenvalues of $\tau^\mu(R_n)$ are calculated as the quotient

$$(32) \qquad \frac{\left| \dfrac{\frac{1}{2}\frac{3}{2}\ldots(k_i-i+j-\frac{1}{2})}{(k_i-i+j)!} \right|}{\left| \dfrac{1}{(k_i-i+j)} \right|}.$$

Factorials of negative numbers are of course infinite. Moreover, the entry in the determinant of the numerator is to be taken to be 1 when $k - i + j = 0$.

For a partition consisting of a single term, this clearly agrees with our previous formula. As an additional confirmation, take the simple partition of n given by $\mu = \{1, 1, \ldots, 1\}$. Then the determinant of (31) is

$$\begin{vmatrix} 1 & \frac{1}{2!} & \frac{1}{3!} & \frac{1}{4!} & \cdots \\ 1 & 1 & \frac{1}{2!} & \frac{1}{3!} & \cdots \\ 0 & 1 & 1 & \frac{1}{2!} & \cdots \\ 0 & 0 & 1 & 1 & \cdots \\ \cdot & \cdot & \cdot & \cdot & \cdots \end{vmatrix}.$$

214   R. P. Langlands

Multiplying this determinant by

$$
1 = \begin{vmatrix}
1 & 0 & 0 & 0 & 0 & \cdots \\
-1 & 1 & 0 & 0 & 0 & \cdots \\
2! & -2! & 1 & 0 & 0 & \cdots \\
-3! & 3! & -\frac{3!}{2} & 1 & 0 & \cdots \\
4! & -4! & \frac{4!}{2!} & -\frac{4!}{3!} & 1 & \cdots \\
& & \cdot & \cdot & \cdot & \cdots
\end{vmatrix},
$$

we obtain the determinant of an upper-diagonal matrix with entries $1, 1/2, 1/3, \ldots$ along the diagonal. So the formula (31) yields 1. The determinant in the numerator of (32) is

$$
\begin{vmatrix}
\frac{1}{2} & \frac{\frac{1}{2}\frac{3}{2}}{2!} & \frac{\frac{1}{2}\frac{3}{2}\frac{5}{2}}{3!} & \frac{\frac{1}{2}\frac{3}{2}\frac{5}{2}\frac{7}{2}}{4!} & \cdots \\
1 & \frac{1}{2} & \frac{\frac{1}{2}\frac{3}{2}}{2!} & \frac{\frac{1}{2}\frac{3}{2}\frac{5}{2}}{3!} & \cdots \\
0 & 1 & \frac{1}{2} & \frac{\frac{1}{2}\frac{3}{2}}{2!} & \cdots \\
0 & 0 & 1 & \frac{1}{2} & \cdots \\
& \cdot & \cdot & & \cdots
\end{vmatrix}.
$$

If it is multiplied by

$$
1 = \begin{vmatrix}
1 & 0 & 0 & 0 & 0 & \cdots \\
-2 & 1 & 0 & 0 & 0 & \cdots \\
-8 & 4 & 1 & 0 & 0 & \cdots \\
-16 & 8 & 2 & 1 & 0 & \cdots \\
-\frac{128}{5} & \frac{64}{5} & \frac{16}{5} & \frac{8}{5} & 1 & 0 \cdots \\
& \cdot & \cdot & \cdot & \cdot & \cdots
\end{vmatrix}
$$

the result is again an upper-diagonal matrix but now with entries $\frac{1}{2}, -\frac{1}{2}\frac{1}{2}, -\frac{3}{2}\frac{1}{3}, -\frac{5}{2}\frac{1}{4}, \ldots$ along the diagonal. To pass from one row of the matrix to the next we multiply the nonzero entries by $2k/(2k-3)$, add one entry equal to 1, and the necessary zeros.

Suppose $\mu = \{k_1, k_2\}$, $k_1 \geq k_2$, $k_1 + k_2 = n$. Then the denominator of (32) is equal to

$$
\frac{1}{k_1! k_2!}\left(1 - \frac{k_2}{k_1 + 1}\right).
$$

The numerator is equal to

$$
\frac{\frac{1}{2} \cdots (k_1 - \frac{1}{2})}{k_1!} \frac{\frac{1}{2} \cdots (k_2 - \frac{1}{2})}{k_2!}\left(1 - \frac{k_2(k_1 + \frac{1}{2})}{(k_2 - \frac{1}{2})(k_1 + 1)}\right).
$$

The quotient is to be divided by (28). This yields

$$\frac{\frac{1}{2}\cdots(k_1-\frac{1}{2})\frac{1}{2}\cdots(k_2-\frac{1}{2})}{\frac{1}{2}\cdots(n-\frac{1}{2})}\left(1-\frac{k_2(k_1+\frac{1}{2})}{(k_2-\frac{1}{2})(k_1+1)}\right)\Big/\left(1-\frac{k_2}{k_1+1}\right).$$

The second term of the numerator is universally bounded and the denominator is at least $(k_1-k_2+1)/k_1$. The first term of the terminator is bounded by a universal constant times

$$\frac{\Gamma(k_1+\frac{1}{2})\Gamma(k_1+\frac{1}{2})}{\Gamma(N+\frac{1}{2})}=O\left(\left(\frac{k_1}{n}\right)^{k_1}\left(\frac{k_2}{n}\right)^{k_2}\right),$$

the constant implicit in this relation being universal. The quotient clearly goes to 0 as n approaches infinity, independently of the ratio of $k_1/k_2 > 1$. I have not yet tried to establish something similar for arbitrary $k_1 \geq \cdots \geq k_s$. Experiments can, however, begin without a full understanding of the eigenvalues of $\tau(R_n)$.

9.6   Final Remarks

Even if the constructions of Section 9.4 have something to offer, it is not clear how to set about convincing oneself that they can indeed be made, thus that the pertinent scaling limits exist, or that the intuitive arguments that we sketched can be rendered effective. For experiments, one might begin with models on the Riemann sphere or, by conformal invariance, with translation-invariant models in the plane or on a cylinder. The construction is also possible for percolation, and for percolation experiments will be easier to perform.

Neither for percolation nor for the Ising model do I see at all clearly what information might be contained in the operators K_Σ. If they can be defined, it is more than likely, indeed almost certain, that they contain not only the expected unitary conformal field theories but also nonunitary ones.

Even if experiments establish that the reflections of this paper are well founded, the problem of proving that the scaling limits exist will remain. There may be some analogue of the finite-model for percolation described in [P3], although its definition will have to be more elaborate, not alone but in part because there is a dependence in the Ising model not present in percolation. Crossing probabilities for regions that do not overlap are independent in percolation. The measures μ_C will, on the contrary, change even when the conditioning data are taken from outside C, although the influence diminishes with increasing distance. It can be taken into account, but whether it can be taken into account effectively so that dynamical systems of manageable size result is another matter. For the construction of a conformal field theory, it may very well be better to work directly with definitions based on (26), and for finite models it may be best to begin with the factorization given by that formula.

Bibliography

[CG] Eric Charpentier and Krzyztof Gawędski, *Wess-Zumino-Witten conformal field theory for simply-laced groups at level one*, Ann. of Phys. 213, (1992),233–294.

[J] G. D. James, The representation theory of the symmetric groups, SLM 682, *Springer Verlag*, 1978

[K] H. Kesten, Percolation theory for mathematicians, *Birkhäuser*, 1982.

[L] O. Lanford, *A computer-assisted proof of the Feigenbaum conjectures*, Bull. Amer. Math. Soc. 6, (1982), 427-434.

[P1] R. P. Langlands, C. Pichet, P. Pouliot and Y. Saint-Aubin, *On the universality of crossing probabilities in two-dimensional percolation*, J. Statist. Phys. 67, (1992), 553–574.

[P2] Robert Langlands, Philippe Pouliot and Yvan Saint-Aubin, *Conformal invariance in two-dimensional percolation*, Bull. A. M. S., (1994),1–61.

[P3] R. P. Langlands and M. A. Lafortune, *Finite models for percolation*, Cont. Math., (1994), 227-246.

[P4] Robert P. Langlands, *Dualität bei endlichen Modellen der Perkolation*, Math. Nachr. 160, (1993), 7–58.

[E] Robert P. Langlands, *An essay on the dynamics and statistics of conformal field theories*, in Canadian Mathematical Society, 1945-1995, Invited papers, 1996, 173–210.

[I] Robert P. Langlands, Marc-André Lewis, and Yvan Saint-Aubin, *Universality and conformal invariance for the Ising model in domains with boundary*, J. Statist. Phys. 98, (2000) 131–244.

[SS] O. Schramm, *Scaling limits of loop-erased random walks and uniform spanning trees*, Israel. J. Math. 118, (2000), 221-288; Stanislav Smirnov, *Critical percolation in the plane: conformal invariance, Cardy's formula, scaling limits*, C. R. Acad. Sci., Paris 333, (2001), 239-244.

[S] Graeme Segal, *Two-dimensional conformal field theories and modular functors*, in IXth International Congress on Mathematical Physics (Swansea,1988), 1989, 22-37; *Geometric aspects of quantum field theory*, in Proc. Int. Cong. Math. (Kyoto, 1990), 1991, 1387-1396.

[W] Kenneth G. Wilson, *Renormalization group and critical phenomena: I. Renormalization group and the Kadanoff scaling picture, II. Phase-space cell analysis of critical behavior*, Phys. Rev. B (1971), 3174-3205.

Chapter 10

A Cohomological Description of Abelian Bundles and Gerbes

Roger Picken[1]

Abstract: We describe the geometrical ladder of equations for Abelian bundles and gerbes, as well as higher generalisations, in terms of the cohomology of an operator that combines de Rham and Čech cohomology.

10.1 Introduction

Gerbes with connection appear in differential geometry as a natural higher-order generalization of abelian bundles with connection, and thus, from the physics standpoint, provide a possible framework in which to generalise Abelian gauge theory. They first appeared in algebraic geometry [Giraud (1971)], and were subsequently developed by Brylinski [Brylinski (1993)], whose motivation was to generalise the geometrical interpretation of the second integral cohomology of a manifold M, $H^2(M, \mathbf{Z})$, in terms of the curvature of a complex line bundle, to $H^3(M, \mathbf{Z})$. There has been renewed interest in the subject recently following a concrete approach due to Hitchin and Chatterjee [Hitchin (1999)], and due to the appearance of possible applications in physics, for instance in anomalies [Carey, Mickelsson and Murray (2000)], new geometrical structures in string theory [Witten (1998)] and Chern-Simons theory [Gomi (2001].

Bundles and gerbes, as well as higher generalisations (n-gerbes), can be understood both in terms of local geometry, i.e. local functions and forms,

[1]Departamento de Matemática and CEMAT, Centro de Matemática e Aplicações, Instituto Superior Técnico, Av. Rovisco Pais,1049-001 Lisboa, Portugal, rpicken@math.ist.utl.pt

and in terms of non-local geometry, i.e. holonomies and parallel transports, and these two viewpoints are equivalent, in a sense made precise by Mackaay and the author in [Mackaay and Picken (2002)], following on from work by Barrett [Barrett (1991)] and Caetano and the author [Caetano and Picken (1994)]. For gerbes, holonomy and parallel transport are along embedded surfaces in the manifold, instead of along loops or paths (as indeed might be expected, since for gerbes everything is "one dimension up", compared to bundles). The non-local viewpoint was further explored by the author in [Picken (2003)], as a special case of a topological quantum field theory framework introduced in [Picken and Semião (2002)]. Here we wish to examine in greater detail the geometrical ladders of local functions and forms that appear in the local geometry perspective. We use a simple cohomological approach that is similar to Deligne cohomology.

The article is organized as follows. In section 2 we recall the equations governing principal $U(1)$ bundles with connection in terms of transition functions and local connection 1-forms. In section 3 we generalise to $U(1)$ gerbes with connection and give a simple example of a gerbe on the 3-sphere. In section 4 we present our cohomological framework for describing the whole ladder of n-gerbes and their equivalences. Section 5 contains some comments.

Acknowledgements

I am very grateful to Marco Mackaay for much stimulating collaboration, and for suggesting the method of dealing with the signs in section 10.4. It is a pleasure to thank the organizers of the XXth Białowieża Workshop for giving me the opportunity to present this material for the first time, and for their remarkable achievement in creating and maintaining the excellent scientific and social traditions of the Białowieża meetings.

This work was supported by *Programa Operacional "Ciência, Tecnologia, Inovação"* (POCTI) of the *Fundação para a Ciência e a Tecnologia* (FCT), cofinanced by the European Community fund FEDER.

10.2 The Local Equations for Bundles with Connection

We start by recalling a few well-known facts about principal bundles. A principal G-bundle P over a manifold M is given by a projection map $\pi : P \to M$, where $\pi^{-1}(x)$ is called the fibre over $x \in M$, P is called the total space, and M is called the base space, together with a (right, effective) G-action on P, written $p.g$ for $p \in P$ and $g \in G$, preserving fibres: $\pi(p.g) = \pi(g)$. Here G is a Lie group, and all manifolds and maps are smooth. The axiom of local triviality says that for any $x \in M$ there exists an open neighbourhood U of x such that $\pi^{-1}(U)$ is isomorphic to $U \times G$. A local trivialization is given by a local section, i.e. $s : U \to P$, satisfying $\pi(s(x)) = x$ for all $x \in U$, via $p = s(x).g \in \pi^{-1}(U) \leftrightarrow (x, g) \in U \times G$.

Given two open sets U_1 and U_2 which intersect, and local sections s_1 and s_2 on U_1 and U_2 respectively, they are related on the overlap $U_1 \cap U_2$ by a transition function g_{12} defined by:

$$s_2(x) = s_1(x).g_{12}(x), \quad \forall x \in U_1 \cap U_2.$$

If we introduce a third local section s_3, defined on an open set U_3 that intersects $U_1 \cap U_2$ non-trivially, then, by writing $s_3(x)$ in two different ways, we obtain the equation

$$g_{12}(x)g_{23}(x) = g_{13}(x), \quad \forall x \in U_1 \cap U_2 \cap U_3$$

on the triple overlap. In particular we have, choosing $U_3 = U_1$,

$$g_{21}(x) = g_{12}^{-1}(x). \tag{10.1}$$

The (Ehresmann) connection is a 1-form ω on P taking values in the Lie algebra of G, and satisfying some extra properties. Using local sections we obtain 1-forms on M by pull-back. Let $A_1 = s_1^*\omega$ and $A_2 = s_2^*\omega$ be two such local 1-forms, defined on U_1 and U_2 which intersect non-trivially. Then they are related by the equation

$$A_2 = g_{12}^{-1}A_1g_{12} + g_{12}^{-1}dg_{12}$$

on $U_1 \cap U_2$, due to the properties of ω.

If we use different local sections $s_i' : U_i \to P$ instead, related to the original local sections s_i by

$$s_i'(x) = s_i(x).h_i(x), \quad \forall x \in U_i,$$

where $h_i : U_i \to G$, we get gauge equivalent transition functions and connection 1-forms:

$$g_{ij}' = h_i^{-1}g_{ij}h_j,$$
$$A_i' = h_i^{-1}A_ih_i + h_i^{-1}dh_i.$$

We now specialize to the case when G is $U(1)$, i.e. Abelian. We will replace functions with values in G by their logarithms, so that all equations take values in $i\mathbf{R}$, the Lie algebra of G, and are modulo integer multiples of $2\pi i$. We will assume that all the open sets, and multiple overlaps $U_{ijk\ldots} := U_i \cap U_j \cap U_j \ldots$ in our cover of M are contractible. A G-bundle with connection is then given by transition functions

$$\ln g_{ij} : U_{ij} \to i\mathbf{R},$$

antisymmetric under exchange of indices because of equation (10.1), and connection 1-forms

$$A_i \in \Lambda^1(U_i),$$

satisfying

$$\ln g_{jk} - \ln g_{ik} + \ln g_{ij} = 0 \tag{10.2}$$

on U_{ijk} (we will see the reason for writing the terms in this order in section 10.4) and

$$i(A_j - A_i) = d\ln g_{ij} \tag{10.3}$$

on U_{ij}. We may also introduce the curvature 2-form F on M defined by

$$F = dA_i \tag{10.4}$$

on each open set U_i (which is indeed globally defined because of equation (10.3)). Finally F satisfies the Bianchi identity

$$dF = 0 \tag{10.5}$$

on M. Transition functions $\ln g'_{ij}$ and connection 1-forms A'_i are gauge-equivalent to transition functions $\ln g_{ij}$ and connection 1-forms A_i, iff there exist functions

$$\ln h_i : U_i \to i\mathbf{R},$$

satisfying

$$\ln g'_{ij} - \ln g_{ij} = \ln h_j - \ln h_i \tag{10.6}$$

on U_{ij} and

$$i(A'_i - A_i) = d\ln h_i \tag{10.7}$$

on U_i.

In this final formulation of the equations describing $U(1)$-bundles, there is no reference any longer to the total space P of the principal bundle—the whole description is "downstairs". This is advantageous for the generalisation to gerbes with connection in the next section, where the notion of a total space, although it exists[2], is less convenient for geometric constructions.

10.3 The Generalisation to Gerbes with Connection and an Example

The main guiding principle for understanding the generalisation from bundles to gerbes, is that for gerbes everything is one step up compared to bundles, in the form degree, the number of open sets in an overlap, or the dimension. Thus gerbes have transition functions defined on triple overlaps, a curvature 3-form, and parallel transport defined along surfaces inside M, instead of paths. A good place to read more about the general background and geometrical applications of gerbes is in Hitchin's lectures [Hitchin (1999)].

The data and equations defining a $U(1)$-gerbe with connection are analogous to the bundle case, except that there are now two separate layers of connections, connection 1-forms and connection 2-forms. A $U(1)$-gerbe with connection is given by transition functions

$$\ln g_{ijk} : U_{ijk} \to i\mathbf{R},$$

completely antisymmetric under exchange of indices, connection 1-forms

$$A_{ij} \in \Lambda^1(U_{ij}),$$

antisymmetric under exchange of indices, and connection 2-forms

$$F_i \in \Lambda^2(U_i),$$

satisfying

$$\ln g_{jkl} - \ln g_{ikl} + \ln g_{ijl} - \ln g_{ijk} = 0 \qquad (10.8)$$

---

[2]In the Dixmier-Douady [Dixmier and Douady (1963)] approach, an equivalence class of gerbes corresponds to a principal G-bundle, where G is the group of projective unitary transformations of a complex Hilbert space, up to equivalence. However, when introducing connections and curvature in this framework, the Lie algebra of G is tricky to handle - see [Brylinski (1993), Chap. 4].

on U_{ijkl},

$$i(A_{jk} - A_{ik} + A_{ij}) = -d\ln g_{ijk} \tag{10.9}$$

on U_{ijk}, and

$$F_j - F_i = dA_{ij} \tag{10.10}$$

on U_{ij}. Again all equations are taken modulo $2\pi i$. As before we may introduce a curvature form on M, this time a curvature 3-form G, defined by

$$G = dF_i \tag{10.11}$$

on each open set U_i (which is again globally defined because of equation (10.10)). Finally G satisfies the "Bianchi" identity

$$dG = 0 \tag{10.12}$$

on M.

The notion of gauge equivalence for gerbes is also a higher notion, since equivalences are specified not just by giving functions, but also 1-forms. More precisely, transition functions $\ln g'_{ijk}$, connection 1-forms A'_{ij} and connection 2-forms F'_i are gauge-equivalent to transition functions $\ln g_{ijk}$, connection 1-forms A_{ij} and connection 2-forms F_i iff there exist functions

$$\ln h_{ij} : U_{ij} \to i\mathbf{R}$$

and 1-forms

$$B_i \in \Lambda^1(U_i)$$

satisfying

$$\ln g'_{ijk} - \ln g_{ijk} = \ln h_{jk} - \ln h_{ik} + \ln h_{ij} \tag{10.13}$$

on U_{ijk},

$$i(A'_{ij} - A_{ij}) = -d\ln h_{ij} + i(B_j - B_i) \tag{10.14}$$

on U_{ij}, and

$$i(F'_i - F_i) = dB_i \tag{10.15}$$

on U_i.

There is clearly a pattern in the above equations, which will be elucidated in the next section. One aspect which is sometime found to be puzzling, is that the connection 1-form A_{ij} for gerbes is not defined everywhere on M, but only on the double overlaps of the cover. It may seem that A_{ii} is defined on U_i, but this is identically zero because of the antisymmetry condition on the indices. However one should really think of A_{ij} as being merely a subsidiary "transition" connection for the genuine gerbe connection 2-form F_i, which is defined on every patch of M.

We will conclude this section with a simple example of a gerbe with connection on the 3-sphere, which is a natural generalisation of the familiar monopole bundle on the 2-sphere[3]. In fact, let us start by considering the monopole bundle on S^2, covered by two patches U_1 and U_2, which intersect in a (not-too-wide) strip around the equator, isomorphic to $S^1 \times]0, 1[$. (This is not contractible, but we could introduce extra patches to get contractible overlaps if necessary.) The transition function $g_{12} : U_{12} \to U(1)$ is chosen to be a winding number 1 map from S^1 to $U(1)$, constant along the transversal direction (which is why we do not want the strip to be too wide). On U_1 we choose the connection 1-form $A_1 = 0$, and on U_2 we choose $iA_2 = \overline{d \ln g_{12}}$, meaning it is equal to $d \ln g_{12}$ on the overlap U_{12} and is continued in some manner to the rest of U_2, which is possible since U_2 is contractible. Then we have the desired equation for a bundle with connection (10.3):

$$i(A_2 - A_1) = d \ln g_{12}$$

on U_{12}. The curvature F of the monopole connection, given by equation (10.4), has support contained in U_2, and integrating $F/2\pi$ over S^2 gives 1.

Now we take the 3-sphere S^3, and cover it with three patches U_1, U_2 and U_3. U_3 covers the equator and one half of S^3, and U_1 and U_2 together cover the equator and the other half of S^3, in such a way that U_{13} and U_{23} are isomorphic to $U_1 \times]0, 1[$ and $U_2 \times]0, 1[$ respectively, where U_i refer to the patches from the monopole bundle case. The intersection between U_3 and the union of U_1 and U_2 is a (not-too-wide) spherical shell isomorphic to $S^2 \times]0, 1[$. The intersection of all three open sets is isomorphic to $S^1 \times]0, 1[^2$. We take the transition function g_{132} on this triple overlap to be the winding number 1 map g_{12} from S^1 to $U(1)$ used previously, taken to be constant along the transversal directions. The 1-form connection is given by: $A_{13} = A_1$ (the 1-form for the monopole, constant along the direction transversal to U_1), $A_{23} = A_2$ (constant in the direction transversal to U_2) and $A_{12} = 0$.

---

[3]Meinrenken [Meinrenken (2002)] has recently obtained gerbes with connection on compact simple Lie groups G, which are G-equivariant. The purpose of our example is different, however, so we do not consider S^3 as a Lie group.

We thus have equation (10.9)

$$i(A_{32} - A_{12} + A_{13}) = -d\ln g_{132}.$$

Now we choose the 2-form connection F_i as follows: $F_1 = F_2 = 0$ and $F_3 = \overline{F}$, meaning it is equal to the monopole curvature F on the overlaps U_{13} and U_{23}, and is continued in some manner to the rest of U_3, which is possible since U_3 is contractible. We have thus also satisfied equation (10.10)

$$F_2 - F_1 = dA_{12} = 0, \quad F_3 - F_1 = dA_{13} = dA_1, \quad F_3 - F_2 = dA_{23} = dA_2.$$

The curvature of the gerbe connection G, given by equation (10.11), has support contained in U_3, and integrating $G/2\pi$ over S^3 gives the integral of $F/2\pi$ over S^2, i.e. 1, after applying Stokes' theorem.

The "gerbopole" we have just described is therefore a natural, higher generalisation of the monopole, and its winding number, or charge, also derives from the winding number 1 map from the circle to the group. In the next section we will see that this map itself also has an interpretation in the dimensional ladder containing bundles and gerbes with connection.

10.4   A Cohomological Formulation

In this section we wish to unify the equations for bundles and gerbes with connection from the previous sections, and generate the whole dimensional ladder of n-gerbes, by using a cohomological formulation. Our approach can be viewed as a variant of Deligne cohomology, in that it blends together de Rham and Čech cohomology. Let us first define the cochain groups $\Lambda^{p,n}$, whose elements are collections of p-forms, valued in $i\mathbf{R}$, defined on each n-fold overlap of open sets of our fixed cover of M. When $n = 1$, a cochain consists of a p-form on each open set of the cover, and when $n = 0$, a cochain is a single, globally-defined p-form on M. We write cochains in the form $C_{ijk\ldots}$ where $ijk\ldots$ ranges over the n-fold overlaps, or C (no index) when $n = 0$.

There are two natural operators acting on these cochain groups, namely the exterior derivative

$$d : \Lambda^{p,n} \to \Lambda^{p+1,n}, \quad C_{ijk\ldots} \mapsto dC_{ijk\ldots}$$

and the Čech coboundary operator

$$\delta : \Lambda^{p,n} \to \Lambda^{p,n+1}, \quad \delta C_{i_1 \ldots i_{n+1}} = \sum_{\alpha=1}^{n+1} (-1)^{\alpha+1} C_{i_1 \ldots \hat{i_\alpha} \ldots i_{n+1}}.$$

Both of these operators are nilpotent, but in order to combine them into a new nilpotent operator, they should anticommute, which is not the case. This can be remedied either by multiplying δ by $(-1)^p$, or by multiplying d by $(-1)^n$. We choose the latter solution, and define

$$\bar{d} : \Lambda^{p,n} \to \Lambda^{p+1,n}, \quad \bar{d} = (-1)^n d,$$

which satisfies $\delta \bar{d} + \bar{d} \delta = 0$.

Now we can define the cohomology that is relevant for describing bundles and gerbes. Let the cochain groups $\Lambda^{(k)}$ be given by

$$\Lambda^{(k)} = \bigoplus_{p+n=k} \Lambda^{p,n},$$

and the operator $D : \Lambda^{(k)} \to \Lambda^{(k+1)}$ be defined by:

$$D = \delta - \bar{d},$$

satisfying $D^2 = 0$. A bundle with connection may now be defined to be a 2-cocycle $\mathcal{B}$:

$$\mathcal{B} = \ln g_{ij} + iA_i - iF \in \Lambda^{(2)}, \quad D\mathcal{B} = 0,$$

which is equivalent to equations (10.2) to (10.5), and a gerbe with connection may be defined to be a 3-cocycle $\mathcal{G}$

$$\mathcal{G} = \ln g_{ijk} + iA_{ij} + iF_i - iG \in \Lambda^{(3)}, \quad D\mathcal{G} = 0,$$

which is equivalent to equations (10.8) to (10.12), as may be easily verified.

Having simplified the equations in this way, we can extend the definition to cocycles of any order. Thus we define an n-gerbe with connection to be:

$$\mathcal{H} \in \Lambda^{(n+2)}, \quad D\mathcal{H} = 0,$$

so that a gerbe is a 1-gerbe, and a bundle is a 0-gerbe, in these terms. For $n \geq 2$, the n-gerbe itself is the $(0, n+2)$ part of $\mathcal{H}$, its multilayered connection consists of the $(1, n+1)$ to $(n+1, 1)$ parts of $\mathcal{H}$, and the n-gerbe curvature is minus the globally-defined $(n+2)$-form part of $\mathcal{H}$.

It is not particularly illuminating to write down the explicit equations for the higher gerbes. However it is amusing to consider a lower case, namely a (-1)-gerbe. This is, by the above definition,

$$\mathcal{H} = \ln f_i - iA \in \Lambda^{(1)}, \quad D\mathcal{H} = 0,$$

i.e. the (-1)-gerbe itself is a collection of functions $\ln f_i$, there is no layer of connections, and the curvature is the 1-form A. Let us give an example of such a (-1)-gerbe on S^1. We cover the circle, parametrised by a 2π-periodic coordinate θ, with three open sets: $U_1 =]0, \pi[$, $U_2 =]2\pi/3, 5\pi/3[$ and $U_3 =]4\pi/3, 7\pi/3[$. The equation $D\mathcal{H} = 0$ implies the following equations:

$$\ln f_j - \ln f_i = 0$$

on U_{ij},

$$d\ln f_i - iA = 0$$

on U_i, and

$$dA = 0$$

on M, which are solved by

$$\ln f_i(\theta) = i\theta, \quad A = d\theta.$$

In terms of the "gerbopole" example at the end of the previous section, the (-1)-gerbe sits on the equator of the monopole bundle, in the same way as the monopole bundle sits on the equator of the gerbopole.

The notion of gauge equivalence between bundles and gerbes in the previous two sections can also be expressed in terms of the language introduced in this section. For bundles the equations (10.6) and (10.7) for gauge equivalence can be expressed as:

$$\mathcal{B}' \sim \mathcal{B} \Leftrightarrow \mathcal{B}' - \mathcal{B} = D\ln h_i, \quad \ln h_i \in \Lambda^{0,1},$$

and for gerbes the equations (10.13) to (10.15) for gauge equivalence can be expressed as:

$$\mathcal{G}' \sim \mathcal{G} \Leftrightarrow \mathcal{G}' - \mathcal{G} = D(\ln h_{ij} + iB_i), \quad \ln h_{ij} + iB_i \in \Lambda^{0,2} \oplus \Lambda^{1,1}.$$

This generalises to gauge equivalence for n-gerbes $\mathcal{H}', \mathcal{H} \in \Lambda^{(n+2)}$:

$$\mathcal{H}' \sim \mathcal{H} \Leftrightarrow \mathcal{H}' - \mathcal{H} = D\mathcal{F}, \quad \mathcal{F} \in \Lambda_0^{(n+1)},$$

where $\Lambda_0^{(n+1)}$ denotes $\Lambda^{(n+1)}$ without the top-degree $(n+1)$-form part.

We conclude this section by remarking that clearly it would be more natural if we could replace $\Lambda_0^{(n+1)}$ simply by $\Lambda^{(n+1)}$ in the above definition of equivalence. For example, in the case of bundles the higher gauge equivalence suggested here modifies equation (10.7) as follows, taking $\mathcal{F} = \ln h_i + iB$,

$$i(A'_i - A_i) = d\ln h_i + iB,$$

but also implies a higher gauge transformation for the curvature F:

$$F' - F = dB.$$

This would not normally be considered as a gauge transformation, but it is an equivalence for some purposes, since e.g. Chern forms for bundles on closed manifolds are preserved under this transformation.

10.5 Comments

Given the importance of abelian gauge theory, and the fact that gerbes generalise abelian bundles with connection in such a natural way, it would be very interesting if a direct, dynamical role could be found for gerbes in physical theories. For this it is necessary to construct couplings to other fields and build actions. A possible coupling arises in so-called twisted vector bundles, where abelian gerbes with connection are coupled to non-abelian bundles with connection (see [Mackaay (2001)], where the holonomy of such objects is also discussed). Actions for gerbes have been studied by Baez [(Baez (2002)], in fact, in the context of non-abelian gerbes. Adapting the equations in section 10.3 to non-abelian gerbes is a challenging task — see [Breen and Messing (2001)] for an algebraic/differential-geometric approach and [Attal (2002)] for a combinatorial approach. Maybe the cohomological approach and example presented here will suggest some way forward in this problem.

Bibliography

Attal, R. (2002). Combinatorics of Non-Abelian Gerbes with Connection and Curvature, math-ph/0203056.

Baez, J. (2002). Higher Yang-Mills Theory, hep-th/0206130.

Barrett, J. W. (1991). Holonomy and path structures in general relativity and Yang-Mills theory, *Int. J. Theor. Phys.*, **30**, 9, pp. 1171–1215.

Breen, L. and Messing, H. (2001). Differential Geometry of Gerbes, math.AG/0106083.

Brylinski, J.-W. (1993). Loop spaces, characteristic classes and geometric quantization, Volume 107 of Progress in Mathematics, Birkhauser.

Caetano, A. and Picken, R. (1994). An axiomatic definition of holonomy, *Int. J. Math.* **5**, 6, pp. 835–848.

Carey, A., Mickelsson, J. and Murray, M. (2000). Bundle gerbes applied to quantum field theory, *Rev. Math. Phys.* **12**, 1, pp. 65–90.

Dixmier, J. and Douady, A. (1963). Champs continus d'espaces hilbertiens et de C^*-algèbres, *Bull. Soc. Math. Fr.* **91**, pp. 227–284.

Giraud, J. (1971). Cohomologie non-abelienne, Volume 179 of Grundl., Springer-Verlag, Berlin.

Gomi, K. (2001). Gerbes in classical Chern-Simons theory, hep-th/0105072.

Hitchin, N. (1999). Lectures on special Langrangian submanifolds, School on Differential Geometry (1999), the Abdus Salam International Centre for Theoretical Physics.

Mackaay, M. (2001). A note on the holonomy of connections in twisted bundles. to appear in *Cahiers de Topologie et Geometrie Differentielle Categoriques* , math.DG/0106019.

Mackaay, M. and Picken, R. (2002). Holonomy and Parallel Transport for Abelian Gerbes, *Adv. Math.* **170**, pp. 287–339.

Meinrenken, E. (2002). The basic gerbe over a compact simple Lie group, math.DG/0209194.

Picken, R. (2003). TQFT's and gerbes, math.DG/0302065.

Picken, R. and Semião, P. (2002). A classical approach to TQFT's, math.QA/0212310.

Witten, E. (1998). D-branes and K-theory, *J. High Energy Phys.*,no.12, pap.19.

Chapter 11

On a Quantum Group of Unitary Operators.
The Quantum '$az + b$' Group

W. Pusz[1] and S. L. Woronowicz[2]

Abstract: The concept of *a quantum group of unitary operators* is relevant for the theory of non-compact locally compact quantum groups. It plays a similar rôle as the concept of *a quantum matrix group* in the compact case. To show the usefulness of this notion we present an approach to a construction of the quantum '$az + b$' group based on this idea. A brief survey of the present status of quantum group theory is also included.

11.1 Introduction

The Conferences in Białowieża were unforgettable events due to a rare combination of a beautiful wild surroundings with its unique old forest and a specific atmosphere. Hopefully for us they started almost at the same time as quantum group theory was initiated and from the very beginning they were open for reports concerning a development of a general quantum group theory as well as to constructions of examples. Due to this we participated in many of them. The present jubilee seems to be a good occasion to give also a brief account of the present status of the theory.

It is not possible in a short review to mention all important approaches and contributions (more than two hundreds of the original papers!). For the discussion of many of them and the introduction to the theory we refer

<hr>

[1]Department of Mathematical Methods in Physics, Faculty of Physics, University of Warsaw, Hoża 74, 00-682 Warszawa, Poland, `wieslaw.pusz@fuw.edu.pl`

[2]Department of Mathematical Methods in Physics, Faculty of Physics, University of Warsaw, Hoża 74, 00-682 Warszawa, Poland, `stanislaw.woronowicz@fuw.edu.pl`

to [7] and [11]. To make the paper self-contained, in our presentation we focus only on some general aspects of the quantum group theory.

Roughly speaking 'a quantum group' is 'a quantum space' endowed with a group structure. One may consider a pure algebraic version of the theory or a topological one.

The first one is known as a Hopf *-algebra approach. Let $\mathcal{A}$ be a unital *-algebra and $\Delta : \mathcal{A} \to \mathcal{A} \otimes_{\text{alg}} \mathcal{A}$ be unital *-homomorphism such that

$$(\Delta \otimes \text{id})\Delta = (\text{id} \otimes \Delta)\Delta \tag{11.1.1}$$

(coassociativity). In this approach $\mathcal{A}$ encodes a quantum space. Now Δ endows a group structure on it if $(\mathcal{A}, \Delta)$ is a Hopf *-algebra.

We recall that $(\mathcal{A}, \Delta)$ is a Hopf *-algebra if there exist linear mappings $e : \mathcal{A} \to \mathbb{C}$ and $\kappa : \mathcal{A} \to \mathcal{A}$ such that

$$(e \otimes \text{id})\Delta(a) = a = (\text{id} \otimes e)\Delta(a)$$

and

$$m(\kappa \otimes \text{id})\Delta(a) = e(a)I = m(\text{id} \otimes \kappa)\Delta(a)$$

for any $a \in \mathcal{A}$, where $m : \mathcal{A} \otimes_{\text{alg}} \mathcal{A} \to \mathcal{A}$ denotes the multiplication map, i.e. $m(a \otimes b) = ab$ for any $a, b \in \mathcal{A}$. It is known that e (called counit) and κ (called coinverse or antipode) are uniquely determined. Moreover e is a unital *-algebra homomorphism, κ is antimultiplicative, anticomultiplicative and

$$\kappa\left(\kappa(a^*)^*\right) = a$$

for any $a \in \mathcal{A}$.

The topological approach uses C^*-algebra language. Any (locally compact) quantum space is encoded by a C^*-algebra A. This is approved by Gelfand-Naimark theorem which says that any commutative C^*-algebra A is isomorphic to the algebra $C_\infty(\Lambda)$ of all complex continuous and vanishing at infinity functions on some locally compact space Λ. The space Λ is unique up to a homeomorphism, i.e. we have a correspondence

$$\Lambda \quad \longleftrightarrow \quad A = C_\infty(\Lambda).$$

Moreover A is unital if and only if Λ is a compact space. No theory of such type is known for noncommutative C^*-algebra A. Formally one can solve the problem (cf. e.g. [23]) by considering the category dual to the category of C^*-algebras. Objects of this dual category are called 'locally compact quantum spaces' and its morphisms - continuous mappings of (locally compact) quantum spaces. The notion of quantum spaces introduces a new language to the C^*-algebra theory. In particular, if A is unital (or non-unital) we say that we deal with compact case (or noncompact, respectively).

Now we pass to the (locally compact) quantum groups. It turned out that the compact case was relatively easy and starting from simple axioms a nice theory (parallel to that for classical compact groups) was build [31] on early stage of the theory.

Definition 11.1.1 Let $G = (A, \Delta)$, where A is a separable C^*-algebra and $\Delta : A \longrightarrow A \otimes A$ is a unital *-algebra homomorphism. We say that G is a compact quantum group if
(1) Δ is coassociative (cf (11.1.1)) and
(2) The sets

$$\{\, (b \otimes I)\Delta(a) : \ a, b \in A \,\}, \qquad \{\, (I \otimes b)\Delta(a) : \ a, b \in A \,\}$$

are linearly dense subsets of $A \otimes A$.

Using these axioms one shows that there are many finite-dimensional representations of G. We recall that a unitary matrix $V = (v_{kl}) \in M_N(A) = M_N(\mathbb{C}) \otimes A$ is a N-dimensional unitary representation of G if

$$\Delta(v_{kl}) = \sum_{r=1}^{N} v_{kr} \otimes v_{rl}$$

or using leg-numbering notation:

$$(\mathrm{id} \otimes \Delta)V = V_{12}V_{13}. \tag{11.1.2}$$

Let $\mathcal{A}$ be the set of all linear combinations of matrix elements of all finite-dimensional unitary representations of G. Then we have the following result [31, Theorem 2.2 and Theorem 2.3]

Theorem 11.1.2 *Let $G = (A, \Delta)$ be a compact quantum group. Then*
*1. $\mathcal{A}$ is dense *-subalgebra of A and $\Delta(\mathcal{A}) \subset \mathcal{A} \otimes_{\mathrm{alg}} \mathcal{A}$.*
*2. $(\mathcal{A}, \Delta|_{\mathcal{A}})$ is a Hopf *-algebra.*
3. There exists unique state (normalized positive linear functional) h on A

such that

$$(h \otimes \mathrm{id})\Delta(a) = h(a)I = (\mathrm{id} \otimes h)\Delta(a).$$

for any $a \in A$.

4. h is faithful on A, i.e. if $a \in A$ and $h(a^*a) = 0$ then $a = 0$.

Statement 1 and Statement 2 give a connection between C^*-algebra approach and Hopf *-algebra one in the compact case. This means that in fact the purely algebraic description of compact quantum groups and the topological one are equivalent (cf. also [5]). In particular all classical compact groups admit a quantum deformation (cf. [6; 18]).

Clearly the functional h in Statement 3 of the above theorem is a Haar state (measure). It is still an open problem how to guarantee the faithfulness of h on the whole algebra A. If this would be the case then a deeper insight into the structure of G could be obtained [31, Theorem 2.]. Let us note that left-invariant state and right-invariant state coincide. Therefore a compact quantum group·is unimodular as in classical case. On the other hand the Haar state needs not be central one. Therefore if h is faithful there is modular structure coming from the Tomita-Takesaki theory.

Definition 11.1.1 is not very useful if one is looking for examples of compact quantum groups. In this aspect a concept of a compact matrix quantum group [25; 26] turns out to be very useful.

Let A be unital C^*-algebra and $V = (v_{kl}) \in M_N(A) = M_N(\mathbb{C}) \otimes A$. We say that V generates A if the smallest *-algebra containing all matrix elements v_{kl}, $k, l = 1, 2, \dots N$ is dense in A.

Definition 11.1.3 Let A be unital C^*-algebra and $V = (v_{kl}) \in M_N(A) = M_N(\mathbb{C}) \otimes A$. Assume that

1. A is generated by V.
2. There exists a unital *-algebra homomorphism $\Delta : A \longrightarrow A \otimes A$ such that

$$(\mathrm{id} \otimes \Delta)V = V_{12}V_{13}. \tag{11.1.3}$$

3. V and its transpose $V^\top$ are invertible in $M_N(A)$.

Then (A, V) is called a compact matrix quantum group.

One can easily show that Δ is uniquely defined by (11.1.3) and when it exists it is automatically coassociative. Moreover, $G = (A, \Delta)$ is a compact group [31, Remark 2] in the sense of Definition 11.1.1. Element V may be treated as a "strongly continuous" quantum family (labeled by the quantum space related to A) of operators acting on the Hilbert space $\mathbb{C}^N$. By (11.1.3) this family is a representation of G. In particular if V is a unitary element of $M_N(\mathbb{C}) \otimes A$ than V is a unitary representation of G (cf.(11.1.2)).

The majority of known examples of compact quantum groups are of this kind. In particular the first nontrivial example, namely the quantum $SU_q(2)$ introduced in [24] is a compact matrix quantum group.

Now we turn to the non-compact case. Then the C^*-algebra A is non-unital. To simplify our presentation we shall mostly deal with concrete C^*-algebras. Let H be a Hilbert space. It is convenient to denote by $C^*(H)$ the set of all non-degenerate separable C^*-algebras of operators acting on H. By definition $A \subset B(H)$ is non-degenerate if AH is dense in H.

In the case of a non-compact locally compact space Λ, $A = C_\infty(\Lambda)$ consists of continuous functions vanishing at infinity on Λ. On the other hand one has to consider also other classes of continuous functions, such as bounded ones or all continuous functions. The counterparts of above notions for general noncommutative C^*-algebra are provided by concepts of a multiplier and an affiliated element. We recall these notions.

The multiplier algebra of A is denoted by $M(A)$:

$$M(A) = \{ a \in B(H) : \ aA \subset A \text{ and } Aa \subset A \}.$$

Clearly $M(A)$ is a unital C^*-subalgebra of $B(H)$. Now we define elements affiliated with A. These should be treated as "unbounded" multipliers. To be more precise, let T be a closed densely defined linear operator acting on H. Its z-transform is by definition

$$z_T = T(I + T^*T)^{-\frac{1}{2}}. \tag{11.1.4}$$

Then z_T is a bounded operator and $\|z_T\| \leq 1$. One should notice that $\|T\| < \infty$ if and only if $\|z_T\| < 1$. The affiliation relation, denoted by η, is introduced as follows.

$$\Big(T \, \eta \, A \Big) \Longleftrightarrow \left(\begin{array}{c} z_T \in M(A) \text{ and} \\[2mm] (I + T^*T)^{-\frac{1}{2}} A \text{ is dense in } A \end{array} \right). \tag{11.1.5}$$

The set of all affiliated elements is denoted by A^η. Let us remark that since $(I + T^*T)^{-\frac{1}{2}}$ is selfadjoint, the density of $(I + T^*T)^{-\frac{1}{2}} A$ is equivalent to the density of $A(I + T^*T)^{-\frac{1}{2}}$. Since $(z_T)^* = z_{T^*}$, $z_T \in M(A)$ if and only if both z_T and z_{T^*} are right multipliers of A.

Clearly $A \subset M(A) \subset A^\eta$. Moreover if $T \in A^\eta$ and T is bounded then $T \in M(A)$. If A is unital then $A = M(A) = A^\eta$. On the other hand for $A = C_\infty(\Lambda)$ we have $M(A) = C_{\text{bounded}}(\Lambda)$ and $A^\eta = C(\Lambda)$. For the algebra $A = \mathcal{K}(H)$ (of all compact operators on H) we get $M(A) = B(H)$ and A^η is the set of all closed densely defined operators acting on H. Therefore in general A^η is not an algebra or even a vector space. Nevertheless the set A^η is closed under $*$-operation: $T^* \, \eta \, A$ and $T^*T \, \eta \, A$ for any $T \in A^\eta$.

Elements of $M(A)$ act on A by left and right multiplications. The natural topology of $M(A)$ is that of pointwise convergence on A. This is called a strict topology of $M(A)$. It is known that $M(A)$ equipped with this topology is a topological $*$-algebra. In turn the strict topology of $M(A)$ induces (by z-transform) a natural topology of A^η.

Let A and B be C^*-algebras and let $B \in C^*(K)$. We say that π is a morphism from A into B if π is a (non-degenerate) representation of A on the Hilbert space K and $\pi(A)B$ is dense subset of B. The set of all morphism from A into B is denoted by $\text{Mor}(A, B)$. In particular $\pi \in \text{Mor}(A, \mathcal{K}(K))$. Clearly π maps A into $M(B)$ and one can easily prove that any $\pi \in \text{Mor}(A, B)$ has unique extension to a C^*-algebra homomorphism from $M(A)$ into $M(B)$ (this allows to make a composition of morphisms possible) and then to a $*$-preserving map from A^η into B^η.

The absence of a good algebraic structure of A^η is the main source of a great discrepancy between Hopf $*$-algebra and C^*-algebra approach to non-compact quantum groups. In latter case generators of Hopf algebra are represented by unbounded (closed and densely defined) operators on a Hilbert space. Since the sum and the product of two such operators are badly defined, no representation of the whole Hopf algebra exists in general. There are also more subtle reasons of the discrepancy coming from the existence of symmetric operators without selfadjoint extensions or due to the phenomenon of commuting but not strongly commuting selfadjoint operators. This diversity of situations not apparent on the Hopf algebra level makes the constructions of topological quantum groups much more sophisticated. On the other hand the nice algebraic structure of $A^\eta = A$ for unital C^*-algebra A explains why Statement 1 and 2 of Theorem 11.1.2 hold for compact quantum groups and fail in non-compact case.

Now we are ready to describe briefly the present status of the theory of locally compact quantum groups. The theory of multiplicative unitary operators plays the central role in the approach. It was developed by S.Baaj and G.Skandalis [3]. Let H be a Hilbert space. A unitary operator W acting on $H \otimes H$ is a called multiplicative unitary if it satisfies the pentagon equation:

$$W_{23}W_{12} = W_{12}W_{13}W_{23}. \tag{11.1.6}$$

Such operators have appeared long ago in the theory of locally compact groups in the context of generalized Pontryagin duality. Let G be a locally compact group and $H = L^2(G, dg)$, where dg is a (right) Haar measure. For any $x \in H$ and $g, g' \in G$ we set

$$(Wx)(g, g') = x(gg', g'). \tag{11.1.7}$$

Then one can easily verify that W is a unitary operator acting on $H \otimes H$. Moreover

$$(W_{23}W_{12}x)(g_1, g_2, g_3) = x(g_1(g_2g_3), g_1g_2, g_2),$$
$$(W_{12}W_{13}W_{23}x)(g_1, g_2, g_3) = x((g_1g_2)g_3, g_1g_2, g_2)$$

for any $x \in H \otimes H \otimes H$ and any $g_1, g_2, g_3 \in G$. So the pentagon equation (11.1.6) is equivalent to the associativity of the group multiplication. The operator W introduced by formula (11.1.7) is called Kac-Takesaki operator. It contains the full information about the group G. Following this idea one may try to assign a quantum group to any multiplicative unitary operator W. It turns out [3] that it is possible if pentagon equation is supplemented by a *regularity condition*:

$$\left\{ (\mathrm{id} \otimes \omega)(\Sigma W) : \omega \in B(H)_* \right\}^{\text{norm closure}} = \mathcal{K}(H) \tag{11.1.8}$$

where $\Sigma \in B(H \otimes H)$ is the flip $\Sigma(x \otimes y) = y \otimes x$ for any $x, y \in H$ and $B(H)_*$ is the space of all normal functionals on $B(H)$.

Unfortunately this theory does not apply to all multiplicative unitaries related to quantum groups [1; 2]. To overcome this difficulty the regularity condition was replaced by condition of manageability [32]. It was shown [9; 10] that any quantum group may be related to a manageable multiplicative unitary. On the other hand it is not easy to verify manageability condition in particular examples. Moreover in specific examples we deal with non-manageable multiplicative unitaries. This is because the correspondence between multiplicative unitaries and quantum groups is not one to one. Different multiplicative unitaries may describe the same quantum object. In fact the manageability condition may be weakened.

A multiplicative unitary W is called *modular* [20] if there exist strictly positive selfadjoint operators $\hat{Q}$ and Q acting on H and a unitary operator $\tilde{W}$ acting on $\overline{H} \otimes H$ such that

$$W(\hat{Q} \otimes Q) = (\hat{Q} \otimes Q)W \tag{11.1.9}$$

and

$$(x \otimes u | W | z \otimes y) = \left(\bar{z} \otimes Qu \middle| \tilde{W} \middle| \bar{x} \otimes Q^{-1}y \right) \tag{11.1.10}$$

for any $x, z \in H$, $u \in \mathcal{D}(Q)$ and $y \in \mathcal{D}(Q^{-1})$. In the above definition $\overline{H}$ is the complex conjugate Hilbert space related to H by the antiunitary mapping $H \ni x \longrightarrow j(x) = \bar{x} \in \overline{H}$. In what follows $\top$ will denote a transposition map

$$B(H) \ni m \longrightarrow m^\top = j \circ m^* \circ j^{-1} \in B(\overline{H}). \tag{11.1.11}$$

Clearly it is antiisomorphisms of the C^*-algebras.

Let A be a C^*-algebra and $\Delta \in \mathrm{Mor}(A, A \otimes A)$. We say that (A, Δ) is a C^*-bialgebra if Δ is coassociative. The following result [20, Theorem 2.3] is a structure theorem for bialgebras (A, Δ) related to a modular multiplicative unitary.

Theorem 11.1.4   *Let $W \in B(H \otimes H)$ be a modular multiplicative unitary. Define*

$$A = \{(\omega \otimes \mathrm{id})W : \omega \in B(H)_*\}^{\mathrm{norm\ closure}}$$
$$\hat{A} = \{(\mathrm{id} \otimes \omega)W^* : \omega \in B(H)_*\}^{\mathrm{norm\ closure}}. \tag{11.1.12}$$

Then

(1) A, $\hat{A} \in C^*(H)$.
(2) $W \in M(\hat{A} \otimes A)$.
(3) There exists a unique $\Delta \in \mathrm{Mor}(A, A \otimes A)$ such that

$$(\mathrm{id} \otimes \Delta)W = W_{12}W_{13}.$$

Moreover Δ is coassociative and

$$\{ (b \otimes I)\Delta(a) : \ a, b \in A \}, \qquad \{ (I \otimes b)\Delta(a) : \ a, b \in A \}$$

are linearly dense subsets of $A \otimes A$.

(4) There exists a unique closed linear operator κ on the Banach space A such that

$\{(\omega \otimes \mathrm{id})W : \omega \in B(H)_*\}$ *is a core for κ and*

$$\kappa\left((\omega \otimes \mathrm{id})W\right) = (\omega \otimes \mathrm{id})W^*$$

for any $\omega \in B(H)_*$. Moreover

  (i) the domain of κ is a subalgebra of A and κ is antimultiplicative: $\kappa(ab) = \kappa(b)\kappa(a)$ for any $a, b \in \mathcal{D}(\kappa)$.

  (ii) $\kappa\left((\mathcal{D}(\kappa))\right) = \mathcal{D}(\kappa)^*$ and $\kappa\left(\kappa(a^*)^*\right) = a$ for all $a \in \mathcal{D}(\kappa)$.

  (iii) the operator κ admits the following polar decomposition

$$\kappa = R {\circ} \tau_{i/2},$$

where R is involutive (normal) antiautomorphism of A and $\tau_{i/2}$ is the analytic generator of a one parameter group of $*$-automorphisms $\{\tau_t\}_{t \in \mathbb{R}}$ of the C^*-algebra A,

  (iv) R commutes with τ_t for any $t \in \mathbb{R}$,

  (v) R and $\{\tau_t\}_{t \in \mathbb{R}}$ are uniquely determined.

(5) We have

  (i) $\Delta {\circ} \tau_t = (\tau_t \otimes \tau_t) {\circ} \Delta$ for all $t \in \mathbb{R}$,

  (ii) $\Delta {\circ} R = \sigma(R \otimes R) {\circ} \Delta$, where σ is a flip automorphisms of $A \otimes A$.

(6) Let $\tilde{W}$ and Q be the operators related to W by modularity condition. Then

  (i) $\tau_t(a) = Q^{2it} a Q^{-2it}$ for any $a \in A$ and $t \in \mathbb{R}$,

  (ii) $W^{\top \otimes R} = \tilde{W}^*$, where a^R denotes $R(a)$ for any $a \in A$ and $\top$ is the transposition map (11.1.11).

The antiautomorphism R which appears in the polar decomposition of the antipode κ is called the unitary antipode and $\{\tau_t\}_{t \in \mathbb{R}}$ is called the scaling group. It is clear that for any locally compact group $\kappa = R$ and the scaling group is trivial.

Now using above Theorem we can say that C^*-bialgebra $G = (A, \Delta)$ is a quantum group if it is related to some modular multiplicative unitary in the way described above (A coincides with the C^*-algebra introduced by he first formula (11.1.12) and Δ as in Statement 3). To verify such definition one has to know a multiplicative unitary W in advance. It is not easy. But if this is the case one has the rich theory of modular multiplicative operators at hand. In particular the operator $\widehat{W} = \Sigma W^* \Sigma$ is a modular multiplicative unitary (operators Q and $\hat{Q}$ exchange their position) and a quantum group related to $\widehat{W}$ is a Pontryagin dual group $\hat{G} = (\hat{A}, \hat{\Delta})$.

As we noticed the above concept of quantum group is very implicit and it would be nice to have much simpler set of axioms that would guarantee that a C^*-bialgebra (A, Δ) is a quantum group in the above sense. At present the situation is much better than a few years ago. There are two approaches in this direction. The first one, proposed by J.Kustermans and S.Vaes [9] is based on the assumption of the existence of a left- and a right-invariant weight. Both weights should be faithful in a certain strong sense. It is very interesting that their theory anticipated non-invariance of Haar weights with respect to the scaling group. In [21] A.Van Daele shows that it really happens in the case of the quantum '$az + b$' group for deformation parameter being a root of unity.

In the second approach presented in [10] we assume the existence of the right faithful (in the strong sense) Haar weight and the existence of antipode coupled to the rest of the structure by the *strong right invariance* condition. It turns out that the two approaches are equivalent.

In both approaches the existence of a Haar weight is postulated. This is very unsatisfactory. In any nice theory the existence of the Haar weight should be derived from simpler, more elementary axioms. This is the case for the theory of compact quantum groups. In our opinion we still have to look for better formulation of the theory of locally compact quantum groups. It is supported by the fact that in particular examples of non-compact locally compact groups a Haar weight can be constructed. We mention

- Pontryagin duals for compact quantum groups [13],

- quantum groups resulting from quantum double group construction over compact quantum groups (in particular a quantum Lorentz group with Iwasawa decomposition property) [13],

- a quantum '$ax + b$' group of affine transformations of the real line [36; 17],

- a quantum '$az + b$' group of affine transformations of the complex plane and its quantum double group (a quantum $GL(2, \mathbb{C})$) [29; 16; 14],

- a quantum $E(2)$ group of motions of the Euclidean plane [28], its Pontryagin dual group [22] and a quantum double group build over quantum $E(2)$ group (a quantum Lorentz group with Gauss decomposition property) [35],

- a quantum $SU(1,1)$ group (in this case 'the non existence on the C^*-

algebra level' theorem was proved in [27] but then the positive result for two-fold covering was obtained by E.Koelink and J.Kustermans [8]).

Moreover, let $G = (A, \Delta)$ be a quantum group produced by a modular multiplicative unitary W acting on $H \otimes H$. It turns out [34] that in many examples the right Haar weight is given by the formula

$$h(a) = \operatorname{Tr}(\hat{Q}a\hat{Q}), \qquad (11.1.13)$$

where $\hat{Q}$ is the operator appearing in (11.1.9). This is the case when this provides finite values for $a = c^*c$ where c runs over a dense subset of A.

A more general method of constructing the Haar weight was elaborated by A.Van Daele [21].

In what follows we shall consider representations of C^*-algebras acting on different Hilbert spaces. If π is such a representation then the corresponding carrier Hilbert space will be denoted by H_π.

When we deal with classical non-compact matrix group G then the matrix elements are unbounded continuous functions. Therefore they are only affiliated with the algebra $C_\infty(G)$. This means that for non-unital C^*-algebra A one has to make precise what it means that A is generated by elements which do not belong to A. This problem was solved in [30]. At first we recall the concept of a C^*-algebra generated by a finite set of affiliated elements [30, Definition 3.1].

Let A be a C^*-algebra and $T_j \in A^\eta$, $j = 1, 2, ...N$. We say that A is generated by $T_1, T_2, ...T_N$ if for any representation π of A and any $B \in C^*(H_\pi)$ we have

$$\begin{pmatrix} \pi(T_j)\,\eta\,B \text{ for any} \\[1mm] j = 1, 2, ...N \end{pmatrix} \Longrightarrow \Big(\pi \in \operatorname{Mor}(A, B) \Big). \qquad (11.1.14)$$

This condition is not easy to verify but we have a nice criterion ([30, Theorem 3.3].

Theorem 11.1.5 *Let A be a C^*-algebra and $T_j \in A^\eta$ for any $j = 1, 2, ...N$ and let*

$$\mathcal{R} = \left\{ (I + T_j^*T_j)^{-1}, \ (I + T_jT_j^*)^{-1} : \ j = 1.2, ...N \right\}.$$

Assume that

1. $T_1, T_2, ...T_N$ separate representations of A : if φ_1, φ_2 are different elements of $\operatorname{Rep}(A, H)$ then $\varphi_1(T_j) \neq \varphi(T_j)$ for some $j \in \{1, 2, ...N\}$.

2. There exist elements $r_1, r_2, ...r_k \in \mathcal{R}$ such that the product $r_1 r_2 ... r_k \in$ A.

Then A is generated by $T_1, T_2, ...T_N$.

For commutative C^*-algebra this criterion simplifies ([30, Exemple 2].

Proposition 11.1.6   *Let Λ be a locally compact space and $f_1, f_2, ...f_N \in C(\Lambda)$. Assume that $f_1, f_2, ...f_N$ separate points of Λ and*

$$\lim_{\lambda \to \infty} \sum_{j=1}^{N} |f_j(\lambda)| = +\infty.$$

Then A is generated by $f_1, f_2, ...f_N$.

To introduce a notion of a quantum group of unitary operators we shall use the concept of a C*-algebra generated by a quantum family of affiliated elements [33, Definition 4.1].

Let C, A be C*-algebras and V be an element affiliated with $C \otimes A$. We may regard V as a family of elements of A^η labeled by the "quantum space" related to C. We say that A is generated by an element $V \, \eta \, (C \otimes A)$ if and only if for any representation π of A and any $B \in C^*(H_\pi)$ we have:

$$\Big((\mathrm{id} \otimes \pi) V \, \eta \, (C \otimes B) \Big) \implies \Big(\pi \in \mathrm{Mor}(A, B) \Big). \qquad (11.1.15)$$

Let us note that if V generates A and B is a C^*-algebra then any morphism $\phi \in \mathrm{Mor}(A, B)$ is completely determined by its value on V. To be more precise, let $\phi_1, \phi_2 \in \mathrm{Mor}(A, B)$ and $B \in C^*(K)$. Then

$$\Big((\mathrm{id} \otimes \phi_1) V = (\mathrm{id} \otimes \phi_2) V \Big) \implies \Big(\phi_1 = \phi_2 \Big). \qquad (11.1.16)$$

Indeed, let $\tilde{\phi} = \phi_1 \oplus \phi_2$. Then $\tilde{\phi} \in \mathrm{Mor}(A, B \oplus B)$. Let $\tilde{B} = \{b \oplus b : b \in B\}$. Clearly $\tilde{B} \in C^*(K \oplus K)$ and one can easily verify that $\tilde{B}^\eta = \{b \oplus b : b \in B^\eta\}$. Now our assumption means that $(\mathrm{id} \otimes \tilde{\phi}) V \eta C \otimes \tilde{B}$. Therefore $\tilde{\phi} \in \mathrm{Mor}(A, \tilde{B})$ and by definition of $\tilde{B}$

$$\tilde{\phi}(c) = \phi_1(c) \oplus \phi_2(c) \in \mathrm{M}(\tilde{B}).$$

for any $c \in A$. Since $\mathrm{M}(\tilde{B}) \subset \tilde{B}^\eta$, the statement is proven.

In this more general situation we also have a useful criterion (cf. [33, Example 10, page 507]):

Proposition 11.1.7 *Let C, A be C^*-algebras and V be a unitary element of $M(C \otimes A)$. Assume that there exists a faithful representation ϕ of C such that:*

1. For any ϕ-normal linear functional ω on C we have: $(\omega \otimes \mathrm{id})V \in A$

2. The smallest $*$-subalgebra of A containing
$\{(\omega \otimes \mathrm{id})V : \omega \text{ is } \phi\text{-normal}\}$ *is dense in A.*

Then A is generated by $V \in M(C \otimes A)$.

Let us remind that a linear functional ω on C is said to be ϕ-normal if there exists a trace-class operator ρ acting on H_ϕ such that $\omega(c) = \mathrm{Tr}(\rho\phi(c))$ for all $c \in C$.

A unitary element $V \in M(\mathcal{K}(K) \otimes A)$ may be treated as a "strongly continuous family" (labeled by the quantum space related to A) of unitary operators acting on the Hilbert space K. Now following the idea of a compact matrix quantum group (cf Definition 11.1.3) we have

Definition 11.1.8 Let A be a C^*-algebra, K be a Hilbert space and let V be a unitary element of $M(\mathcal{K}(K) \otimes A)$. Assume that
1. A is generated by V.
2. There exists a morphism $\Delta \in \mathrm{Mor}(A, A \otimes A)$ such that

$$(\mathrm{id} \otimes \Delta)V = V_{12}V_{13}. \tag{11.1.17}$$

Then we say that (A, V) is a quantum group of unitary operators.

By previous considerations there is at most one $\Delta \in \mathrm{Mor}(A, A \otimes A)$ satisfying (11.1.17). On the other hand if Δ exists then it is co-associative. Indeed, $\Phi_1 = (\mathrm{id} \otimes \Delta)\Delta$ and $\Phi_2 = (\Delta \otimes \mathrm{id})\Delta$ are both elements of $\mathrm{Mor}(A, A \otimes A \otimes A)$ and

$$(\mathrm{id} \otimes \Phi_1)V = V_{12}V_{13}V_{14} = (\mathrm{id} \otimes \Phi_2)V$$

Since they coincide on V, they are equal. Therefore (A, Δ) is a C^*-bialgebra and V is a co-representation. Now one can study whether $G = (A, \Delta)$ is a quantum group in the sense described in Section 11.1.

Recently (cf [17]) an approach based on above concepts was used for construction of new deformations of quantum '$ax + b$' group. In this approach a role of generating aspects is more transparent. To demonstrate these ideas we consider the construction of quantum '$az + b$' group introduced in [29] from this point of view. This is the content of next sections. There are no new results concerning the theory of quantum '$az + b$' group. Nevertheless, with respect to the methodology and to the tools involved in the approach this presentation may be interesting.

11.2   Group Γ, Related Special Functions and Generating Algebras

In this section we recall the basic facts concerning the construction of quantum '$az + b$' group for real values of deformation parameter (cf [33], [29], [15]). To this end for a fixed value of a real parameter q, $0 < q < 1$ we consider a multiplicative subgroup Γ of nonzero complex numbers,

$$\Gamma = \left\{ z \in \mathbb{C} : |z| \in q^{\mathbb{Z}} \right\}.$$

Then Γ is an abelian locally compact group. Denote by $d\gamma$ the Haar measure:

$$\int_{\Gamma} x(\gamma) d\gamma = \sum_{n \in \mathbb{Z}} \frac{1}{2\pi} \int_{0}^{2\pi} x(q^n e^{i\varphi}) d\varphi.$$

Clearly any $\gamma \in \Gamma$ is of the form $\gamma = q^{i\varphi + n}$ for unique $n \in \mathbb{Z}$ and $\varphi \in \left[0, -\frac{2\pi}{\log q} \right[$. For any $\gamma, \gamma' \in \Gamma$ we set

$$\chi(\gamma, \gamma') = \chi(q^{i\varphi + n}, q^{i\varphi' + n'}) = q^{i(\varphi n' + \varphi' n)}. \tag{11.2.1}$$

Then $\chi \colon \Gamma \times \Gamma \to S^1$ and χ is a symmetric function. One can easily check that

$$\chi(\gamma, q) = \text{Phase}\, \gamma, \qquad \chi(\gamma, q^{it}) = |\gamma|^{it} \tag{11.2.2}$$

for any $\gamma \in \Gamma$. Moreover χ is a nondegenerate bicharacter on Γ. Therefore we may identify Γ with its Pontryagin dual $\widehat{\Gamma}$.

Let $\overline{\Gamma}$ denote the closure of Γ. Clearly $\overline{\Gamma} = \Gamma \cup \{0\}$. The C^*- algebras $C_\infty(\Gamma)$ and $C_\infty(\overline{\Gamma})$ play a key role in further consideration. We consider C^*-algebra $C_\infty(\Gamma)$ first.

Let

$$f_1(\gamma) = \gamma, \qquad f_2(\gamma) = \gamma^{-1} \tag{11.2.3}$$

for any $\gamma \in \Gamma$. Then f_1, $f_2 \in C(\Gamma) = C_\infty(\Gamma)^\eta$. Moreover f_1 and f_2 separates points of Γ and $|f_1(\gamma)| + |f_2(\gamma)| \longrightarrow +\infty$ whenever γ tends to infinity in Γ, i.e. $|\gamma| \longrightarrow 0$ or $|\gamma| \longrightarrow +\infty$. Therefore by Proposition 11.1.6, f_1, f_2 generates $C_\infty(\Gamma)$.

Let X be a normal operator acting on the Hilbert space K. Assume that X is invertible and $\operatorname{Sp} X \subset \overline{\Gamma}$. Then the mapping

$$C_\infty(\Gamma) \ni f \longrightarrow \pi(f) = f(X) \in B(K) \tag{11.2.4}$$

is a representation of $C_\infty(\Gamma)$ acting on K. Operators X and X^{-1} are determined by π. Indeed, $X = \pi(f_1)$ and $X^{-1} = \pi(f_2)$, where f_1, f_2 are given by (11.2.3). Recall that f_1, f_2 generate $C_\infty(\Gamma)$. Therefore for any representation π of $C_\infty(\Gamma)$ and any $B \in C^*(H_\pi)$ we have:

$$\Big(\pi(f_1), \pi(f_2)\,\eta\,B\Big) \Longrightarrow \Big(\pi \in \operatorname{Mor}(C_\infty(\Gamma), B)\Big)$$

$$\Longrightarrow \Big(\pi(f)\,\eta\,B \text{ for any } f \in C(\Gamma)\Big)$$

In particular for π introduced by (11.2.4) and $B \in C^*(K)$ we get:

$$\begin{pmatrix} X, X^{-1}\,\eta\,B \\ f \in C(\Gamma) \end{pmatrix} \Longrightarrow \Big(f(X)\,\eta\,B\Big). \tag{11.2.5}$$

Let $f_\gamma(\gamma') = \chi(\gamma', \gamma)$ for any $\gamma, \gamma' \in \Gamma$. Then $f_\gamma \in C(\Gamma)$ and $f_\gamma(X) = \chi(X, \gamma)$ is a unitary element of $B(K)$. Let us note that X is completely determined by $\chi(X, \gamma)$. Indeed, using (11.2.2) one can easily show

Proposition 11.2.1   *Let X_k $(k = 1, 2)$ be a normal invertible operator acting on a Hilbert space K and such that $\operatorname{Sp} X_k \subset \overline{\Gamma}$. Then*

$$\begin{pmatrix} \chi(X_1, \gamma) = \chi(X_2, \gamma) \\ \text{for all } \gamma \in \Gamma \end{pmatrix} \Longleftrightarrow \Big(X_1 = X_2\Big).$$

Assume that X is a normal invertible operator and $\operatorname{Sp} X \subset \Gamma$. Then the mapping

$$\Gamma \ni \gamma \longrightarrow \chi(X, \gamma) \in B(K) \tag{11.2.6}$$

is strongly continuous. By the general theory strongly continuous mappings from Γ into the set of unitary operators acting on K correspond to unitary multipliers of $\mathcal{K}(K) \otimes C_\infty(\Gamma)$.

Proposition 11.2.2   *Let X be a normal invertible operator acting on a Hilbert space K and $\mathfrak{X} \in M(\mathcal{K}(K) \otimes C_\infty(\Gamma))$ be the unitary corresponding to the mapping (11.2.6). Assume that the spectral measure of X is absolutely continuous with respect to the Haar measure on Γ.*
Then $\mathfrak{X}$ generates $C_\infty(\Gamma)$.

Proof.   We use Proposition 11.1.7. For any normal linear functional ω on $B(K)$ we set $f_\omega = (\omega \otimes \mathrm{id})\mathfrak{X}$. Then $f_\omega \in M(C_\infty(\Gamma)) = C_{\mathrm{bounded}}(\Gamma)$. Clearly

$$f_\omega(\gamma) = \omega\left(\chi(X, \gamma)\right)$$

for any $\gamma \in \Gamma$. Since the spectral measure of X is absolutely continuous with respect to the Haar measure, $f_\omega \in C_\infty(\Gamma)$ by the Riemann-Lebesgue lemma.

     We shall show that f_ω separates points of Γ. To this end let $\gamma, \gamma' \in \Gamma$, $\gamma \neq \gamma'$. Suppose that $f_\omega(\gamma) = f_\omega(\gamma')$ for all ω. Then $\chi(X, \gamma) = \chi(X, \gamma')$ and $\chi(X, \gamma_0) = I$ where $\gamma_0 = \gamma'\gamma^{-1}$. This means that the spectral measure of X is supported by the set $\{z \in \Gamma : \chi(z, \gamma_0) = 1\}$. Inspecting formula (11.2.1) we find that this is a discrete subset of Γ. This is in contradiction with the assumption of absolute continuity with respect to the Haar measure. Therefore f_ω separates points of Γ. Now by the Stone-Weierstrass theorem the smallest *-subalgebra of $C_\infty(\Gamma)$ containing all f_ω is dense in $C_\infty(\Gamma)$. This ends the proof. $\square$

As a conclusion we formulate the following Proposition which will be very useful in further considerations.

Proposition 11.2.3   *Let X be a normal invertible operator acting on a Hilbert space K. Assume that $\mathrm{Sp}\, X \subset \overline{\Gamma}$ and the spectral measure of X is absolutely continuous with respect to the Haar measure. Let Z be a normal invertible operator acting on a Hilbert space H. Assume that $\mathrm{Sp}\, Z \subset \overline{\Gamma}$.*
Then for any $A \in C^*(H)$ we have:

$$\left(\chi(X \otimes I, I \otimes Z) \in M(\mathcal{K}(K) \otimes A)\right) \implies \left(Z, Z^{-1}\, \eta\, A\right)$$

Proof.   For any $f \in C_\infty(\Gamma)$ we set $\pi(f) = f(Z)$. Then π is a representation of $C_\infty(\Gamma)$ acting on the Hilbert space $H_\pi = H$. Let $\mathfrak{X} \in M(\mathcal{K}(K) \otimes C_\infty(\Gamma))$ be unitary introduced in Proposition 11.2.2. A moment of reflection shows that $(\mathrm{id} \otimes \pi)\mathfrak{X} = \chi(X \otimes I, I \otimes Z)$. If $\chi(X \otimes I, I \otimes Z)$ is affiliated with $\mathcal{K}(K) \otimes A$ then $\pi \in \mathrm{Mor}(C_\infty(\Gamma), A)$ and π maps continuous functions on Γ into elements affiliated with A. Applying this rule to the functions f_1, f_2 introduced by (11.2.3) we obtain $Z = \pi(f_1)\,\eta\,A$, $Z^{-1} = \pi(f_2)\,\eta\,A$. $\square$

We shall need an operator version of Proposition 11.2.1.

Proposition 11.2.4 *Let X be a normal invertible operator acting on a Hilbert spaces K. Assume that $\operatorname{Sp} X \subset \overline{\Gamma}$ and the spectral measure of X is absolutely continuous with respect to the Haar measure. Let Z_k $(k = 1, 2)$ be a normal invertible operator acting on a Hilbert space H. Assume that $\operatorname{Sp} Z_k \subset \overline{\Gamma}$. Then*

$$\Big(\chi(X \otimes I, I \otimes Z_1) = \chi(X \otimes I, I \otimes Z_2)\Big) \implies \Big(Z_1 = Z_2\Big).$$

Proof. Let $A = \{m \oplus m : m \in \mathcal{K}(H)\}$ and $Z = Z_1 \oplus Z_2$. Then $\chi(X \otimes I, I \otimes Z) \in M(\mathcal{K}(K) \otimes A)$ due to the assumption. Therefore $Z \in A^{\eta}$ by Proposition 11.2.3. This means that $Z_1 = Z_2$. $\square$

Now we pass to the set $\overline{\Gamma}$ and the C^*-algebra $C_\infty(\overline{\Gamma})$.
Let

$$f_0(\gamma) = \gamma \tag{11.2.7}$$

for any $\gamma \in \overline{\Gamma}$. Then $f_0 \in C(\overline{\Gamma}) = C_\infty(\overline{\Gamma})^{\eta}$. Using Proposition 11.1.6 one easily verifies that $C_\infty(\overline{\Gamma})$ is generated by f_0.

Let Y be a normal operator acting on a Hilbert space K and $\operatorname{Sp} Y \subset \overline{\Gamma}$. Then the mapping

$$C_\infty(\overline{\Gamma}) \ni f \longrightarrow \pi(f) = f(Y) \in B(K) \tag{11.2.8}$$

is a representation of $C_\infty(\overline{\Gamma})$ acting on K. The operator Y is determined by π, $Y = \pi(f_0)$ (cf (11.2.7)). Since f_0 generate $C_\infty(\overline{\Gamma})$,

$$\Big(\pi(f_0)\,\eta\,B\Big) \implies \Big(\pi \in \operatorname{Mor}(C_\infty(\overline{\Gamma}), B)\Big) \implies \Big(\pi(f)\,\eta\,B \text{ for any } f \in C(\overline{\Gamma})\Big)$$

for any representation π of $C_\infty(\overline{\Gamma})$ and any $B \in C^*(H_\pi)$. In particular for π introduced by (11.2.8) and $B \in C^*(K)$ we get

$$\binom{Y\,\eta\,B}{f \in C(\overline{\Gamma})} \implies \Big(f(Y)\,\eta\,B\Big). \tag{11.2.9}$$

Now consider a special function $F_q : \overline{\Gamma} \to \mathbb{C}$. This is a *quantum exponential function* introduced in [33] by the formula

$$F_q(\gamma) = \prod_{k=0}^{\infty} \frac{1 + q^{2k}\overline{\gamma}}{1 + q^{2k}\gamma}$$

for $\gamma \in \overline{\Gamma} \setminus \{-1, -q^{-2}, -q^{-4}, \ldots\}$. Setting $F_q(\gamma) = -1$ for $\gamma \in \{-1, -q^{-2}, -q^{-4}, \ldots\}$ one gets a continuous function on $\overline{\Gamma}$. In addition $F_q(0) = 1$.

Let $z, \gamma \in \Gamma$. Due to [33, p. 427] the asymptotic behaviour of $F_q(z\gamma)$ for large $|z\gamma|$ is described by the formula

$$F_q(z\gamma) \sim \alpha(z)\alpha(\gamma)\chi(z, \gamma) \qquad (11.2.10)$$

where $\alpha(z) = (\text{Phase } z)^{\log_q |z| - 1}$ and '$\sim$' means that the difference goes to 0 when $|z\gamma| \longrightarrow \infty$.

We know that $F_q \in C(\overline{\Gamma})$ and assumes values of modulus one. Therefore if Y is a normal operator acting on a Hilbert space H and $\text{Sp}\, Y \subset \overline{\Gamma}$ then $F_q(Y)$ is unitary. Moreover we have

Proposition 11.2.5   *Let Y_k $(k = 1, 2)$ be normal operator acting on a Hilbert space H and such that $\text{Sp}\, Y_k \subset \overline{\Gamma}$. Then*

$$\left(\begin{array}{c} F_q(zY_1) = F_q(zY_2) \\ \\ \text{for all } z \in \Gamma \end{array} \right) \Longleftrightarrow \left(Y_1 = Y_2 \right).$$

Proof.   One may proceed as in the proof of [16, Lemma 3.1] but here we present another proof. It is known (cf [33, p.425]) that asymptotic behavior of $F_q(\gamma)$ for small γ is described by the formula

$$F_q(\gamma) = 1 - \frac{\gamma}{1 - q^2} + \frac{\overline{\gamma}}{1 - q^2} + o(|\gamma|). \qquad (11.2.11)$$

Let

$$f_n(\gamma) = \frac{1}{2\pi i} \int_{|z|=q^n} \frac{F_q(z\gamma)}{z} \frac{dz}{z} = \frac{1}{2\pi} \int_0^{2\pi} \frac{F_q(q^n e^{i\varphi}\gamma)}{q^n e^{i\varphi}} d\varphi. \qquad (11.2.12)$$

where n is an integer. Then $f_n \in C_{\text{bounded}}(\overline{\Gamma})$. Set $n \longrightarrow +\infty$. Then $f_n(\gamma) \longrightarrow -\frac{1}{1-q^2}\gamma$ for all $\gamma \in \overline{\Gamma}$ due to (11.2.11). The convergence is almost uniform. Therefore if Y is a normal operators acting on a Hilbert space H and $\operatorname{Sp} Y \subset \bar{\Gamma}$ we have

$$\lim_{n \to +\infty} f_n(Y) = -\frac{1}{1 - q^2} Y$$

in a natural topology (cf. [30]) on the set of affiliated elements $\mathcal{K}(H)^\eta$. We know that $F_q(zY_1) = F_q(zY_2)$. Therefore $f_n(Y_1) = f_n(Y_2)$ and $Y_1 = Y_2$ (the limit is unique). $\qquad\square$

To reveal the usefulness of F_q we need a notion of a G-pair. This notion involves a pair (X, Y) of normal operators and assigns a precise meaning to the relations of the form

$$XY = q^2 YX, \qquad XY^* = Y^*X. \tag{11.2.13}$$

They were investigated in [33],[29].

Definition 11.2.6 Let X and Y be closed densely defined operators acting on a Hilbert space H. We say that (X, Y) is a G-pair on H if X and Y are normal, $\operatorname{Sp} X$, $\operatorname{Sp} Y \subset \bar{\Gamma}$, $\ker X = \{0\}$ and

$$\chi(X, \gamma) Y \chi(X, \gamma)^* = \gamma Y \tag{11.2.14}$$

for all $\gamma \in \Gamma$.

Setting $\gamma = q$ and $\gamma = q^{it}$ in the above formula we have (cf (11.2.2))

$$(\operatorname{Phase} X) Y (\operatorname{Phase} X)^* = qY \quad \text{and} \quad |X|^{it} Y |X|^{-it} = q^{it} Y.$$

respectively. In particular $|X|$ and $|Y|$ strongly commute and $(\operatorname{Phase} X) |Y| = q |Y| (\operatorname{Phase} X)$.

Remark 11.2.7 It is known that if (X, Y) is a G-pair on H then (Y^*, X^*) and (XY, Y) are G-pairs on H as well. If in addition Y is an invertible operator then formula (11.2.14) takes the form of Weyl relation:

$$\chi(X, \gamma) \chi(Y, \gamma') = \chi(\gamma, \gamma') \chi(Y, \gamma') \chi(X, \gamma) \tag{11.2.15}$$

for any $\gamma, \gamma' \in \Gamma$. Then one can show that $(Y, X^{-1}), (Y^{-1}, X)$ and $(Y^{-1}, Y^{-1}X)$ are G-pairs on H.

We shall need the following result [33, Theorem 2.1, Theorem 2.2 and Theorem 3.1].

Theorem 11.2.8   *Let (X, Y) be a G-pair on a Hilbert space H. Then the sum $Y + X$ is a densely defined closeable operator and its closure $Y \dotplus X$ is a normal operator and $\mathrm{Sp}\,(Y \dotplus X) \subset \overline{\Gamma}$. Moreover*

$$F_q(Y \dotplus X) = F_q(Y)F_q(X). \qquad (11.2.16)$$

If in addition $\ker Y = \{0\}$ *then*

$$Y \dotplus X = F_q(Y^{-1}X)Y F_q(Y^{-1}X)^*. \qquad (11.2.17)$$

The reader should notice that the last formula combined with (11.2.16) leads to

$$F_q(Y)F_q(X) = F_q(Y^{-1}X)F_q(Y)F_q(Y^{-1}X)^*. \qquad (11.2.18)$$

Formula (11.2.16) justifies the name "quantum exponential function" assigned to the function F_q.

Now we shall introduce a generating element for $C_\infty(\overline{\Gamma})$ associated with F_q. To this end for any $z \in \Gamma$ and $\gamma \in \overline{\Gamma}$ we set:

$$\Phi(z, \gamma) = \overline{F_q(\gamma)}F_q(z\gamma). \qquad (11.2.19)$$

Then $|\Phi(z, \gamma)| = 1$ and Φ is a continuous function on $\Gamma \times \overline{\Gamma}$. Therefore it may be treated as a unitary element of $\mathrm{M}\left(C_\infty(\Gamma) \otimes C_\infty(\overline{\Gamma})\right)$. We have

Proposition 11.2.9   *The C^*-algebra $C_\infty(\overline{\Gamma})$ is generated by $\Phi \in \mathrm{M}\left(C_\infty(\Gamma) \otimes C_\infty(\overline{\Gamma})\right)$.*

Proof.   We shall use Proposition 11.1.7 setting $C = C_\infty(\Gamma)$, $A = C_\infty(\overline{\Gamma})$ and $V = \Phi$. Denote by dz the Haar measure on Γ and let ϕ be the natural representation of $C_\infty(\Gamma)$ acting on $L^2(\Gamma, dz)$: $\phi(h)$ is the multiplication by h for any $h \in C_\infty(\Gamma)$. Then ϕ is faithful representation and a linear functional ω on $C_\infty(\Gamma)$ is ϕ-normal if and only if it is of the form

$$\omega(h) = \int_\Gamma h(z)\varphi_\omega(z)\, dz,$$

where $\varphi_\omega \in L^1(\Gamma, dz)$.

Let $f_\omega = (\omega \otimes \mathrm{id})\Phi$. Then $f_\omega \in \mathrm{M}\left(C_\infty(\overline{\Gamma})\right)$ i.e. f_ω is a bounded continuous function on $\overline{\Gamma}$. Clearly for any $\gamma \in \overline{\Gamma}$ we have

$$f_\omega(\gamma) = \int_\Gamma \Phi(z, \gamma)\varphi_\omega(z)\, dz = \overline{F_q(\gamma)} \int_\Gamma F_q(z\gamma)\varphi_\omega(z)\, dz. \qquad (11.2.20)$$

Now using the asymptotic behaviour (11.2.10) and the Riemann–Lebesgue lemma one can show that the integral on the right hand side vanishes when $|\gamma| \to +\infty$. This means that $f_\omega \in C_\infty(\overline{\Gamma})$.

To prove that the smallest *-algebra containing all functions of the form (11.2.20) is dense in $C_\infty(\overline{\Gamma})$ we apply the Stone-Weierstrass theorem to the one point compactification of $\overline{\Gamma}$. Clearly for any $\gamma \in \overline{\Gamma}$ one can find a functional ω such that $f_\omega(\gamma) \neq 0$. It remains to show that functions f_ω separate points of $\overline{\Gamma}$. Let $\gamma, \gamma' \in \overline{\Gamma}$ and assume that $f_\omega(\gamma) = f_\omega(\gamma')$ for all ϕ-normal functionals ω. Then $\overline{F_q(\gamma)}F_q(z\gamma) = \overline{F_q(\gamma')}F_q(z\gamma')$ for all $z \in \Gamma$. Recall that F_q is a continuous function and $F_q(0) = 1$. Therefore taking the limit $z \to 0$ we get $\overline{F_q(\gamma)} = \overline{F_q(\gamma')}$. This formula combined with the previous one imply that $F_q(z\gamma) = F_q(z\gamma')$ for all $z \in \Gamma$. Then for any integer n the function f_n introduced by formula (11.2.12) attains the same value on γ and γ', $f_n(\gamma) = f_n(\gamma')$. Remembering that $\lim_{n\to\infty} f_n(\gamma) = -\frac{1}{1-q^2}\gamma$ for any $\gamma \in \overline{\Gamma}$ we conclude that $\gamma = \gamma'$. The statement is proved. $\square$

To solve some technical problems which appear in further considerations we need the following result.

Proposition 11.2.10   *Let Y, U and X be operators acting on a Hilbert space H and $C \in C^*(H)$. Assume that:*

1. *X and Y are normal and (X, Y) is a G-pair on H,*

2. *U is unitary and commutes with X,*

3. *Operators X, X^{-1}, Y and U are affiliated with C.*

Then $F_q(Y) \in M(C)$ and

1. *For any representation ρ of C and any $B \in C^*(H_\rho)$ we have:*

$$\left(\begin{array}{c} \rho(X),\ \rho(X^{-1}),\ \rho\left(F_q(Y)U\right) \\ \text{are affiliated with } B \end{array} \right) \implies \left(\begin{array}{c} \rho(Y),\ \rho(U) \\ \text{are affiliated with } B \end{array} \right)$$

2. *For any representations ρ_1 and ρ_2 of C acting on the same Hilbert space $H_{\rho_1} = H_{\rho_2}$ we have:*

$$\left(\begin{array}{c} \rho_1(X) = \rho_2(X), \\ \rho_1\left(F_q(Y)U\right) = \rho_2\left(F_q(Y)U\right) \end{array} \right) \implies \left(\begin{array}{c} \rho_1(Y) = \rho_2(Y), \\ \rho_1(U) = \rho_2(U) \end{array} \right)$$

Proof. Relation $F_q(Y) \in \mathrm{M}(C)$ follows immediately from (11.2.9).
Ad 1. Let $z \in \Gamma$. Using the commutation relations satisfied by operators X, Y and U we have:

$$\chi(X, z)\, F_q(Y)U\, \chi(X, z)^* = F_q(zY)U.$$

Passing to a representation ρ of C we get

$$\chi\left(\rho(X), z\right) \rho\left(F_q(Y)U\right) \chi\left(\rho(X), z\right)^* = \rho\left(F_q(zY)U\right).$$

If $\rho(X)$, $\rho(X^{-1})$, $\rho\left(F_q(Y)U\right) \eta\, B$, then all factors on the left hand side of the above equation belong to $\mathrm{M}(B)$ and depend continuously on z in the strict topology of $\mathrm{M}(B)$ (cf [29, Theorem 5.2]). Therefore $\rho\left(F_q(zY)U\right) \in \mathrm{M}(B)$ for any $\gamma \in \Gamma$ and the mapping

$$\Gamma \ni z \longmapsto \rho\left(F_q(zY)U\right) \in \mathrm{M}(B)$$

is strictly continuous. Multiplying from the right by the hermitian conjugation of $\rho\left(F_q(Y)U\right)$ we get

$$\rho\left(F_q(zY)F_q(Y)^*\right) = \rho\left(F_q(Y)^*F_q(zY)\right) = \rho\left(\Phi(z, Y)\right) \in \mathrm{M}(B)$$

where Φ is the function introduced by formula (11.2.19). Moreover the mapping

$$\Gamma \ni z \longmapsto \rho\left(\Phi(z, Y)\right) \in \mathrm{M}(B) \tag{11.2.21}$$

is strictly continuous. By general theory (cf [30]) such mappings from Γ into $\mathrm{M}(B)$ correspond to elements of $\mathrm{M}(C_\infty(\Gamma) \otimes B)$. A moment of reflection shows that the mapping (11.2.21) corresponds to the element $(\mathrm{id} \otimes \rho \circ \pi)\Phi$, where π is the representation of $C_\infty(\overline{\Gamma})$ introduced by (11.2.8). Therefore $(\mathrm{id} \otimes \rho \circ \pi)\Phi \in \mathrm{M}(C_\infty(\Gamma) \otimes B)$. Now using Proposition 11.2.9 we conclude that $\rho \circ \pi \in \mathrm{Mor}(C_\infty(\overline{\Gamma}), B)$. In consequence $\rho \circ \pi$ maps continuous functions on $\overline{\Gamma}$ into elements affiliated with B. Applying this rule to function f_0 (cf. (11.2.7)) and F_q we obtain that $\rho(Y)$ is affiliated with B and $\rho(F_q(Y)) \in \mathrm{M}(B)$. By passing to adjoint $\rho(F_q(Y)^*) \in \mathrm{M}(B)$. We have assumed that $\rho\left(F_q(Y)U\right) \in \mathrm{M}(B)$. Therefore $\rho(U) \in \mathrm{M}(B)$ and Statement 1 is proved.

Ad 2. Let $\rho = \rho_1 \oplus \rho_2$. Then $H_\rho = H_{\rho_1} \oplus H_{\rho_2}$ and $\rho(c) = \rho_1(c) \oplus \rho_2(c)$. In our case $H_{\rho_1} = H_{\rho_2}$. We set: $B = \{m \oplus m : m \in \mathcal{K}(H_{\rho_1})\}$. Then

$B \in C^*(H_\rho)$. One can easily verify that for any $c \,\eta\, C$ we have:

$$\Big(\rho(c)\,\eta\,B\Big) \iff \Big(\rho_1(c) = \rho_2(c)\Big).$$

Now Statement 2 follows immediately from Statement 1. $\qquad\square$

We shall use slightly different version of Statement 2 of the above Proposition.

Proposition 11.2.11 *Let Y_1, U_1, Y_2, U_2, X be operators acting on a Hilbert space H. Assume that for each $k = 1,2$ the operators Y_k, U_k, X satisfy the assumptions 1-3 of the previous Proposition. Then*

$$\Big(F_q(Y_1)U_1 = F_q(Y_2)U_2\Big) \implies \begin{pmatrix} Y_1 = Y_2, \\ U_1 = U_2. \end{pmatrix} \tag{11.2.22}$$

Remark 11.2.12 *Since $F_q(Y)U = UF_q(U^*YU)$ the same result holds under assumption that we have $U_1F_q(Y_1) = U_2F_q(Y_2)$.*

Proof. Let $C = \mathcal{K}(H)\oplus\mathcal{K}(H)$ and for any $m_1, m_2 \in \mathcal{K}(H)$ we set $\rho_k(m_1\oplus m_2) = m_k$ $(k = 1,2)$. We use Proposition 11.2.10 with Y, U and X replaced by $Y_1 \oplus Y_2$, $U_1 \oplus U_2$ and $X \oplus X$, Now (11.2.22) follows immediately from Statement 2 of Proposition 11.2.10. $\qquad\square$

11.3 Construction of Quantum '$az + b$' Group

The quantum '$az + b$' group considered in this paper was introduced in [29, Appendix A]. Following the idea of [17] in this section we shall present it as a quantum group of unitary operators. In this approach one considers a C^*-algebra A and a Hilbert space K endowed with a certain additional structure. The main object is a pair (A, V) where V is a unitary element of $M(\mathcal{K}(K)\otimes A)$. It may be treated as a quantum family of unitary operators acting on K 'labeled by elements' of quantum space related to the C^*-algebra A.

At first we define A. To this end we consider two operators a and b acting on the Hilbert space $H = L^2(\Gamma, d\gamma)$. For any $\gamma \in \Gamma$ let u_γ denote the shift operator:

$$(u_\gamma x)(\gamma') = x(\gamma\gamma')$$

for any $x \in H$. Clearly $\Gamma \ni \gamma \longrightarrow u_\gamma \in B(H)$ is a unitary representation of Γ. Therefore by SNAG theorem [4, Chap. 6, §2, Theorem 1] there exists a spectral measure $dE(\gamma)$ on $\widehat{\Gamma} = \Gamma$ such that

$$u_\gamma = \int_\Gamma \chi(\gamma', \gamma) dE(\gamma')$$

for all $\gamma \in \Gamma$. Let

$$a = \int_\Gamma \gamma' dE(\gamma').$$

Then a is a normal operator, $\ker a = \{0\}$ and $\operatorname{Sp} a \subset \overline{\Gamma}$. Moreover $u_\gamma = \chi(a, \gamma)$. By b we denote the multiplication operator:

$$(bx)(\gamma') = \gamma' x(\gamma').$$

By definition a domain $\mathcal{D}(b)$ consists of all $x \in H$ such that the right hand side is square integrable. Clearly b is normal and $\operatorname{Sp} b \subset \overline{\Gamma}$. Moreover $\ker b = \{0\}$. Now one can easily check that

$$\chi(a, \gamma) b \chi(a, \gamma)^* = u_\gamma b u_\gamma^* = \gamma b. \tag{11.3.1}$$

This means (cf (11.2.14)) that (a, b) is a G-pair on H. We refer to it as a Schrödinger pair.

Theorem 11.3.1   *Let*

$$A = \left\{ f(b)g(a) : f \in C_\infty(\overline{\Gamma}), g \in C_\infty(\Gamma) \right\}^{\substack{\text{norm closed}\\ \text{linear envelope}}}. \tag{11.3.2}$$

Then: 1. A is a nondegenerate C^*-algebra of operators acting on $L^2(\Gamma, d\gamma)$,

     2. a, a^{-1} and b are affiliated with A: $a, a^{-1}, b \, \eta \, A$,

     3. a, a^{-1} and b generate A.

Proof.   Ad 1. Operator b is normal and $\operatorname{Sp} b \subset \overline{\Gamma}$. Therefore the mapping $C_\infty(\overline{\Gamma}) \ni f \longrightarrow f(b) \in B(H)$ is a representation of the C^*-algebra $C_\infty(\overline{\Gamma})$ on the Hilbert space H. Let

$$B = \left\{ f(b) : f \in C_\infty(\overline{\Gamma}) \right\}. \tag{11.3.3}$$

Then B is a non-degenerate C*-subalgebra of $B(H)$. Let $C_0(\Gamma, B)$ denote the set of all continuous mappings from Γ into B with compact support. Then

$$A = \left\{ \int_\Gamma h(\gamma) \chi(a, \gamma) d\gamma : h \in C_0(\Gamma, B) \right\}^{\text{norm closure}}. \tag{11.3.4}$$

Indeed, for $h(\gamma) = f(b)\hat{g}(\gamma)$, where $\gamma \in \Gamma$ and $\hat{g} \in C_0(\Gamma)$ we have

$$\int_\Gamma h(\gamma)\chi(a,\gamma)d\gamma = f(b)g(a),$$

where $g(\gamma') = \int_\Gamma \hat{g}(\gamma)\chi(\gamma',\gamma)d\gamma$ $(\gamma' \in \Gamma)$. By the Riemann-Lebesgue lemma (e.g. [19, Theorem 1.2.4]), $g \in C_\infty(\Gamma)$ and the set consisting of functions of such form is dense in $C_\infty(\Gamma)$. This proves formula (11.3.4). Now (11.3.1) shows that the unitaries $\chi(a,\gamma)$ $(\gamma \in \Gamma)$ implement a one parameter group of automorphisms of B. Using the standard technique of the theory of crossed products (cf. [12, Section 7.6]) one can show that (11.3.4) is a (non-degenerate) C*-algebra of operators acting on $L^2(\Gamma, d\gamma)$.

Ad 2. The affiliation relation was introduced in (11.1.5). We consider the operator a first. We know that a is a normal invertible operator and $\mathrm{Sp}\, a \subset \Gamma$. Let

$$g_1(\gamma) = \frac{1}{\sqrt{1 + |\gamma|^2}}, \qquad g_2(\gamma) = \frac{\gamma}{\sqrt{1 + |\gamma|^2}}, \tag{11.3.5}$$

for any $\gamma \in \Gamma$. For $T = a$ and $T = a^*$ we have $z_a = g_2(a)$ and $z_{a^*} = \bar{g}_2(a)$. In both cases $(I + T^*T)^{-\frac{1}{2}} = g_1(a)$. Clearly $g_1, g_2 \in M(C_\infty(\Gamma))$. Now inspecting definition (11.3.2) one can easily show that $Ag_1(a)$ is dense in A and z_a, z_{a^*} are right multipliers of A. This means (cf the comment after (11.1.5)) that z_a is a multiplier of A and a is affiliated with A. In the same manner we prove that $a^{-1} \eta A$.

Now consider the operator b. Let g_1 and g_2 be given by the expression (11.3.5) again but now for any $\gamma \in \overline{\Gamma}$. Then $z_b = g_2(b)$, $z_{b^*} = \bar{g}_2(b)$ and in both cases $(I + T^*T)^{-\frac{1}{2}} = g_1(b)$. Now $g_1, g_2 \in M(C_\infty(\overline{\Gamma}))$. Therefore $g_1(b)A$ is dense in A and z_b, z_{b^*} are left multipliers. In consequence z_b is a multiplier of A and $b \eta A$.

Ad 3. Let $c \in A$ be of the form $c = f(b)g(a)$, where $f \in C_\infty(\overline{\Gamma})$ and $g \in C_\infty(\Gamma)$. By definition (11.3.2) the set of such elements is total in A. Let π be a non-degenerate representation of A. Then $\pi(a)$ is invertible and $\pi(c) = f(\pi(b))g(\pi(a))$. Therefore π is completely determined by $\pi(a)$ and $\pi(b)$. This means that a, a^{-1} and b separate representations of A.

Let $r_1 = (I+b^*b)^{-1}$, $r_2 = (I+a^*a)^{-1}$ and $r_3 = [I+(a^{-1})^*a^{-1}]^{-1}$. To end the proof it is sufficient (cf Theorem 11.1.5) to show that $r_1 r_2 r_3 \in A$. Since $r_1 r_2 r_3 = f(b)g(a)$ where $f(\gamma) = (1 + |\gamma|^2)^{-1}$, and $g(\gamma) = |\gamma|^2 (1 + |\gamma|^2)^{-2}$, the result follows from (11.3.2). $\qquad\square$

Now we describe the Hilbert space K. The structure of K is determined by two normal operators $\widehat{a}$ and $\widehat{b}$ such that

$$(\widehat{a},\,\widehat{b}) \text{ is a } G\text{-pair on } K \text{ and } \ker\widehat{b} = \{0\}.$$

It is known that any such pair is unitary equivalent to the direct sum of copies of the Schrödinger pair. In particular spectral measures of $\widehat{a}$ and $\widehat{b}$ are absolutely continuous with respect to the Haar measure on Γ.

Let

$$V = F_q(\widehat{b}\otimes b)\,\chi(\widehat{a}\otimes I, I\otimes a). \qquad (11.3.6)$$

It is the basic object considered in this Section. We shall prove

Theorem 11.3.2

1. V is a unitary operator and $V \in \mathrm{M}(\mathcal{K}(K)\otimes A)$

2. A is generated by $V \in \mathrm{M}(\mathcal{K}(K)\otimes A)$.

Proof.　Let $Y = \widehat{b}\otimes b$, $U = \chi(\widehat{a}\otimes I, I\otimes a)$, $X = \widehat{a}\otimes I$ and $C = \mathcal{K}(K)\otimes A$. Then all the assumptions of Proposition 11.2.10 are satisfied. Therefore $V = F_q(Y)U \in \mathrm{M}(C)$ and Statement 1 is proved.

Let π be a representation of A and $B \in C^*(H_\pi)$. Then $\mathrm{id}\otimes\pi$ is a representation of C acting on $K\otimes H_\pi$. Let us note that $(\mathrm{id}\otimes\pi)X = \widehat{a}\otimes I$ is affiliated with $\mathcal{K}(K)\otimes B$. Assume that $(\mathrm{id}\otimes\pi)V \in \mathrm{M}(\mathcal{K}(K)\otimes B)$. By Proposition 11.2.10, operators: $(\mathrm{id}\otimes\pi)Y = \widehat{b}\otimes\pi(b)$ and $(\mathrm{id}\otimes\pi)U = \chi(\widehat{a}\otimes I, I\otimes\pi(a))$ are affiliated with $\mathcal{K}(K)\otimes B$. By Proposition A.1 of [36] operator $\pi(b)$ is affiliated with B. On the other hand operators $\widehat{a}$ and $\pi(a)$ satisfy the assumptions of Proposition 11.2.3. Therefore $\pi(a)$ and $\pi(a)^{-1}$ are affiliated with B.

According to Statement 3 of Theorem 11.3.1, a, a^{-1} and b generate A. Therefore $\pi \in \mathrm{Mor}(A, B)$. This way we showed that $(\mathrm{id}\otimes\pi)V \in \mathrm{M}(\mathcal{K}(K)\otimes B)$ implies $\pi \in \mathrm{Mor}(A, B)$. It means that A is generated by $V \in \mathrm{M}(\mathcal{K}(K)\otimes A)$. $\qquad\square$

Now we formulate the main result of this Section:

Theorem 11.3.3　*There exists $\Delta \in \mathrm{Mor}(A, A\otimes A)$ such that*

$$(\mathrm{id}\otimes\Delta)V = V_{12}V_{13} \qquad (11.3.7)$$

Proof.　Let us recall that b is an invertible operator. Therefore (b^{-1}, a) and $(b^{-1}, b^{-1}a)$ are G-pairs on H by Remark 11.2.7. In particular $b^{-1}a$ is normal and $\mathrm{Sp}\,(b^{-1}a) \subset \overline{\Gamma}$. Let

$$W = F_q(b^{-1}a \otimes b)\,\chi(b^{-1} \otimes I, I \otimes a). \tag{11.3.8}$$

Clearly W is a unitary operator acting on $H \otimes H$. We shall prove that

$$V_{12}V_{13} = W_{23}V_{12}W_{23}^*. \tag{11.3.9}$$

To deal with shorter formulae we set

$$U = \chi(\widehat{a} \otimes I, I \otimes a), \qquad Z = \chi(b^{-1} \otimes I, I \otimes a).$$

Applying formula (11.2.14) for the G-pairs $(\widehat{a}, \widehat{b})$ and (b^{-1}, a) one can easily verify that

$$U(\widehat{b} \otimes I)U^* = \widehat{b} \otimes a \tag{11.3.10}$$

and

$$Z(a \otimes I)Z^* = a \otimes a. \tag{11.3.11}$$

With the above notation $V = F_q(\widehat{b} \otimes b)U$ and

$$V_{12}V_{13} = F_q(\widehat{b} \otimes b \otimes I)\,U_{12}\,F_q(\widehat{b} \otimes I \otimes b)\,U_{13}. \tag{11.3.12}$$

By the relation (11.3.10) we get

$$U_{12}\,F_q(\widehat{b} \otimes I \otimes b) = F_q(\widehat{b} \otimes a \otimes b)U_{12} \tag{11.3.13}$$

and

$$V_{12}V_{13} = F_q(\widehat{b} \otimes b \otimes I)F_q(\widehat{b} \otimes a \otimes b)\,U_{12}U_{13}. \tag{11.3.14}$$

Let us consider the first factor in (11.3.14). We apply formula (11.2.18) with $X = \widehat{b} \otimes a \otimes b$ and $Y = \widehat{b} \otimes b \otimes I$. Then

$$F_q(Y^{-1}X) = F_q(I \otimes b^{-1}a \otimes b).$$

Now (11.3.14) takes the form

$$V_{12}V_{13} = F_q(I \otimes b^{-1}a \otimes b)F_q(\widehat{b} \otimes b \otimes I)F_q(I \otimes b^{-1}a \otimes b)^*\,U_{12}U_{13}. \tag{11.3.15}$$

Since χ is a bicharacter, $U_{12}U_{13} = \chi(\widehat{a} \otimes I \otimes I, I \otimes a \otimes a)$. Since $a \otimes a$ commutes with $b^{-1}a \otimes b$,

$$F_q(I \otimes b^{-1}a \otimes b)^*\,U_{12}U_{13} = U_{12}U_{13}\,F_q(I \otimes b^{-1}a \otimes b)^*.$$

The relation (11.3.11) implies that $Z_{23}U_{12}Z_{23}^* = U_{12}U_{13}$ and the formula (11.3.15) takes now the form

$$V_{12}V_{13} = F_q(I \otimes b^{-1}a \otimes b)F_q(\widehat{b} \otimes b \otimes I)Z_{23}U_{12}Z_{23}^*F_q(I \otimes b^{-1}a \otimes b)^* \tag{11.3.16}$$

Finally $b \otimes I$ commutes with Z. Therefore $F_q(\widehat{b} \otimes b \otimes I)$ commutes with Z_{23}. Clearly $F_q(\widehat{b} \otimes b \otimes I)U_{12} = [F_q(\widehat{b} \otimes b) \otimes I]U_{12} = V_{12}$ and $F_q(I \otimes b^{-1}a \otimes b)Z_{23} = [I \otimes F_q(b^{-1}a \otimes b)]Z_{23} = W_{23}$. Now (11.3.9) follows immediately from (11.3.16).

Now we prove the main statement. For any $c \in A$ we set

$$\Delta(c) = W(c \otimes I)W^*. \tag{11.3.17}$$

Then Δ is a representation of A acting on $L^2(\Gamma, d\gamma) \otimes L^2(\Gamma, d\gamma)$. We know that $V \in \mathrm{M}(\mathcal{K}(K) \otimes A)$. Formula (11.3.9) shows that

$$(\mathrm{id} \otimes \Delta)V = V_{12}V_{13}.$$

Clearly $V_{12}, V_{13} \in \mathrm{M}(\mathcal{K}(K) \otimes A \otimes A)$. Therefore $(\mathrm{id} \otimes \Delta)V = V_{12}V_{13} \in \mathrm{M}(\mathcal{K}(K) \otimes A \otimes A)$. Remembering that A is generated by V we conclude that $\Delta \in \mathrm{Mor}(A, A \otimes A)$. $\qquad\square$

We conclude the section by discussion to what extent C^*-bialgebra (A, Δ) is a quantum group.

Using formula (11.3.17) one can calculate $\Delta(c)$ for any $c \in A$. The same is true for any c affiliated with A. We shall show that

$$\begin{aligned} \Delta(a) &= a \otimes a, \\ \Delta(b) &= a \otimes b \dotplus b \otimes I. \end{aligned} \tag{11.3.18}$$

Since $b^{-1}a \otimes b$ commutes with $a \otimes a$, formula for $\Delta(a)$ follows immediately from (11.3.11). To prove formula for $\Delta(b)$ we notice that Z and $b \otimes I$ commute. Therefore

$$W(b \otimes I)W^* = F_q(b^{-1}a \otimes b)(b \otimes I)F_q(b^{-1}a \otimes b)^* \tag{11.3.19}$$

Now we use formula (11.2.17) with $X = a \otimes b$ and $Y = b \otimes I$. Then $Y^{-1}X = b^{-1}a \otimes b$ and the right hand side of (11.3.19) coincides with $X \dotplus Y$. The formula for $\Delta(b)$ is proved.

Formula (11.3.18) shows that (A, Δ) does not depend on the particular choice of a Hilbert space K nor on operators $\widehat{a}$ and $\widehat{b}$. One can choose $K = L^2(\Gamma, d\gamma)$ and $(\widehat{a}, \widehat{b}) = (a, b)$. However it turns out that $K = L^2(\Gamma, d\gamma)$ and

$$(\widehat{a}, \widehat{b}) = (b^{-1}, b^{-1}a). \tag{11.3.20}$$

is a more interesting choice. If this is the case then operator (11.3.6) coincides with (11.3.8): $V = W$. Relation (11.3.9) takes the form:

$$W_{23}W_{12} = W_{12}W_{13}W_{23}.$$

This is a pentagon equation (11.1.6). It means that W is a multiplicative unitary. It turns out [29],[20] that W is modular with (cf formulae (11.1.9) and (11.1.10))

$$\widehat{Q} = |b|, \qquad Q = |a|$$

and

$$\widetilde{W} = F_q\left((b^{-1}a)^{\top} \otimes (-qa^{-1}b)\right)^* \chi\left((b^{-1})^{\top} \otimes I, I \otimes a\right). \tag{11.3.21}$$

One can easily verify that (A, Δ) is related to W in the sense explained after Theorem 11.1.4. Therefore (A, Δ) is a quantum group. Its structure is described by Theorem 11.1.4. In particular there exists an antipode admitting a polar decomposition. We shall show that in this case

$$a^R = a^{-1}, \qquad b^R = -qa^{-1}b \tag{11.3.22}$$

where R is a unitary antipode (cf Statement 4(iii) of Theorem 11.1.4). We use Statement 6(ii) of this theorem to prove these formulae. Since $\top \otimes R$ is an antiisomorphism of $B(H) \otimes A$ into $B(\overline{H}) \otimes A$, it is antimultiplicative. We get

$$W^{\top \otimes R} = \chi\left((b^{-1})^{\top} \otimes I, I \otimes a^R\right) F_q\left((b^{-1}a)^{\top} \otimes b^R\right).$$

On the other hand (cf (11.3.21))

$$\widetilde{W}^* = \chi\left((b^{-1})^{\top} \otimes I, I \otimes a\right)^* F_q\left((b^{-1}a)^{\top} \otimes (-qa^{-1}b)\right).$$

Clearly $\overline{\chi(\gamma', \gamma)} = \chi(\gamma', \gamma^{-1})$ for any $\gamma', \gamma \in \Gamma$. Therefore

$$\chi\left((b^{-1})^{\top} \otimes I, I \otimes a\right)^* = \chi\left((b^{-1})^{\top} \otimes I, I \otimes a^{-1}\right).$$

Now formula $W^{\top \otimes R} = \tilde{W}^*$ may be written as

$$\chi\left((b^{-1})^\top \otimes I, I \otimes a^R\right) F_q\left((b^{-1}a)^\top \otimes b^R\right)$$

$$= \chi\left((b^{-1})^\top \otimes I, I \otimes a^{-1}\right) F_q\left((b^{-1}a)^\top \otimes (-qa^{-1}b)\right). \tag{11.3.23}$$

We apply Proposition 11.2.11 with

$$Y_1 = (b^{-1}a)^\top \otimes b^R, \qquad U_1 = \chi\left((b^{-1})^\top \otimes I, I \otimes a^R\right),$$

$$Y_2 = (b^{-1}a)^\top \otimes (-qa^{-1}b), \quad U_2 = \chi\left((b^{-1})^\top \otimes I, I \otimes a^{-1}\right),$$

$X = b^\top \otimes I$ and $C = \mathcal{K}(\overline{H}) \otimes A$. Taking into account that $(b^\top, a^\top)$ is a G-pair on $\overline{H}$ one can easily check that all assumptions of Proposition 11.2.11 are satisfied. Then (cf Remark 11.2.12) we get $Y_1 = Y_2$ (this proves the second formula in (11.3.22)) and $U_1 = U_2$. Now the first formula in (11.3.22) follows by Proposition 11.2.4.

Now consider formula (11.1.13). In this case it takes the form

$$h(c) = \mathrm{Tr}(|b|\, c\, |b|). \tag{11.3.24}$$

We shall show that h is locally finite (cf [34]), i.e. the set $\{c \in A : h(c^*c) < +\infty\}$ is dense in A. By Statement 1 of Theorem 11.3.1 we know that a set of elements of the form $c = g(a)f(b)$, where $g \in C_\infty(\Gamma)$ and $f \in C_\infty(\overline{\Gamma})$ is total in A. The same is true if g is of the form $g(\gamma) = \int_\Gamma \hat{g}(\gamma')\chi(\gamma, \gamma')d\gamma'$, where $\hat{g} \in C_0(\Gamma)$. Clearly $h(c^*c) = \mathrm{Tr}((c\,|b|)^*c\,|b|)$ and since $[\chi(a, \gamma')x](\gamma) = x(\gamma\gamma')$, $c\,|b|$ is an integral operator

$$[c\,|b|\,x](\gamma) = \int_\Gamma \hat{g}(\gamma')f(\gamma\gamma')\,|\gamma\gamma'|\,x(\gamma\gamma')d\gamma' = \int_\Gamma \hat{g}(\gamma^{-1}\gamma')f(\gamma')\,|\gamma'|\,x(\gamma')d\gamma'$$

with a kernel $K_c(\gamma, \gamma') = \hat{g}(\gamma^{-1}\gamma')f(\gamma')\,|\gamma'|$. Therefore

$$h(c^*c) = \int_{\Gamma \times \Gamma} |K_c(\gamma, \gamma')|^2\, d\gamma d\gamma' = \left(\int_\Gamma |f(\gamma)|^2\,|\gamma|^2\, d\gamma\right)\left(\int_\Gamma |\hat{g}(\gamma)|^2\, d\gamma\right)$$

where we used the invariance of Haar measure on Γ. Now by Plancherel formula

$$h(c^*c) = \left(\int_{\overline{\Gamma}} |f(\gamma)|^2\,|\gamma|^2\, d\gamma\right)\left(\int_\Gamma |g(\gamma)|^2\, d\gamma\right).$$

Clearly $L^2(\Gamma, d\gamma) \cap C_\infty(\Gamma)$ is dense in $C_\infty(\Gamma)$ and since the measure $d\mu(\gamma) = |\gamma|^2\, d\gamma$ is locally finite on $\overline{\Gamma}$, $L^2(\overline{\Gamma}, d\mu) \cap C_\infty(\overline{\Gamma})$ is dense in $C_\infty(\overline{\Gamma})$. This means that h is finite on a dense subset of A. Therefore h is a (right) Haar weight.

We recall that (cf Statement 6, (ii) of Theorem 11.1.4) the scaling group acts in the following way

$$\tau_t(c) = Q^{2it} c Q^{-2it} = |a|^{2it} c |a|^{-2it} .$$

Remembering that $|a|$ and $|b|$ commute we conclude (cf (11.3.24)) that in this case (in contrast to that of the quantum '$az+b$' group at roots of unity) the Haar weight is invariant with respect to the scaling group, $h \circ \tau_t = h$.

Acknowledgement

The authors are grateful to Komitet Badań Naukowych (grant No 2 P03A 040 22) for financial support.

Bibliography

[1] S. Baaj, *Représentationes régulière de groupe quantique* $E_\mu(2)$, C.R.Acad.Sci., Paris, Sér. I, **314** (1992), 1021–1026.

[2] S. Baaj, *Représentationes régulière de groupe quantique de déplacements de Woronowicz*, Asterisque, **232** (1995), 11–48.

[3] S. Baaj et G. Skandalis, *Unitaries multiplicatifs et dualité pour les produits croisé de C^*-algèbres*, Ann. Sci. Ec. Norm. Sup., 4e série, **26** (1993), 425–488.

[4] A.O. Barut and R. Rączka, *Theory of Group Representations and Applications*, Warszawa PWN – Polish Scientific Publishers, 1977.

[5] M.S. Dijkhuizen and T.H. Koornwinder: *CQG algebras: a direct algebraic approach to compact quantum groups.* Lett. Math. Phys. **32** (1994), 315–330.

[6] V.G. Drinfel'd: *Quantum groups.* Proceedings ICM Berkeley, 1986, 798–820.

[7] A. Klimyk and K. Schmüdgen: *Quantum groups and their representations.* Springer-Verlag, 1997.

[8] E. Koelink and J. Kustermans: *A locally compact quantum group analogue of the normalizer of $SU(1,1)$ in $SL(2,\mathbb{C})$.* Commun. Math. Phys. **232** (2003), 231–296.

[9] J. Kustermans and S. Vaes: *Locally compact quantum groups.* Annales Scientifiques de l'Ecole Normale Superieure. **33**, No. 6 (2000), 837–934.

[10] T. Masuda, Y. Nakagami and S.L. Woronowicz: *A C^*-algebraic framework for the quantum groups,* Int. J. Math. **14**, No. 9 (2003), 903–1001.

[11] S. Majid: *Foundations of Quantum Group Theory.* Cambridge University Press, 1995.

[12]   Gert K. Pedersen: *C*-algebras and their Automorphism Groups*, Academic Press, London, New York, San Francisco 1979.

[13]   P. Podleś and S.L. Woronowicz: *Quantum deformation of Lorentz group*, Commun. Math. Phys. **130** (1990), 381–431.

[14]   W. Pusz: *Quantum GL(2, ℂ) group as a double group over 'az+b' quantum group*, Rep. Math. Phys. **49** (2002), 113–122.

[15]   W. Pusz and P.M. Sołtan: *Functional form of unitary representations of the quantum 'az + b' group*, Rep. Math. Phys. **52** (2003), 309–319.

[16]   W. Pusz and S.L. Woronowicz: *A quantum GL(2, ℂ) group at roots of unity*, Rep. Math. Phys. **47** (2001), 431–462.

[17]   W. Pusz and S.L. Woronowicz, *A new quantum deformation of 'ax + b' group*, in preparation.

[18]   N.Yu. Reshetikhin, L.A. Takhtadjan and L.D. Faddeev: *Quantization of Lie groups and Lie algebras.* Leningrad Math. J. **1** (1990), 193–225.

[19]   Walter Rudin: *Fourier Analysis on Groups*, Interscience Publishers, New York - London 1962.

[20]   P.M. Sołtan and S.L. Woronowicz, *A remark on manageable multiplicative unitaries*, Lett. Math. Phys., **57** (2001), 239–252.

[21]   A. Van Daele, *The Haar measure on some locally compact group*, Preprint of K.U.Leuven (2000), arXiv:math.OA/0109004v1 (2001).

[22]   A. Van Daele and S.L. Woronowicz, *Duality for the quantum E(2) group* Pac. J. Math., **173**, No 2, (1996), 375–385.

[23]   S.L. Woronowicz, *Pseudospaces, pseudogroups and Pontriagin duality* Proceedings of the International Conference on Mathematical Physics, Lausanne, 1979, *Lecture Notes in Physics* **116** (1980), 407–412.

[24]   S.L. Woronowicz, *Twisted SU(2) group. An example of a non-commutative differential calculus*, Publ. RIMS, Kyoto Univ. **23** (1987), 117–181.

[25]   S.L. Woronowicz, *Compact matrix pseudogroups*, Commun.Math.Phys. **111** (1987), 613–635.

[26]   S.L. Woronowicz, *A remark on compact matrix quantum groups*, Lett. Math. Phys, **21** (1991), 35–39.

[27]   S.L. Woronowicz, *Unbounded elements affiliated with C*-algebras and non-compact quantum groups*, Commun. Math. Phys. **136** (1991), 399–432.

[28]   S.L. Woronowicz: *Quantum E(2) group and its Pontryagin dual*, Lett. Math. Phys., **23** (1991), 251–263.

[29]   S.L. Woronowicz, *Quantum 'az + b' group on complex plane*, Commun. Math. Phys. **144** (1992), 417-428.

[30]   S.L. Woronowicz: *C*-algebras generated by unbounded elements*, Rev. Math. Phys., Vol. 7, No. 3, (1995), 481–521.

[31]   S.L. Woronowicz: *Compact quantum groups*, in "Symétries quantiques" (Les Houches, Session LXIV, 1995), North-Holland, Amsterdam (1998), 845–884. Rev. Math. Phys., Vol. 7, No. 3, (1995), 481–521.

[32]   S.L. Woronowicz, *From multiplicative unitaries to quantum groups*, International J. Math., **7** (1996), 127–149.

[33]   S.L. Woronowicz, *Operator equalities related to the quantum E(2) group*, International J. Math., **12** (2001), 461–503.

[34] S.L. Woronowicz, *Haar weights on some quantum groups,* in GROUP 24. Physical and Mathematical Aspects of Symmetries, eds. J-P.Gazeau, R.Kerner, J-P.Antoine, S.Métens and J-Y.Thibon, IOP Publishing Ltd., Inst. Phys. Conf. Series **No. 173** (2003), 763–772.

[35] S.L. Woronowicz and S. Zakrzewski: *Quantum Lorentz group having Gauss decomposition property,* Publ. RIMS, Kyoto Univ. **28** (1992), 809–824.

[36] S.L. Woronowicz and S. Zakrzewski, *Quantum 'ax + b' group,* Rev. Math. Phys., **14**, Nos 7 & 8 (2002), 797–828.